Oncogene and Transgenics Correlates of Cancer Risk Assessments

NATO ASI Series

Advanced Science Institutes Series

A series presenting the results of activities sponsored by the NATO Science Committee, which aims at the dissemination of advanced scientific and technological knowledge, with a view to strengthening links between scientific communities.

The series is published by an international board of publishers in conjunction with the NATO Scientific Affairs Division

A	Life Sciences	Plenum Publishing Corporation
B	Physics	New York and London
C	Mathematical and Physical Sciences	Kluwer Academic Publishers
D	Behavioral and Social Sciences	Dordrecht, Boston, and London
E	Applied Sciences	
F	Computer and Systems Sciences	Springer-Verlag
G	Ecological Sciences	Berlin, Heidelberg, New York, London,
H	Cell Biology	Paris, Tokyo, Hong Kong, and Barcelona
I	Global Environmental Change	

Recent Volumes in this Series

Volume 226—Regulation of Chloroplast Biogenesis
    edited by Joan H. Argyroudi-Akoyunoglou

Volume 227—Angiogenesis in Health and Disease
    edited by Michael E. Maragoudakis, Pietro Gullino, and Peter I. Lelkes

Volume 228—Playback and Studies of Animal Communication
    edited by Peter K. McGregor

Volume 229—Asthma Treatment—A Multidisciplinary Approach
    edited by D. Olivieri, P. J. Barnes, S. S. Hurd, and G. C. Folco

Volume 230—Biological Control of Plant Diseases: Progress and Challenges
    for the Future
    edited by E. C. Tjamos, G. C. Papavizas, and R. J. Cook

Volume 231—Formation and Differentiation of Early Embryonic Mesoderm
    edited by Ruth Bellairs, Esmond J. Sanders, and James W. Lash

Volume 232—Oncogene and Transgenics Correlates of Cancer Risk Assessments
    edited by Constantine Zervos

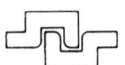

Series A: Life Sciences

Oncogene and Transgenics Correlates of Cancer Risk Assessments

Edited by
Constantine Zervos
Consulting in Health and Environmental Protection
Simpsonville, Maryland

Springer Science+Business Media, LLC

Proceedings of a NATO Advanced Research Workshop on
Oncogene and Transgenics Correlates of Cancer Risk Assessments,
held October 7-11, 1991,
in Vouliagmeni Beach, Attiki, Greece

NATO-PCO-DATA BASE

The electronic index to the NATO ASI Series provides full bibliographical references (with keywords and/or abstracts) to more than 30,000 contributions from international scientists published in all sections of the NATO ASI Series. Access to the NATO-PCO-DATA BASE is possible in two ways:

—via online FILE 128 (NATO-PCO-DATA BASE) hosted by ESRIN, Via Galileo Galilei, I-00044 Frascati, Italy.

Additional material to this book can be downloaded from http://extra.springer.com.


```
                Library of Congress Cataloging-in-Publication Data

Oncogene and transgenics correlates of cancer risk assessments /
  edited by Constantine Zervos.
       p.    cm.  -- (NATO ASI series. Series A, Life sciences ; v.
  232)
     "Proceedings of a NATO Advanced Research Workshop on Oncogene and
  Transgenics Correlates of Cancer Risk Assessments, held October
  7-11, 1991, in Vouliagmeni Beach, Attiki, Greece"--T.p. verso.
     "Published in cooperation with NATO Scientific Affairs Division."
     Includes bibliographical references and index.
     ISBN 978-1-4613-6333-0      ISBN 978-1-4615-3056-5 (eBook)
     DOI 10.1007/978-1-4615-3056-5
     1. Oncogenes--Congresses.  2. Cancer--Risk factors--Congresses.
  3. Health risk assessment--Congresses.  4. Transgenic animals-
  -Congresses.  5. Cancer--Animal models--Congresses.  I. Zervos,
  Constantine.  II. North Atlantic Treaty Organization.  Scientific
  Affairs Division.  III. NATO Advanced Research Workshop on Oncogene
  and Transgenics Correlates of Cancer Risk Assessments (1991 :
  Attike, Greece)  IV. Series.
     [DNLM: 1. Mice, Transgenic--genetics--congresses.  2. Neoplasms-
  -genetics--congresses.  3. Oncogenes--congresses.  4. Risk Factors-
  -congresses.   QZ 202 05413]
  RC268.42.05  1992
  616.99'4071--dc20
  DNLM/DLC
  for Library of Congress                                   92-21812
                                                                CIP
```

ISBN 978-1-4613-6333-0

© 1992 Springer Science+Business Media New York
Originally published by Plenum Press in 1992

All rights reserved

No part of this book may be reproduced, stored in a retrieval system, or transmitted in any form or by any means, electronic, mechanical, photocopying, microfilming, recording, or otherwise, without written permission from the Publisher

PREFACE

The data, the information, and even the overarching knowledge necessary for risk assessments of economically important environmental carcinogens come, for the most part, from the applied biological disciplines, e.g., toxicology, epidemiology, biostatistics, etc. The more fundamental biological disciplines, e.g., biochemistry, cell biology, molecular biology, molecular genetics of cancer, etc., have enormous but unrealized potential to improve current cancer risk assessment methods. The objective of this advanced research workshop ARW was to advance the state of the art of cancer risk assessment methods by identifying potential short and long term contributions to such methods from the more fundamental disciplines. Attention was paid to short and long term contributions from research advances in the biochemistry and physiology of oncogenes (oncogenes research) and in the construction and utilization of transgenic animals (transgenics research).

In the last 20 years, researchers in the fundamental biological disciplines, i.e., biochemists, geneticists, molecular and cell biologists, etc., have, *inter alia*, advanced spectacularly our understanding of the nature of neoplastic diseases. Their phenomenal progress is the combined result of both advances and refinements of the techniques available to them and of new fundamental discoveries. Among the latter the most significant are the discoveries of oncogenes and of the feasibility of creating transgenic animals, i.e., of transferring well defined and expressible genes from the cells of one species of organisms to the embryonic cells of another.

Oncogene research efforts have taken two major directions: (a) to identify, isolate, and study the structure, properties, functions and interrelations of an increasing number of members of the proto-oncogene families; and (b) to elucidated the role proto-oncogenes play in the development and homeostasis of normal eukariotic cells and in the initiation promotion and maintenance of the malignant state.

Thus far, research efforts in transgenics have been directed toward multiple objectives mostly in diagnostics and therapeutics. The potential of transgenics in safety testing is just beginning to be recognized.

The same 20 years of great progress in the fundamental biological disciplines have also witnessed the expanded uses of cancer risk assessments (CRA) by individuals and institutions concerned with public health and environmental protection. Risk assessment methods are now an integral part of various strategies of cancer avoidance and CRA documents for natural or man-made substances in the environment are yearly produced and published in increasing numbers.

CRAt methods have been improved some since the assessments were first introduced back in the early 70s. However, incorporation of new information and knowledge has been slow. This observation rings particularly true for information and knowledge from the fundamental disciplines of biology (biochemistry, molecular biology, genetics, etc.)

The reasons are probably many. Most likely, they include the absence of an acceptable overarching theory that credibly bridges the knowledge gap that separates fundamental biology from CRA methods. It is also likely, however, that a major reason is the absence of facile interdisciplinary communications between the practitioners in the major areas.

There is a need to bring closer and to bridge fundamental biology and CRA method research. This ARW brought together for in-depth discussions internationally known researchers with experience in the molecular biology and molecular genetics of cancer and CRA practitioners and theoreticians from the academe, government and industry. It was a 5-day, work-intensive gathering of a small number of scientists and experts. The invited participants prepared background papers in specified topics the majority of which were distributed for study to all participants before the Workshop.

During the Workshop, speakers present their papers and lively discussion followed. At the appropriate points round table discussions were held to propose, discuss, and formulate recommendations.

The ARW featured the following specific themes.

A. A review of the current state of development of CRA methods. This review focused on identifying and describing the knowledge gaps that force risk assessors to make sweeping assumptions about the mechanisms of spontaneous and environmentally induced neoplasia.

B. A review of the current state of development and the likely future research directions of molecular biology and cancer genetics. This review focused specifically on the following two areas:
   1. The production and characteristics of transgenic animals and their potential utility in CRA; and
   2. The role of oncogenes in the processes of spontaneous and induced neoplasia.

C. Case studies to identify currently possible contributions to CRA methods from current and immediately forthcoming knowledge about transgenics and oncogenes. Recent cancer risk assessments for methylene chloride and benzene were taken up as study cases.

D. Attempts to formulate recommendations for long term approaches to research both in transgenics/oncogenes and in CRA methods. Recommendations were sought likely to bring closer the work of the fundamental and applied research groups and to increase the contributions of each group to the advances and the utility of the other.

A significant portion of the workshop time was taken up by discussions and by efforts to formulate consensus recommendations for research pursuant to the workshop's basic objective.

This volume is the product of the Workshop. In addition to the invited papers, it includes a summary, findings, and recommendations for future work.

The editor of this volume wishes to thank the North Atlantic Treaty Organization and the U.S. National Science Foundation for their support. He also gratefully acknowledges the support of the Society for Risk Analysis and the International Life Sciences Institute.

C. Zervos
Washington, D.C.

CONTENTS

CANCER RISK ASSESSMENTS: AN OVERVIEW 1
    Elizabeth L. Anderson

PHARMACODYNAMIC MODELS FOR CANCER RISK ASSESSMENT 21
    Suresh H. Moolgavkar

PREDICTIVE VALUE FOR CANCER RISK ASSESSMENT OF
    CELL- AND TISSUE-SPECIFIC FORMATION OF
    CARCINOGEN-DNA ADDUCTS 31
    Ewalt Scherer

STUDIES ON THE ROLE OF PROTEIN KINASE C IN MULTISTAGE
    CARCINOGENESIS AND THEIR RELEVANCE TO RISK
    EXTRAPOLATION 43
    Kevin R. O'Driscoll, Scott M. Kahn, Christoph M. Borner,
    Wei Jiang, and I. Bernard Weinstein

ONCOGENE ACTIVATION AND HUMAN CANCER 61
    Demetrios A. Spandidos and Margaret L.M. Anderson

GENETICS OF HEPATOCARCINOGENESIS IN MOUSE AND MAN 67
    Tommaso A. Dragani, Giacomo Manenti, Manuela Gariboldi,
    F. Stefania Falvella, Marco A. Pierotti, and
    Giuseppe Della Porta

PROTEIN KINASE C IN CARCINOGENESIS: COMPARATIVE
    EXPRESSION OF ONCOGENES AND PROTEIN KINASE
    C IN RHABDOMYOSARCOMA MODELS 81
    Monique Castagna, Jacqueline Robert Lezenes,
    Xupei Huang, and Nicole Hanania

ONCOGENES AND HUMAN CANCERS 91
    J.S. Rhim, J.H. Yang, I.H. Lee, M.S. Lee, and J.B. Park

'IN VIVO' MODEL SYSTEMS TO STUDY ras ONCOGENE
    INVOLVEMENT IN CARCINOGENESIS 111
    Ramon Mangues and Angel Pellicer

K-*ras* ONCOGENE ACTIVATION IN NEOPLASIAS OF PATIENTS
  WITH FAMILIAL ADENOMATOUS POLYPOSIS OR
  KIDNEY TRANSPLANT 127
 A. Haliassos and D.A. Spandidos

AN OVERVIEW OF TRANSGENIC MICE AS FUTURE MODELS
  OF HUMAN DISEASES, IN DRUG DEVELOPMENT,
  AND FOR GENOTOXICITY AND
  CARCINOGENESIS STUDIES 133
 J.C. Cordaro

REGULATING GENE EXPRESSION IN MAMMALIAN CELL
  CULTURE AND TRANSGENIC MICE WITH YEAST
  GAL4/UAS CONTROL ELEMENTS 155
 David M. Ornitz, Radek Skoda, Randall W. Moreadith,
 and Philip Leder

GENES AND ONCOGENES OF THE *ras* SUPERFAMILY 173
 Armand Tavitian

ONCOGENIC TRANSGENIC MICE IN THE STUDY
  OF CARCINOGENESIS 185
 Chang-Ho Ahn and Won-Chul Choi

LEUKEMIA, LYMPHOMA, EMBRYONIC CARCINOMA AND
  HEPATOMA INDUCED IN TRANSGENIC RABBITS
  BY THE c-*myc* ONCOGENE FUSED WITH
  IMMUNOGLOBULIN ENHANCERS 199
 Helga Spieker-Polet, Periannan Sethupathi, Herman Polet,
 Pi-Chen Yam, and Katherine L. Knight

AN OVERVIEW OF THE OUTSTANDING ISSUES IN THE RISK
  ASSESSMENT OF METHYLENE CHLORIDE 217
 Rory B. Conolly, Kannan Krishnan, and Melvin E. Andersen

CANCER DOSE-RESPONSE MODELING AND
  METHYLENE CHLORIDE 231
 Gail Charnley

RISK ASSESSMENTS FOR BENZENE-INDUCED LEUKEMIA
  — A REVIEW 241
 Kenny S. Crump

COMPARATIVE METABOLISM AND GENOTOXICITY DATA
  ON BENZENE: THEIR ROLE IN CANCER
  RISK ASSESSMENT 263
 Sandro Grilli, Silvio Parodi, Maurizio Taningher,
 and Annamaria Colacci

DOSE-RESPONSE RELATIONSHIPS FOR BENZENE: HUMAN
    AND EXPERIMENTAL CARCINOGENICITY DATA 293
  Silvio Parodi, Davide Malacarne, Sandro Grilli,
    Annamaria Colacci, and Maurizio Taningher

A THRESHOLD FOR BENZENE LEUKEMOGENESIS 305
  Bruce Molholt

SOME POINTS FOR DISCUSSION FROM THE
    ONCOGENE WORKSHOP 313
  Bruce Molholt

CONCLUDING REMARKS, FINDINGS AND
    RECOMMENDATIONS 319
  C. Zervos

Index 333

CANCER RISK ASSESSMENTS: AN OVERVIEW

Elizabeth L. Anderson

President
Clement International Corporation

INTRODUCTION

The process of risk assessment and risk management is widely recognized in the United States as a tool for making policy decisions to control risk associated with toxic chemical exposures. This two-step process, first to evaluate risk and then to decide

---

(1) VHS model using simple equation, EPA fixed default values

$$\frac{C}{C_o} = \text{erf}\left(\frac{Z}{2(DY)^{0.5}}\right) \text{erf}\left(\frac{X}{4(DY)^{0.5}}\right)$$

$$= 0.34$$

(2) VHS model using complex equation, measured site values

$$\frac{C}{C_o} = \frac{1}{4}\left(\text{erf}\left[\frac{z+Z}{2(D_zY)^{0.5}}\right] - \text{erf}\left[\frac{z-Z}{2(D_zY)^{0.5}}\right]\right)$$

$$\times \left(\text{erf}\left[\frac{x+X/2}{2(D_xY)^{0.5}}\right] - \text{erf}\left[\frac{x-X/2}{2(D_xY)^{0.5}}\right]\right)$$

$$= 0.06$$

Figure 1.

---

what, if anything, should be done to reduce exposures, was adopted by the U.S. Environmental Protection Agency (EPA) in 1976 its announcement of guidelines for assessing cancer risk (EPA 1976, Albert et al. 1977). This approach was later endorsed by committees of the National Academy of Sciences (NAS) as the most appropriate process for informed public policy decisions to protect public health from exposure to toxic chemicals (NAS 1982, 1983). Other United States interagency committees reviewed the basis for cancer risk assessment and published background documents that are largely consistent with the earlier and much briefer statement of risk assessment guidance published by EPA in 1976 (IRLG 1979, OSTP 1984). EPA has published updates of its original cancer risk assessment guidelines and guidelines for other health effects that take into account its decade of experience in assessing cancer risks for hundreds of chemicals (EPA 1984a,b,c,d, 1985a, 1986a,b,c,d,e).

RISK ASSESSMENT

Risk assessment is an organized approach to evaluating scientific data that involve considerable uncertainties. Risk assessment attempts to answer two questions: (1) How likely is an event to occur? and (2) If it does occur, how bad could it be in quantitative terms? This approach has long been used to estimate risks associated with a variety of activities, including economic forecasting, transportation safety, engineering safety (e.g., nuclear power plants), and radiation exposures. More recently, risk assessment approaches have been used to estimate risk associated with toxic chemical exposures (EPA 1976; Albert et al. 1977; NAS 1982, 1983; IRLG 1979; OSTP 1984; EPA 1984a,b,c,d, 1985a, 1986a,b,c,d,e).

Risk assessment of toxic chemical exposures may be defined as having four steps: (1) hazard identification, (2) dose-response modeling, (3) exposure assessment, and (4) risk characterization. The hazard identification step (or qualitative risk assessment) (EPA 1976, 1984a,b,c,d, 1985a, 1986a,b,c,d,e) answers the first risk assessment question: How likely is a risk to occur? The next two steps, dose-response modeling and exposure assessment, are combined to quantify the risk associated with current and anticipated exposures. Finally, the risk characterization step presents both the qualitative likelihood that the hazard will occur and, on the assumption of hazard occurrence, the quantitative estimates of risk.

Hazard identification for toxic chemicals is based on the evaluation of all available data (e.g., epidemiology, animal bioassay studies, and *in vivo* and *in vitro* studies) to characterize the strength of evidence indicating potential health effects that might occur in exposed human populations.

In the absence of human data to describe low-dose effects, two different approaches for dose-response characterization are most frequently used: one for "threshold" effects and the other for "nonthreshold" effects. The doses associated with human diseases that are thought to occur by a "threshold dose mechanism," i.e., no significant health risks are likely to occur until a certain dose is reached, are most frequently defined by applying safety factors to the no-observed-effect levels in animal bioassay studies. However, human data are used whenever possible. These so-called safe exposure levels are labeled acceptable daily intakes (ADIs) or reference doses. For nonthreshold effects (e.g., cancer), meaning that there may be some risks associated with any exposure, dose-response extrapolation models are used to estimate cancer risks associated with low-dose exposure based on observed incidence in humans and animals exposed at high doses.

Since 1976, quantitative risk assessment has received widespread recognition and acceptance for describing health risks associated with toxic chemical exposure. The model most often relied upon for estimating cancer risk at exposure levels far below the

TABLE 1. GROUNDWATER MONITORING DATA

Well Number	Sampling Date	TCE mg/L		Well Average TCE mg/L
1	1/ 1/85	0.100		
	1/20/85	0.144	0.124	
	2/ 2/85	0.127		
2	1/1/85	<.005		
	1/20/85	1.10		
	2/2/85	1.98	1.27	
	2/8/85	2.00		
3	1/1/85	<0.005		
	2/2/85	<0.001	0.003	
4	1/1/85	0.005		
	2/2/85	0.080	0.068	0.36
5	1/20/85	0.210		
	2/2/85	0.177	0.194	
6	1/1/85	0.510		
	2/8/85	1.40	0.96	
7	1/20/85	0.160		
	2/2/85	0.305	0.23	
8	1/1/85	0.070		
	2/8/85	<0.005	0.04	

observed range has been the linear nonthreshold model, which provides a plausible upper-bound cancer risk estimate (Anderson 1983, EPA 1984a, IRLG 1979, OSTP 1984). The conservative nature of the linear, nonthreshold approach, as well as its biological inflexibility, have stimulated scientists, regulators, and economists to search for methods that can incorporate a more biological understanding of the cancer process. A number of chemicals have been identified that most likely exert their carcinogenic effects in manners for which a linear nonthreshold relationship would be invalid at low doses. As a result of these concerns, scientists have sought to develop dose-response models that account for varying biological mechanisms.

Focus on dose-response modeling probably represents the most dramatic departure from practices of the last 15 years. There is a clear effort by regulatory agencies to seek a biological basis for the development of more accurate estimates of risks expected to occur at environmental exposure levels. This effort represents a substantially different approach from applying empirical formulas to estimate low-dose responses from high-dose data; the attention is instead focused on the importance of research data that may guide low-dose modeling efforts. Such an approach provides, at a minimum, an indication of the extent to which the "plausible upper bounds" may be overestimating risk for particular chemicals. Early efforts to define more accurate estimates of risk began at EPA in early 1985 and have culminated in the development of a generic approach using a two-stage model. This model adapts the clinical observations of Moolgavkar and Knudson (1981) to parameters involving exposure to toxic chemicals. The effort was first undertaken by EPA's Risk Assessment Forum and was ultimately

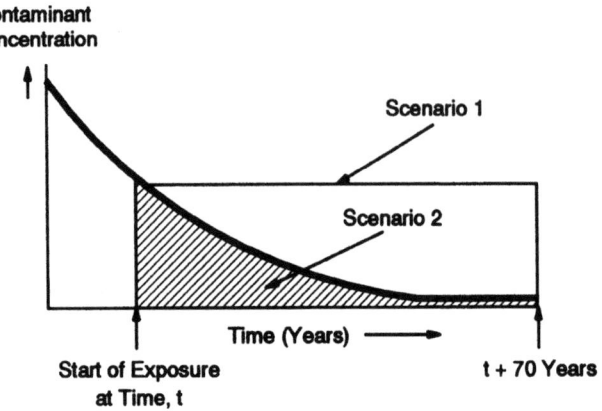

Scenario 1 assumes decay in contaminant levels until start of consumption and then a lifetime exposure to a constant concentration of the contaminant.
Scenario 2 assumes decay in contaminant levels both before and during the period of exposure.

Figure 2.

published in the *Journal of Risk Analysis* in early 1987 (Thorslund et al. 1987).

EPA has proposed two important decisions in line with the trend toward less conservatism in dose-response modeling. For example, the Risk Assessment Forum has recommended lowering the arsenic ingestion potency by approximately an order of magnitude (Levine et al. 1987, Moore 1987) based on modifications in dose-response calculation methodology and better estimates of the exposure involved in the epidemiology studies that were the basis for the evaluation. There has been further consideration of reducing the potency of arsenic by ingestion by still another order of magnitude to reflect the fact that skin cancer caused by arsenic ingestion is less likely to lead to death than is lung cancer induced by inhalation. This raises the issue of whether or not treatability, survival, and severity should be routinely considered as a part of the risk assessment process, and in particular the potency evaluation.

EPA is preparing to reassess its approach to evaluating the toxic effects of dioxin. A public meeting was held in October 1991 on EPA's Scientific Reassessment of Dioxin (EPA 1991). One of the principal tasks in the reassessment is the development of a new biologically based dose-response model based on current knowledge of dioxin toxicity.

The two-stage model of carcinogenesis has also been applied to several other chemicals with similar outcomes. For example, the model has also been applied to chlordane and heptachlor, and methylene chloride. While the mechanisms in each case differ, the outcomes of the model are to indicate most often several orders of magnitude lower potency at low dose than predicted by the linear nonthreshold model at the "plausible upper bounds."

Additional applications of the biological model have involved the polycyclic organic compounds. Past practices have used the potency of benzo(a)pyrene as a unit equivalency to all other potentially carcinogenic polycyclic organic compounds, greatly overestimating risk. This practice has continued in spite of the fact that comparative potency methods have been developed for other chemical classes, such as the dioxins. When assembled in the aggregate, several laboratory studies provide a more substantial basis for developing a comparative potency approach for PAHs (Thorslund et al. 1986). In addition, the shape of the dose-response curve for benzo(a)pyrene itself has been reevaluated. Benzo(a)pyrene is a genotoxic agent as indicated by a linear rate of DNA adduct formation that parallels exposure. The tumor dose-response data do not parallel DNA adduct formation, however, but appear to fit a quadratic equation, indicating that two events are probably necessary to induce the response. EPA's initial cancer potency estimate for benzo(a)pyrene does not reflect this relationship. The comparative potency approach for other polycyclic compounds, together with a revised dose-response curve for benzo(a)pyrene, has been used to predict tumor outcomes accurately in bioassays of chemical mixtures, which is not possible using upper-bound estimates (Thorslund et al. 1986).

Benzene is another chemical that may require two events to produce a cancer outcome. Current investigations of the mechanistic data indicate that benzene causes chromosome damage, which is thought to be responsible for the chromosomal deletions and rearrangements observed in leukemia patients. This relationship implies that, although linearity may establish a plausible upper bound on human leukemia risk from benzene exposure, a quadratic relationship may be more appropriate to estimate the actual risk. If this turns out to be the case, the risk from low-dose exposure to benzene would be considerably lower than previously estimated (Voytek and Thorslund 1991).

Other important efforts are attempting to model pharmacokinetic data that describe the pathway that a toxic chemical follows through the body to deliver the effective carcinogen at the target organ that develops the cancer response. Earlier cancer models assumed that exposure or some adjustment of actual exposure according to uptake or absorption by the skin, gastrointestinal tract, or lungs constituted the effective dose. Current efforts have contributed to important progress toward describing the actual dose of the effective chemical carcinogen that is delivered to the target tissue (Andersen et al. 1980, 1984; Andersen 1981a,b; Crump 1981, 1987).

Of equal importance, trends in exposure assessment research are also leading to improved estimates of population exposures, which provide a better foundation for current and projected exposures. Traditional practices have relied heavily on generic models to describe exposure to human populations. EPA has developed generalized dispersion models for describing air transport and similar generalized dispersion models for surface and groundwater. The overall impact of these dispersion models has been to provide conservative estimates of exposure.

TABLE 2.   RISK COMPARISON

Time of Exposure Initiation (yrs)	TCE at Start of Exposure (mg/L)	Scenario 1	Scenario 2
	0.36	1.1×10^{-4}	3.5×10^{-6}
1	0.23	7.1×10^{-5}	2.2×10^{-6}
2	0.14	4.5×10^{-5}	1.4×10^{-6}
3	0.09	2.8×10^{-5}	8.8×10^{-7}
4	0.06	1.8×10^{-5}	
5	0.04	1.1×10^{-5}	
6	0.02	7.2×10^{-6}	
7	0.01	4.6×10^{-6}	
8	0.009	2.9×10^{-6}	
9	0.006	1.8×10^{-6}	
10	0.004	1.1×10^{-6}	

The use of generalized models provides a practical approach for widespread exposure estimation by regulatory agencies because it would be highly impractical for a national agency to evaluate site-specific parameters for every source. For important cases, however, it is possible to estimate actual parameters that may refine the estimates obtained by generic modeling. An example is EPA's risk assessment of the ASARCO smelter in Tacoma, Washington (Patrick and Peters 1985). The use of generalized dispersion modeling using the human exposure model (HEM) (which assumes a flat terrain, an immobile population, and uses meteorological data from the closest weather station), when coupled with the dose-response curve, estimated a maximum individual risk of about 1×10^{-1} for populations living near the smelter. Subsequently, a local study was conducted that permitted the use of several site-specific assumptions including a more accurate description of the actual terrain, local meteorological data, and better emissions information. The outcome was to lower the exposure assessment and the overall risk about an order of magnitude. This brought the risk into closer alignment with the limited monitoring data available for the ambient air.

The same phenomenon has been observed when comparing estimates using generalized dispersion models for groundwater with estimates that rely on site specific parameters. For example, in Figure 1, the generalized dispersion model, the vertical horizontal spread (VHS) model using EPA default values, overestimates the risk by a factor of 5.7 when compared to the results from the more complex equation, which incorporates measured site values (Domenico and Palciauskas 1982, EPA 1985). Considerations of bioavailability have also sharpened exposure estimates and lowered the outcome from exposure assessment by several orders of magnitude and, thus the quantitative risk assessment. For example, dioxin was originally assumed to be 100% biologically available in soil. Recent studies, however, have demonstrated that dioxin is only partially available, 0.5-85% depending on soil type (Umbreit et al. 1986). In practice, it has been our experience that dioxin is mostly available in the range of 15-

TABLE 3.   CHEMICALS PROPOSED BY EPA'S OFFICE OF AIR PROGRAMS FOR UNIT RISK ASSESSMENT[a]

Group I	Group II	Group III
Acrylonitrile	Beryllium	Acetaldehyde
Carbon tetrachloride	Cresols (o-, m-, p-)	Acetylene tetrachloride
Chloroform	Formaldehyde	Acrolein
Ethylene dibromide	Maleic anhydride	Allyl chloride
Ethylene dichloride	Manganese	Benzyl chloride
Nitrosamines (4)	Methylene chloride	Bischloromethyl ether (BCME)
Perchloroethylene	Nickel	Chlorobenzene
Trichloroethylene	Toluene	Chloromethylmethyl ethyl
Vinylidene chloride	Xylenes (o-, m-, p-)	Chloroprene
		Dichlorobenzene (o-, p-)
		Dioxane
		Epichlorohydrin
		Hexachlorocyclopentadiene
		Methyl iodine
		Naphthylamine (1- and 2-)
		2-Nitropropane
		Phenol
		Phosgene
		Polychlorinated biphenyls (PCBs)
		Propylene oxide

[a] Unit risk is the incremental excess lifetime risk associated with breathing 1 $\mu g/m^3$ of the chemical over a 70-year life span for a 70-kg person.

50%. Dioxin in fly ash was also originally assumed to be up to 100% available.

Recent studies have found that this is not correct but rather that dioxin in fly ash is biologically available between 0.1% and 0.001% (van den Berg et al. 1986). The bioavailability issue is now being investigated in many different situations where the availability in soil and fly ash is important to the outcome of the risk assessment. Other considerations that tend to lower exposure assessments may come from knowledge of the hydrogeology of an area. For example, work at a site in California indicated an "upper-bound" lifetime risk associated with ingestion of water containing trichloroethylene (TCE) in an aquifer to be as high as a 10^{-3} risk (Figure 2). This level is associated with a 70-year lifetime exposure via exposure to drinking water from the contaminated aquifer. Scenario 2 in Figure 2 describes the decline in risk associated when hydrogeology models are applied to the site; the model assumes that the source of contamination has been removed. In Tables 1 and 2, the monitoring well data are displayed and likewise the risk comparison over time given the ability to model the area. In this particular circumstance, the remedial action being considered would cost in the

TABLE 4.   UPPER-BOUND UNIT CALCULATIONS FOR SUSPECTED CARCINOGENIC AIR POLLUTANTS

Chemical	Upper-Bound Unit Risk Estimates[a]	Reference
Acrylonitrile	7×10^{-5}	IRIS 1991
Arsenic	4×10^{-3}	IRIS 1991
Benzene	8×10^{-6}	IRIS 1991
Beryllium	2×10^{-3}	IRIS 1991
Diethylnitrosamine (DEN)	4×10^{-2}	IRIS 1991
Dimethylnitrosamine (DMN)	1×10^{-2}	IRIS 1991
Ethylene dibromide	2×10^{-4}	IRIS 1991
Ethylene dichloride	3×10^{-5}	IRIS 1991
Ethylene oxide	1×10^{-4}	HEAST 1991
Formaldehyde	1×10^{-5}	HEAST 1991
Nickel	2×10^{-4}	IRIS 1991
Perchloroethylene	5×10^{-7}	HEAST 1991
Vinyl chloride	8×10^{-5}	HEAST 1991
Vinylidene chloride	5×10^{-5}	IRIS 1991

[a] Unit risk is the incremental excess lifetime risk associated with breathing 1 $\mu g/m^3$ of the chemical over a 70-year life span for a 70-kg person.

million dollar range and require a number of months to install. If the hydrogeology models are correct, the theoretical risk could be lowered considerably over the first 18-month period given the natural ability of the hydrogeology of the area to remove the contamination. Caution, however, should still be exercised in assuming that the source has been removed because recent publications indicate that in some circumstances some chemicals may remain entrapped in soil micropores and thereby provide a slow, diffuse release (Sawhney et al. 1988).

Although improving the scientific information available for site-specific exposure assessment tends to lower the overall outcome of the exposure assessment and thus the risk assessment, there are important exceptions. For example, in a paper that addressed the risk associated with inhaling volatile organic chemicals from contaminated drinking water during shower activity, Foster and Chrostowski (1987) indicated that as much as half or more of the total body risk could be associated with the shower exposure rather than with the drinking water exposure. In addition, recent improved methods for modeling the actual deposition of particulate matter from stationary sources tends to raise the risk, compared to the earlier EPA air transport models, which assumed that both large and small particles bounced from the surface of the earth in very similar ways and were carried from the site by air transport. The more recent models take into account that the small particles deposit on the surface and are not so readily transported (Sehmel and Hodgson 1979). Also, closer attention to chemical conversions may tend

TABLE 5. UPPER-BOUND LIFETIME CANCER RISK FOR ARSENIC EXPOSURE[a,b]

Source	Exposed[c]	Exporsure Levels×10^4 [d]	Cancer Risk[e]	Cases/yr[e]
Copper smelters	43,800	2.7-1.5	2-1x10^{-3}	1.5-0.821
Lead smelters	3,400	0.69-0.27	6-2x10^{-4}	0.029-0.017
Zinc smelters	37,000	0.69-0.27	6-2x10^{-4}	0.32-0.13
Cotton gins	32	15.4-6.9	13-6x10^{-3}	0.0061-0.0027
Pesticide manufacturing	1,480	0.026-0.014	2-1x10^{-5}	0.0004-0.00025
Glass manufacturing	11,580	0.69-0.014	6-2x10^{-4}	0.099-0.040

[a]From EPA, Carcinogenic Assessment Group's Risk Assessment on Arsenic, May 2, 1980. National Technical Information Service, PB-81-20613. [b]The significant figures presented do not indicate precision or accuracy; they are included to make it easier to trace the deviation of these numbers through the various extrapolation and mathematical calculations. [c]Populations (numbers) of exposed to the highest two ambient levels of arsenic from the sources listed. [d] The highest two; e.g., for copper smelters the highest is 2.7x10^{-4} mg/kg/d. [e]Upper bounds.

---

to raise or lower the risk; for example, trichloroethylene is converted under anaerobic conditions to vinyl chloride, which has a higher potency value by ingestion than does trichloroethylene (Cline and Viste 1984, Parsons et al. 1984). Recognition of this conversion raises the overall risk assessment for circumstances evaluated by these methods.

Numerous other refinements in exposure assessment are also being incorporated into the risk assessment process--for example, the use of human biological data to assist in exposure estimation, better descriptions of life style for subpopulation groups, the use of statistical methods to describe likely exposure below detectable limits, and the use of pharmacokinetic data to describe the actual dose to target tissue. These developments rely on advancing research in multiple disciplines for use in the practical consideration of human exposure.

RISK MANAGEMENT

Performing the risk assessment as a separate step from the risk management decision ensures that the assumptions necessary for dealing with scientific uncertainty in the risk assessment process are not biased by the desired risk management outcome (EPA 1976; NAS 1982, 1983).

Factors considered in deciding how much health risk to accept inevitably include considerations of social benefits and costs, economic cost associated with control measures, and available technology for control. The relative weight assigned to each

TABLE 6. COMPARISON OF UPPER-BOUND RISKS ASSOCIATED WITH AMBIENT EXPOSURE TO CARCINOGENIC AIR POLLUTANTS[a,b]

Chemical	Probability [c]	Exposed[d]	Deaths/Year[f]
Relative Individual Risks			
Arsenic	2×10^{-3}	44,000	1
Benzene	2×10^{-4}	55,000	0.1
Coke ovens	6×10^{-3}	1,800	0.2
Vinyl chloride[e]			
Before regulation	4×10^{-3}	34,000	1.9
After regulation	2×10^{-4}	34,000	0.1
Lifetime Population Risks			
Arsenic	4×10^{-5}	25 million	16
Benzene	3×10^{-5}	220 million	78
Coke ovens	7×10^{-4}	15 million	150
Vinyl chloride[e]			
Before regulation	2×10^{-4}	5 million	20
After regulation	1×10^{-5}	5 million	1

[a] From EPA, Carcinogen Assessment Group Reports 1976-1981. These estimates may change as additional data become available.
[b] All risks are before regulation unless otherwise indicated.
[c] Upper-bound lifetime probability of cancer death from maximum exposure near stationary sources. The significant figures presented do not indicate precision or accuracy; they are included to make it easier to trace the derivation of these numbers through the various extrapolation and mathematical calculations.
[d] Total number in populations exposed to ambient levels of chemical listed. Exposure is from stationary sources.
[e] If risks were based on the incidence of mammary tumors in the animal bioassay studies, the results would be four times higher.
[f] Cancer deaths. Total Number at Upper Bound in U.S. from Chemical in Air

of the factors varies under different U.S. environmental regulations. For example, the Clean Air Act calls for setting national ambient air quality standards based on health protection alone, without regard to social or economic considerations (e.g., standards for sulfur oxides, nitrogen oxides, and lead). The Federal Insecticide, Fungicide, and Rodenticide Act (FIFRA) calls for balancing the risks of use versus the benefits, and the Drinking Water Act calls for basing final standards, maximum contaminant levels (MCLs), on the feasibility of control. The role of risk assessment in the risk management process, therefore, varies according to different legal constraints and practical circumstances.

APPLICATIONS OF RISK ASSESSMENT TO PUBLIC POLICY DECISIONS

Common to all risk assessments is the adoption of an organized approach in the face of uncertainty and the formulation of the most credible projections possible as a basis for management decisions. Most experience with risk assessment approaches in public health protection has been in the areas of radiation and chemical carcinogenesis. Certainly, these same approaches are amenable to, and are being applied to, other kinds of health effects data. For example, for acutely toxic effects that require rapid response in the case of an accidental release, risk assessment approaches may be used to identify the health effects of concern and the dose of concern. By coupling this information with physical and chemical parameters, the area surrounding the release that requires evacuation may be defined (EPA 1985b).

In the ecological area, consistent guidance for assessing risk is badly needed to achieve more organized approaches to evaluating these data to guide both national and international decision making for a host of problems, such as acid rain (NAS 1986).

Applications with Emphasis on Carcinogens

Quantitative risk assessment, together with qualitative assessment of the biomedical evidence, has been used in five distinct situations in the United States--to decide public policy to set priorities, to review residual risk after application of the best available technology to see if anything more should be done, to balance risks against benefits, to set standards and target levels of risk, and to provide information about the urgency of situations where population subgroups are inadvertently exposed to toxic agents, e.g., populations near uncontrolled waste sites. Several examples are discussed below.

Setting Priorities

Under provisions of the Clear Air Act, EPA must list hazardous air pollutants and regulate sources of such air pollutants as necessary. In order to set priorities for reviewing hundreds of agents that may be potential air pollutants, EPA's Office of Air Programs identified three groups of potentially toxic chemicals suspected of being present in the ambient air at levels of concern because of their use patterns (Table 3). The highest priority for performing in-depth health evaluations was given to Group I, then to Group II, and finally to Group III. These priorities reflected judgments from the Office of Air Programs regarding those chemicals that, based on preliminary information for likely exposure and possible toxicity, might present the greatest hazard to humans from air pollution. The Carcinogen Assessment Group (CAG), one of the health subgroups in EPA's Office of Health and Environmental Assessment, provided a qualitative weight-of-evidence statement and an index of potency for chemicals selected from Groups I and II. The index of potency is expressed as an upper-bound unit risk estimate, with the unit risk estimate defined as the incremental excess lifetime risk for a 70-kg individual breathing air containing 1 $\mu g/m^3$ of the chemical of interest, for a 70-yr life span. The data are presented in Table 4. Notice that the potency index, expressed as unit risk, ranges a millionfold, and that chemicals having the strongest biomedical evidence for carcinogenicity based on responses in humans may have relatively low potencies. For example, vinyl chloride has a unit risk of 10^{-6} and benzene has a unit risk of 10^{-6}. Strong evidence of carcinogenicity need not also mean high potency. In the absence of information regarding potency, regulators are inclined to regulate human carcinogens more severely than animal carcinogens, even though some human carcinogens appear to be relatively much less potent than some chemicals whose carcinogenic effect has only been demonstrated in animal studies.

While there are many chemicals of concern, some pose a greater risk than others. The weight-of-evidence for carcinogenicity, the unit risk estimate as a measure of potency, and information concerning exposure levels have all provided a basis for selecting the most hazardous air pollutants for further study and possible regulation, and for disregarding others.

TABLE 7.   UPPER-BOUND RISK ESTIMATES FOR POPULATION EXPOSURE TO SUSPECTED CARCINOGENIC PESTICIDES[a]

Pesticide	Exposed Population	Probability[g]	Death/yr[h]
Chlorobenzilate	220 million--citrus consumption	2×10^{-6}	7
	--citrus applicators[c]	4×10^{-4}	-
		1×10^{-3}	-
Amitraz (BAAM)	220 million--apple consumption	3×10^{-6}	8
	--pear consumption	2×10^{-6}	6
	1,400 applicators--spraying apples	1×10^{-4}	0.002
	1,550 applicators--spraying pears	6×10^{-5}	0.001
	1,600 applicators--spraying pears	1×10^{-4}	0.003
Chlordane/ heptachlor	220 million	2×10^{-4} [d]	500 [b]
		5×10^{-5} [e]	150 [d]

[a]From the EPA Carcinogen Assessment Group Reports 1976-1981. These estimates may change as additional data become available. [c]The significant figures presented do not indicate precision or accuracy; they are included to make it easier to trace the derivation of these numbers through the various extrapolation and mathematical calculations. [d]The total number of applicators was not included in the study. [e]Based on total tumors. [f]Based on large carcinomas. [g]Upper bound probability of cancer death from exposure. [h]Number of expected cancer deaths per year at the upper bound.

After an agent has been listed as a hazardous air pollutant, EPA must decide which sources to regulate first, and indeed, whether regulation is warranted at all. Table 5 presents a comparison of data for different source categories contributing arsenic to the ambient air. The upper-bound risk estimates to population subgroups and the related upper-bound nationwide impacts always rely on estimates of exposures, which also have great uncertainties. Uncertainties must always be included in the exposure assessment and taken into account when using risk assessment information. For example, where estimates of exposure are highly uncertain it may be possible to present a range for the exposure. Risk estimates based on this range can be instructive, particularly when either the upper end of the range provides low estimates or, conversely, when the lower end of the range still suggests possible associated high risks.

Residual Risk

The next example of the use of risk management is in examining residual risk. Quantitative risk assessment was used to compare residual risk, after application of the best available technology to control ambient levels of vinyl chloride monomer, with the risk associated with other potentially hazardous air pollutants that had not yet been regulated (see Table 6). The risk assessment information indicated that reductions in risk had been considerable after regulation for vinyl chloride, and that the remaining risk was low relative to risks associated with the other air pollutants, namely, the individual risks for arsenic and benzene and the nationwide impacts for arsenic and benzene (Table 4. Relative individual risks and lifetime population risks are generally both considered in making regulatory decisions. Based on such analysis, EPA's Office of Air Quality Programs allocated agency resources to consider other air pollutants and not to consider additional reductions in risks associated with vinyl chloride emissions. To date, vinyl chloride has not been further regulated. The agency has agreed, however, to review the regulation of these emissions periodically.

Balancing Risk and Benefits

Many decisions involving the balancing of risks and benefits under EPA's pesticide registration authorities have relied on risk assessment. Table 7 presents the quantitative risk estimates associated with three pesticides on which registration decisions have been made. In the case of chlorobenzilate, which is used on citrus fruit, the weight-of-evidence for carcinogenic potential is based on responses in the liver of both male and female mice; studies in rats were negative (EPA 1978). There is considerable disagreement among some scientists regarding the appropriate weight to be given to such responses. Nevertheless, based on the assumption that chlorobenzilate is a human carcinogen, quantitative risk estimates indicate that risk associated with exposure to the general population is relatively low--on the order of one chance in a million of increased risk--and the annual cancer rate on a nationwide basis is also relatively low. However, the risk to applicators of the pesticide was higher by two orders of magnitude. Since FIFRA requires the balancing of risks and benefits, the presence of increased risk to applicators was evaluated in view of the fact that no substitute exists for chlorobenzilate on citrus. EPA decided that the risks did not outweigh the benefits and therefore retained the registration of chlorobenzilate for use on citrus. The agency added stringent requirements for the label on chlorobenzilate, urging applicators to take further protective measures.

The next example in Table 7 involved the application of risk assessment to the registration of the new pesticide BAAM for use on pears and apples. The only carcinogenesis bioassay performed on BAAM provided very weak evidence of carcinogenic activity. In the absence of additional data but on the assumption that this one test reflected true biological activity, a quantitative risk assessment was performed. The calculated upper-bound risk estimates indicated on the order of one chance in a million of increased risk, a relatively low projected risk to the U.S. population as a whole. Balancing risks against benefits, EPA made a decision to permit a three-year temporary registration of BAAM for use on pears but not on apples, because substitutes were not available in the former case but were available in the latter, and require submission of

TABLE 8. UPPER-BOUND PROJECTED LIFETIME CANCER RISK BASED ON ONE-HIT MODEL FROM NTA EXPOSURE-RESPONSE[a,b]

Exposure	Number	Levels[c]	Risk[e]	Cases/yr[e]
Public drinking water	220 million	8×10^{-5}	4×10^{-7}	1
Range		7×10^{-4}	3×10^{-6}	10
Mean		4×10^{-5}	2×10^{-7}	1
Private wells	66 million	up to 0.1^d	$4 \times 10^{-4\ d}$	370
General consumers				
Laundry	125 million	2×10^{-4}	1×10^{-6}	2
Dishwashing	125 million	2×10^{-4}	1×10^{-6}	2
Residue on unwashed dishes	2 million	0.01	6×10^{-5}	2
Workers				
Manufacture	100	1×10^{-3}	6×10^{-6}	
		7×10^{-3}	3×10^{-5}	
Formulations	1,750	5×10^{-3}	2×10^{-5}	0.001
		5×10^{-2}	5×10^{-4}	

----

[a]From the EPA Office of Toxic Substances, Draft Report, 1979. [b]The significant figures presented do not indicate precision or accuracy; they are included to make it easier to trace the derivation of these numbers through the various extrapolation and mathematical calculations. [c]Projected U.S. exposures based on Canadian monitoring data; mg/kg/day. [d]Insufficient data, based on only 21 samples, of which only one was significantly contaminated. [e]Upper bound.

more definitive data before granting a permanent registration for any uses. This example demonstrates how time and effort can be put to good use when guided by quantitative risk assessment.

In the third example in Table 7, risk assessment was used to balance risks and benefits for registered uses of chlordane/heptachlor. These chemicals are classified as probable human carcinogens (B2), based on liver carcinomas in a series of bioassay studies in the mouse and rat (IRIS 1991). These chemicals bioaccumulate, and most humans carry a body burden of the chemicals in adipose tissue. Application of quantitative risk assessment indicated risks at least an order of magnitude higher than the previous two cases presented in Table 7; considerable potential nationwide impacts were also projected. The decision in this case was to cancel most uses of chlordane/heptachlor with the exception of underground applications for termite control, because good substitutes were not available at the time and potential exposures to humans were estimated to be less.

In the last example of the use of risk assessment to balance risks and benefits, data are presented in Table 8 for projected risks associated with the resumed manufacture of nitrilotriacetic acid (NTA) in the United States. The manufacture of NTA was voluntarily suspended in the early 1970s because of the early indications of animal bioassay studies that NTA might be a carcinogen. This risk assessment was done because the manufacturer asked EPA for guidance as to whether EPA would regulate NTA if manufacturing resumed. NTA can be used in detergents to replace phosphates, which contribute to the eutrophication of bodies of water. The benefits associated with using NTA would be a potential reduction of the concentration of phosphates in water. The risk estimates presented in Table 9 are based on monitoring data from Canada, where NTA has been in continuous use for a number of years. With the exception of private wells, where only 21 samples have been monitored, potential cancer risks from the Canadian data indicated low projected risks for populations in the United States. Although questions were raised about the applicability of the Canadian exposure data to projected exposures in the United States, the decision not to regulate the resumed manufacture of NTA cited these relatively low risk estimates as the reason.

<u>Setting Target Levels of Risk</u>

In this example (Table 9), EPA was obligated to recommend nationwide water quality criteria for a large number of chemicals, including suspected carcinogens (EPA 1980a). The statute under which these criteria were issued, the Federal Water Pollution Control Act, required that water quality criteria be published by the Agency to protect the public health; no provisions were included in this section of the statute to incorporate social and economic factors in setting water quality criteria. Since threshold concentrations could not be established for suspected carcinogens, quantitative risk assessment was used to recommend concentrations in water associated with lifetime risks of 10^{-7} to 10^{-5}. Concentrations that would result in such risks were calculated by assuming an ingestion of 2 liters of drinking water per day and an average consumption of fish of 6.5 grams per day (edible portion). The upper bound slope of the dose-response curve for each chemical was calculated, as was the concentration in water for each chemical that would correspond to a cancer risk level of 10^{-5} at the upper bound. In calculating these data, an extrapolation must be used to estimate the dose of carcinogen needed to cause cancer. In the proposed criteria, the data were calculated using the one-hit model. In response to public comments, the agency reviewed alternative models and decided to adopt the Crump linearized, multistage model in order to make full use of all of the data points (EPA 1980a; and see Armitage 1985, Moolgavkar 1986, Thorslund 1987, and Thorslund et al. 1987a,b for discussions of biological models of carcinogenesis). The upper-bound slope and the individual chemical concentrations are presented in Table 10 using the multistage model and one-hit model (EPA 1979a-c). Both values have been included so that the relative slopes and concentrations can be compared. From these comparisons it is evident that the data derived from the one-hit model and from the multistage model are very close for most cases. The weight of the biomedical evidence varies enormously for the chemicals presented in Table 10; this information should be taken into consideration in applying these target concentrations during the risk management process.

CONCLUSIONS AND FUTURE DIRECTIONS

These examples illustrate the applications of quantitative risk assessment in a variety of practical circumstances to provide information regarding risk as a basis for making public health policy decisions in the United States. These policy decisions did not hinge

TABLE 9. GUIDANCE FOR WATER QUALITY CRITERIA UPPER BOUND CALCULATIONS WITH A LOWER BOUND APPROACHING ZERO[a].

	B_H [b]	Water Concentrations [b,c]
Acrylonitrile	0.6 (2.0)	0.6 (0.08)
Aldrin	11.4 (6.3)	7.4×10^{-4} (5×10^{-5})
Arsenic[d]	14.0	0.02
Asbestos	--	300,000 (fibers/L) (0.05)
Benzene[d]	0.1	7
Benzidine[d]	234.1	1×10^{-3}
Beryllium	4.9 (3.4)	0.1 (0.1)
Carbon tetrachloride	0.1 (0.1)	4 (3)
Chloroform	0.2 (0.2)	2 (2)
Chlordane	1.6 (5.4)	5×10^{-3} (1×10^{-3})
Chloroalkyl ethers		
BCME	9,300 (13,600)	4×10^{-5} (2×10^{-5})
BCEE	1.1 (0.7)	0.3 (0.4)
Chlorinated benzenes (HCB)	1.7 (2.5)	7×10^{-3} (1×10^{-3})
Chlorinated ethanes		
1,2-di-	0.04 (0.05)	9 (7)
1,1,2-tri-	0.1 (0.1)	6 (3)
1,1,2,2-tetra-	0.2 (0.2)	2 (2)
Hexa-	0.01 (0.02)	19 (6)
Dichlorobenzidine	2 (2)	0.1 (0.02)
DDT	8 (18)	2×10^{-4} (4×10^{-4})
Dichloroethylenes (1,1-)	1 (0.3)	0.3 (1)
Dieldrin	30 (180)	7×10^{-4} (4×10^{-4})
Dinitrotoluene	0.3 (0.4)	1 (0.1)
Dioxins (2,3,7,8-tetrachloro-)	4×10^{5} (1×10^{6})	2×10^{-9} (5×10^{-7})
Diphenylhydrazine dioxin	0.8 (0.7)	0.4 (0.4)
Halomethanes	Same as chloroform	
Heptachlor	3 (30)	3×10^{-3} (2×10^{-4})
Hexachlorobutadiene	0.008 (0.05)	5 (1)
Hexachlorocyclohexane		
Technical grade	5 (2) (0.05)	0.1 (0.02)
Alpha isomer	11 (3)	0.02 (0.02)
Beta isomer	2 (2)	0.2 (0.03)
Gamma isomer	1 (1)	0.2 (0.1)
Nitrosamines		
DMNA	26 (13)	1×10^{-2} (3×10^{-2})
DENA	44 (38)	8×10^{-3} (9×10^{-3})
DBNA	5 (27)	0.1 (0.01)
N-N-P	2 (4)	0.2 (0.1)
PAH	12 (28)	3×10^{-2} (10×10^{-3})
PCBs	4 (3)	8×10^{-4} (3×10^{-4})
Tetrachloroethylene	0.04 (0.1)	8x(2.0)
Trichlorethylene	0.01 (0.01)	27 (21)
Toxaphene	1 (4)	7×10^{-3} (5×10^{-4})
Vinyl chloride[d]	0.02	20

[a] *Fed. Reg.*, Vol. 45, No. 231, November 1980. [b] The values in parentheses were calculated on the one-hit model. Other values were calculated based on the multistage model. [c] Assuming a lifetime daily consumption of 2 liters of water and 0.0065 kg fish. Note that a daily consumption of 0.0187 kg fish was assumed in the calculation using the one-hit model and that some of the bioconcentration factors are different in the calculation using the multistage model. [d] Scope determined from epidemiological data. [e] Upper-bound slope in $(mg/kg/day)^{-1}$. [f] Water concentration in $\mu g/L$ corresponding to 10^{-5} increment of risk at the upper bound.

on any "acceptable level" of risk; each decision reflected achievability in some measure. Nevertheless, most risk management decisions regulated exposures so that individual lifetime cancer risks were near 10^{-5} at the upper bound. There were some circumstances in which this level of risk was not achievable, such as in setting haloform standards for drinking water (EPA 1978). Such decisions, in which risks higher than 10^{-5} were accepted, generally were justified on grounds of social and economic tradeoffs, such as the protective value of chlorination of the drinking water supply to prevent infection. Risks to the population as a whole that were lower than 10^{-5} were generally unregulated, as exemplified by the risk management decisions for NTA, vinyl chloride, and chlorobenzilate. Exceptions for risks lower than 10^{-5} include the voluntary cancellation of safrole as a dog repellent (risk of 10^{-7}) and the recommendation of water quality criteria associated with risks ranging from 10^{-7} to 10^{-5} (EPA 1980). In a large sample of risk assessments for different chemicals in various exposure situations, the upper-bound risks fell into a relatively low risk range of 10^{-5} for about 80-90% of the cases studied.

Uncertainties in exposure estimates, and other uncertainties inherent in the extrapolation process, need to be taken into account on a case-by-case basis. Despite thesedeficiencies, the use of upper-bound estimates to identify those cases where the risks may be so low, even at the upper bound, as to fall into a low-priority category for regulatory consideration, has helped regulators focus attention on more compelling public health problems.

The process of risk assessment is undergoing continual refinement and improvement. Considerable effort is now concentrated on improving the methods of dose and exposure assessment, centering on advances in molecular biology and our increasing ability to detect low levels of toxic chemicals and/or their metabolic products in human tissues and fluids. Biologically motivated mathematical models are under development to reduce the uncertainty concerning the upper-bound cancer risk at low doses. Such efforts are necessary so that we do not "over regulate" chemicals where the true risks are significantly below the upper bound (Thorslund et al. 1987a).

The risk analysis process begins with the generation of data on which to base the scientific analysis of health risk estimation, both in qualitative and quantitative terms. The management of these projected risks involves considerations of legal and general social and economic factors in deciding how to manage risk. Finally, risk communication and risk acceptance are essential concerns in the risk analysis process because the affected public is increasingly playing an active role in public health decisions. The complex area of risk communication is an interdisciplinary area that involves a spectrum of elements, including communication of scientific data, the effects of news communication, risk perceptions, risk and compensation tradeoffs, psychological factors of risk and control, and communication and arbitration processes. Historically these areas have been pursued by researchers working in the specialized areas mentioned above.

REFERENCES

Albert, R. E., Train, R. E., and Anderson, E. L. 1977. Rationale developed by the Environmental Protection Agency for the assessment of carcinogenic risks. J. Natl. Cancer Inst. 58:1537-1541.

Andersen, M. E., Gargas, M. L., Jones, R. A., and Jenkins, L. J., Jr. 1980. Determination of the kinetic constants for metabolism of inhaled toxicants *in vivo* using gas uptake measurements. Toxicol. Appl. Pharmacol. 54:100-116.

Andersen, M. E. 1981a. A physiologically based toxicokinetic description of the metabolism of inhaled gases and vapors: Analysis at steady state. Toxicol. Appl. Pharmacol. 60:509-526.

Andersen, M. E. 1981b. Saturable metabolism and its relationship to toxicity. CRC Crit. Rev. Toxicol. 9:105-150.

Andersen, M. E., Gargas, M. L., and Ramsey, J. C. 1984. Inhalation pharmacokinetics: Evaluating systemic extraction, total *in vivo* metabolism, and the time course of enzyme induction for inhaled styrene in rats based on arterial blood: Inhaled air concentration ratios. Toxicol. Appl. Pharmacol. 73:176-187.

Anderson, E. L., and Carcinogen Assessment Group (CAG) of the Environmental Protection Agency. 1983. Quantitative approaches in use to assess cancer risk. Risk Anal. 3:277-295.

Armitage, P. 1985. Multistage models of carcinogenesis. Environ. Health Perspect. 63:195-201.

Cline, P.V., and D.R. Viste. 1984. Migration and Degradation Patterns of Volatile Organic Compounds. National Conference of Uncontrolled Hazardous Waste Sites Proceedings. P. 217.

Crump, K. S. 1981. Investigation of the Use of Pharmacokinetic Data in Carcinogenic Risk Assessment. (Submitted to Environmental Protection Agency).

Crump, K. S. 1987. Investigation of the Incorporation of Pharmacokinetic Data for Selected Volatile Organic Chemicals into Quantitative Risk Assessment Methodology, (Submitted to State of New Jersey).

Domenico, P.A., and V.V. Palciauskas. 1982. Alternative Boundaries in Solid Waste Management. Groundwater 20:303.

Environmental Protection Agency (EPA). 1976. Interim procedures and guidelines for health risks and economic impact assessments of suspected carcinogens. Fed. Reg. 41:21402.

Environmental Protection Agency (EPA). 1978. Summary and Conclusions for Assessment of Carcinogenic Risk of Chlorobenzilate. Carcinogen Assessment Group (CAG). (Unpublished).

Environmental Protection Agency (EPA). 1979a. Water quality criteria: Request for comments (proposed). Fed. Reg. 44:15926-15981.

Environmental Protection Agency (EPA). 1979b. Water quality criteria: Availability (proposed). *Fed. Reg.* 44:43660-43697.

Environmental Protection Agency (EPA). 1979c. Water quality criteria: Availability (proposed). *Fed. Reg.* 44:56628-56656.

Environmental Protection Agency (EPA). 1980. Water quality criteria documents: Availability. Fed. Reg. 45:79316-79379. Environmental Protection Agency (EPA). 1984a. Proposed guidelines for carcinogen risk assessment. Fed. Reg. 49(227):46294-46301.

Environmental Protection Agency (EPA). 1984b. Proposed guidelines for the health assessment of suspect developmental toxicants. Fed. Reg. 49:46324-46331.

Environmental Protection Agency (EPA). 1984c. Proposed guidelines for mutagenicity risk assessment. Fed. Reg. 49:46312-46321.

Environmental Protection Agency (EPA). 1984d. Proposed guidelines for exposure assessment. Fed. Reg. 49:46304-46312.

Environmental Protection Agency (EPA). 1985. Final Exclusions and Final Vertical and Horizontal Spread Model (VHS). Fed. Reg. 50:229.

Environmental Protection Agency (EPA). 1985a. Proposed guidelines for the health risk assessment of chemical mixtures. *Fed. Reg.* 50:1170-1176.

Environmental Protection Agency (EPA). 1985b. Chemical Emergency Preparedness Program, Interim Guidance, Washington, D.C.

Environmental Protection Agency (EPA). 1986a. Guidelines for carcinogen risk assessment, Fed. Reg.51:33992.

Environmental Protection Agency (EPA). 1986b. Guidelines for health assessment of suspect developmental toxicants. Fed. Reg. 51:34028.

Environmental Protection Agency (EPA). 1986c. Guidelines for mutagenicity risk assessment. *Fed. Reg.*51:34006.

Environmental Protection Agency (EPA). 1986d. Guidelines for estimating exposure. Fed. Reg. 51:34042.

Environmental Protection Agency (EPA). 1986e. Guidelines for the health risk assessment of chemical mixtures. Fed. Reg. 51:34014.

Environmental Protection Agency (EPA). 1991. Public meeting on EPA's scientific assessment of dioxin. U.S. Environmental Protection Agency. Fed. Reg. 56:50903-50904.

Foster, S., and P. Chrostowski. 1987. Inhalation exposures to volatile organic contaminants in the shower. Presented at 80th annual meeting of the Air Pollution Control Association, New York, June 21.

HEAST. 1991. Health Effects Assessment Summary Tables. U.S. Environmental Protection Agency. Office of Research and Development, Washington, D.C.

Interagency Regulatory Liaison Group (IRLG). 1979. Scientific bases for identification of potential carcinogens and estimation of risks. *J. Natl. Cancer Inst.* 63:244-268.

IRIS. 1991. Integrated Risk Information System. U.S. Environmental Protection Agency. Office of Health and Environmental Assessment, Environmental Criteria and Assessment Office, Cincinnati, Ohio.

Levine, T., A. Rispin, C.S. Scott, W. Marcus, C. Chen, and H. Libb. 1987. Special Report on Ingested Inorganic Arsenic: Skin Cancer; Nutritional Essentiality. Draft for U.S. Environmental Protection Agency Science Advisory Board Review.

Moolgavkar, S. H. 1986. Carcinogenesis modeling: From molecular biology to epidemiology. Ann. Rev. Pub. Health 7:151-169.

Moolgavkar, S.H., and A.G. Knudson. 1981. Mutation and cancer: A model for human carcinogenesis. J. Natl. Cancer Inst. 66:1037.

Moore, J.A. 1987. Recommended Agency Policy on the Carcinogenicity Risk Associated with the Ingestion of Inorganic Arsenic--Action Memorandum. U.S. Environmental Protection Agency. Office of Pesticides and Toxic Substances.

National Academy of Sciences (NAS). 1982. Risk and Decision Making: Perspectives and Research, National Research Council Committee on Risk and Decision Making, National Academy of Sciences Press, Washington, D.C.

National Academy of Sciences/National Research Council (NAS/NRC). 1983. Risk Assessment in the Federal Government: Managing the Process. Prepared by the Committee on the Institutional Means for Assessment of Risk to Public Health, Commission on Life Sciences (National Academy Press, Washington, D.C.).

National Academy of Sciences (NAS). 1986. Ecological Knowledge and Environmental Problem Solving: Concepts and Case Studies, National Research Council Committee on Risk and Decision Making, National Academy of Sciences Press, Washington, D.C.

Office of Science and Technology Policy (OSTP). 1984. Chemical carcinogens: Notice of review of the science and its associated principles. *Fed. Reg* 49:21594-21661.

Parsons, F., Wood, P.R., and DeMarco, J. 1984. Transformation of tetrachloroethene and trichloroethane in microcosms and groundwater. Res. Technol. P. 56.

Patrick, D., and Peters, W.D. 1985. Exposure assessment in setting air pollution regulations: ASARCO, Tacoma, a case study. Presented at the Society for Risk Analysis annual meeting. Washington, D.C.

Sawhney, B.L., Pignatello, J.J., and Steinberg, S.M. 1988. Determination of 1,2-dibromoethane (EDB) in field soils: Implications for volatile organic compounds. J. Environ. Qual. 17:149.

Sehmel, G.A., and Hodgson, W.H. 1979. A Model for Predicting Dry Deposition of Particles and Gases to Environmental Surfaces. Prepared for the U.S. Department of Energy by Pacific Northwest Laboratory. October. PNL-SA-6271-REV 1.

Thorslund, T. W. 1987. Quantitative Dose-Response Model for the Tumor Promoting Activity of TCDD. (Submitted to Environmental Protection Agency).

Thorslund, T. W., Brown, C. C., and Charnley, G. 1987a. The use of biologically motivated mathematical models to predict the actual cancer risk associated with environmental exposure to a carcinogen. J. Risk Anal. 7:109-119.

Thorslund, T. W., Charnley, G., Bayard, S., and Brown, R. 1987b. Quantitative Model for the Tumor Promoting Activity of 2,3,7,8-TCDD. (Presented at the Seventh International Symposium on Chlorinated Dioxins and Related Compounds, Las Vegas, Nevada, October 4-9, 1987.)

van den Berg, M., van Greevenbroek, M., Olie, K., Hatzinder, O. 1986. Bioavailability of PCDDs and PCDFs on fly ash after semi-chronic ingestion by rat. Chemosphere 15:509.

Voytek, P.E., and Thorslund, T.W. 1991. Benzene risk assessment: Status of quantifying the leukemogenic risk associated with the low-dose inhalation of benzene. Risk Anal. 11:355-357.

PHARMACODYNAMIC MODELS FOR CANCER RISK ASSESSMENT

Suresh H. Moolgavkar

Fred Hutchinson Cancer Research Center Division of Public Health
Sciences 1124 Columbia Street, MP-665 Seattle, Washington 98104

INTRODUCTION

Dose-response models for cancer risk assessment can be classified broadly into two categories. The first class of models to be considered included statistical models that were constructed as empirical descriptions of the dose-response curve in the observed range of data. While these models have good statistical properties and are generally easy to use, they are not based on knowledge of cancer mechanisms and are, thus, unsuited to addressing two of the fundamental problems of cancer risk assessment, namely the problems of low-dose and inter-species extrapolation. It appears reasonable to assume that incorporation of knowledge of cancer mechanisms into dose-response models will lead to more meaningful extrapolations. To that end, various attempts have been made to develop biologically-based dose-response models for cancer risk assessment. Success has been modest, at best. The problem is that, even if the broad features of the model of carcinogenesis are correct, slight mis-specification of the model can lead to large errors in the extrapolation. Thus, in my opinion, models of carcinogenesis have been more useful from the scientific point of view, i.e., in providing a framework within which the process of carcinogenesis can be viewed, than for risk assessment. Nevertheless, their use in risk assessment must be explored thoroughly because only through incorporation of sound biological principles into dose-response modelling is there any hope of a rational approach to the extrapolation problems.

Considerable evidence now indicates that the somatic mutation theory of carcinogenesis is correct. Further, it is being increasingly realized that cell replication and differentiation also play important roles in carcinogenesis (Cohen and Ellwein, 1990, Ames and Gold, 1990; Preston-Martin et al 1990; Moolgavkar and Knudson, 1981) - increase in cell division rates leads to an increase in mutation rates, and an increase in division rates relative to differentiation or death rates leads as well to an increase in the population of cells susceptible to malignant transformation. It is now widely believed that many of the so-called non-genotoxic carcinogens act by affecting cell division and differentiation rates. Thus, any biologically realistic model of carcinogenesis must consider not only mutations but also cell replication and differentiation. One such model is the two-mutation model for retinoblastoma proposed by Knudson in 1971. In fact, when cell replication kinetics are taken into consideration, a two-mutation model is the most parsimonious model consistent with both epidemiologic and experimental data (Moolgavkar and Knudson, 1981; Moolgavkar and Luebeck, 1990).

In this paper I would like to discuss two examples of the use of biologically- based models for data analysis. In the first example, I will apply various models to the analysis of colon cancer incidence in the general population and among subjects with Familial Adenomatous Polyposis (FAP), a dominantly inherited condition that leads to greatly increased risk of colon cancer. One of the questions I will address is whether the lessons learnt from retinoblastoma can be applied to colon cancer. In the second example, I will apply a two-mutation model to the analysis of experimental data on radon-induced lung tumors in rats. One of the intriguing features of exposure to high LET radiation, such as radon, is that fractionation of dose appears to confer an increased life-time risk of tumor. The two mutation model provides a biological explanation of this phenomenon, which I shall discuss. An immediate implication for risk assessment is that, when considering dose-response, it is not sufficient to consider total dose: pattern of exposure may be important as well. Both the examples that I discuss will emphasize the importance of cell proliferation kinetics in carcinogenesis.

A MODEL FOR RETINOBLASTOMA

Retinoblastoma is a rare tumor of the retina in children. It occurs in two forms in the population. One form is sporadic and occurs at the rate of about 3 per 100,000 children. The other form clusters in families, and appears to segregate in an autosomal dominant fashion on pedigree analysis. Inheritance of the gene for retinoblastoma increases the risk of the disease some 100,000 fold. This makes the gene for retinoblastoma the strongest known risk factor for cancer in humans. Some 95% of children who inherit the gene develop the disease by the age of 5 years. In 1971, based on a statistical study, Knudson proposed a two mutation model for retinoblastoma. According to this model, retinoblastoma results from two mutations. The first of these mutations is either germinal (in inherited cases of the disease) or somatic (in sporadic cases of the disease), whereas the second mutation is always somatic. Thus, in Knudson's model for retinoblastoma, inheritance of the susceptibility gene is equivalent to inheritance of one of the steps in multistage carcinogenesis. Can the same be said for carcinoma of the colon among polyposis cases? I will argue below that the answer is no, and that the gene for FAP increases the risk for colon cancer by a mechanism that is fundamentally different.

FAMILIAL ADENOMATOUS POLYPOSIS (FAP) AND COLON CANCER

Like retinoblastoma, colon cancer occurs in sporadic and dominantly inherited forms. One of the dominantly inherited conditions in which virtually 100% of affected individuals develop colon cancer is FAP. The gene for FAP has recently been mapped and cloned (Kinzler et al, 1991; Nishisho et al, 1991). How does the gene for FAP confer this high risk for colon cancer? One obvious possibility is that, as is the case for retinoblastoma, inheritance of the gene is equivalent to inheritance of one of the steps in multistage carcinogenesis. A careful examination of the incidence data does not support this hypothesis, however.

Data on the incidence of colon cancer among polyposis subjects is difficult to come by because a prophylactic colectomy is performed as soon as polyps first begin to appear. To my knowledge, there is only one data set, that compiled by Veale (Veale, 1965, Ashley, 1969), that reports incidence of colon cancer among polyposis subjects. I will present an analysis of this data set. I will also present an analysis of the incidence of sporadic colon cancer in Birmingham, U.K. I chose this population for the analysis because Veale's data were collected in London, and I wanted the populations in which the sporadic and inherited cases were studied to be as comparable as possible.

A simple computation, detailed elsewhere (Moolgavkar and Luebeck, 1992), shows that there are approximately 10^8-10^9 stem cells in the adult human colon. Further, each stem cell divides at most 100 times annually. I first fit the Armitage-Doll (1954) model to the age-specific incidence of colon cancer in the male population of Birmingham. This model predicts an approximate age-specific incidence curve of the form $I(t)=ct^k$, where c and k are constants, and t is age. The fit of this model to the age-specific incidence data in Birmingham leads to estimates of the parameters, $\hat{c}=4 \cdot 10^{-2}$ and $k=4.54$. Now, in the Armitage-Doll model $k=n-1$, where n is the number of stages required for malignant transformation. Thus, according to the Armitage-doll model, either 5 or 6 rate-limiting steps are required for the genesis of colon cancer. From the estimate of c, and assuming that the number of stem cells is 10^8 and that the mutation rates are equal, it is possible to estimate the common mutation rate. This estimate is $2.5 \cdot 10^{-3}$ per cell per year if there are 5 mutations on the pathway to colon cancer and $8.85 \cdot 10^{-3}$ per cell per year if there are 6 mutations on this pathway. Assuming that each stem cell divides 100 times annually leads to an estimate of $2.5 \cdot 10^{-5}$ per cell division if there are 5 mutations and $8.85 \cdot 10^{-5}$ if there are 6 mutations. In any case, these mutation rates appear to be far too high. Thus, if 5 or 6 mutations are involved in the genesis of colon cancer, then an early step must confer a mutator phenotype on the cell. Further, since the version of the Armitage-Doll model used here does not make allowance for differential cell growth in the intermediate compartments, the number of mutations derived from the model can be viewed as an upper bound on the actual number of mutations required for malignant transformation.

I fit two- and three-mutation models incorporating cell kinetics to the colon cancer incidence data as well. The two-mutation model that I used here has been widely discussed elsewhere (Moolgavkar and Knudson, 1981; Moolgavkar and Luebeck, 1990). The three-mutation model is a simple extension of the two-mutation model. Both models are shown in Figure 1, and both models described the data equally well (figure 2). Using the assumptions above, namely, that the mutation rates are equal and that each stem cell divides 100 times annually leads to an estimate, for the two-mutation model, of $3 \cdot 10^{-10}$ per cell generation for the common mutation rate. This mutation rate is much too low unless mutation at one of a few specific codons in the relevant gene is required. Such might be the case, for example, if gene activation were involved in colon carcinogenesis. However, there is little evidence of this in the laboratory data, and thus a two-mutation model appears to be implausible for colon cancer. The common estimate of the mutation rates for the three-mutation model, with the assumptions made above, is $4.7 \cdot 10^{-8}$ per cell generation. This estimate is in line with mutation rates measured in the

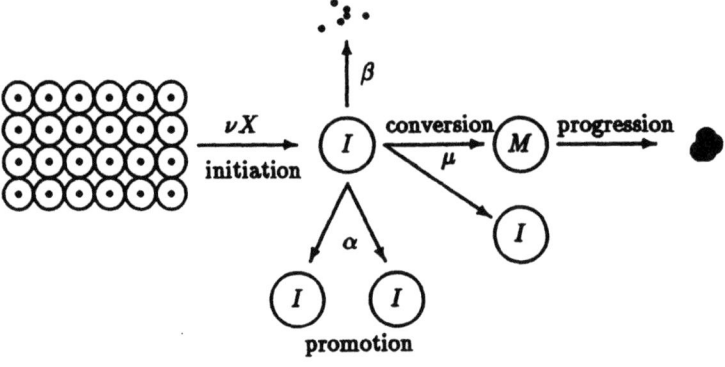

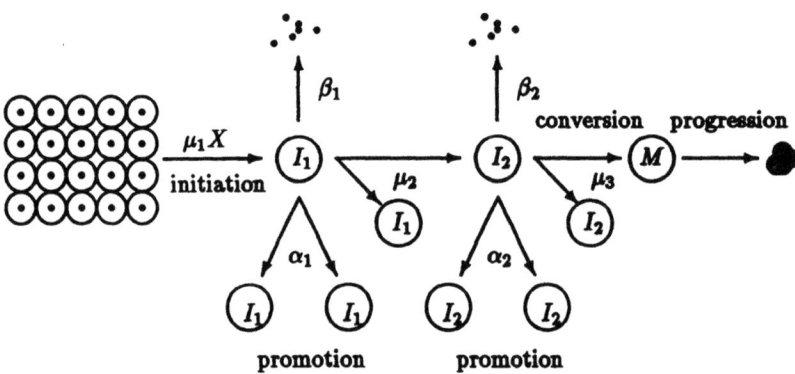

Fig. 1.<u>Upper Panel:</u> Two-mutation model for carcinogenesis. Rate of initiation depends upon the number of normal susceptible cells X and upon mutation rate u. Promotion results in clonal expansion of initiated cells with division rate a (per cell per year) and cell death (or differentiation) rate b (per cell per year). During cell division of initiated cells, one of the daughters may sustain the second mutation (with rate m per cell per year) and become malignant. Subsequent changes are rapid and lead to tumor progression and metastasis. <u>Lower Panel:</u> Three-mutation model for carcinogenesis. Note that, in this model, promotion results in clonal expansion of cells in both intermediate compartments. Rate of initiation depends upon X and the first mutation rate ml . Rates of cell division in the two intermediate compartments I_1 and I_2 are denoted by a_1 nd a_2. The rates of cell death or differentiation are b_1 and b_2. An I_1 cell may divide into one I_1 and one I_2 cell (i.e., sustain the second mutation) with rate m_2. Similarly an I_2 cell may sustain the third mutation during cell division with rate m_3, leading to a malignant daughter. Subsequent changes are rapid leading to tumor progression and metastasis. All rates are per cell per year as in the two-mutation model.

laboratory. Thus, the three-mutation model appears to be the most reasonable description of colon cancer incidence.

The difference, a-b, between the cell division and cell death rates is the net rate of cell division of intermediate cells. The estimate of this net rate of cell division is 0.107 per cell per year for cells in the intermediate compartment for the two-mutation model. With the three-mutation model, most of the differential growth appears to take place in the second intermediate compartment, and the estimate of $a_2 - b_2$ is 0.1 per cell per year.

I next fit the two- and three-mutation models to the incidence of colon cancer among subjects who had inherited the gene for FAP. Both models described the data well (table 1). I did not analyze the data using the Armitage-Doll model because cell replication kinetics are clearly important in FAP, and the Armitage-Doll model does not consider these. As can be seen from the parameter estimates in table 1, both the rates of mutation and the effective rates of intermediate cell division (a-b) are increased among polyposis subjects. The rate of cell division is known to be elevated in the colons of polyposis subjects (Lipkin, 1988), and the increase in mutation rates may be attributable entirely to this fact. Further, the estimated mutation rates with the two-mutation model still appear to be too low. The major effect of inheriting the gene for polyposis appears to be a substantial increase in the rate of effective intermediate cell division. For example, the effective intermediate cell division rate estimated from the three mutation model is approximately 3 times as high in polyposis subjects as in individuals without the polyposis gene. This increase in effective cell division rate confers a substantially increased risk of colon cancer among polyposis subjects.

How does inheritance of the polyposis gene increase the risk of colon cancer? I have argued above that the polyposis gene probably acts like a 'promoter' gene, i.e., increases the rate of effective intermediate cell division, and thus leads to clonal expansion of intermediate cell populations. What about the retinoblastoma paradigm? Is it possible that the inheritance of the polyposis gene is equivalent to inheriting one of the steps on the pathway to colon cancer?

I believe that it is possible to make a plausible argument against this being the case. The first thing to note is that the polyposis gene is much weaker than the retinoblastoma gene in conferring the cancer phenotype. Whereas 95% of retinoblastoma gene carriers develop the disease by the age of 5 years, 95% of polyposis subjects develop colon cancer by the age of 75 years (figure 2). This is all the more striking when one considers that the colon is a much larger organ than the retina, and that there are probably many more cell divisions in the colon in a single year than there are in the retina in an entire life-time (retinoblasts undergo terminal differentiation and stop dividing at about the age of 5 years). Another way to view this is to ask the following question. If inheritance of the polyposis gene is equivalent to inheriting one of the essential mutations on the pathway to colon cancer, then what would the age-specific prevalence of colon cancer be among polyposis subjects? The curve labelled 'P' in figure 2 is the hypothetical prevalence curve that would be expected if inheritance of the polyposis gene were equivalent to inheriting one of the mutations in a three-mutation model for colon cancer. Virtually 100% of subjects would be expected to develop colon cancer by the age of 30 years. A similar argument could be made if colon cancer were a two-mutation or a four-mutation tumor.

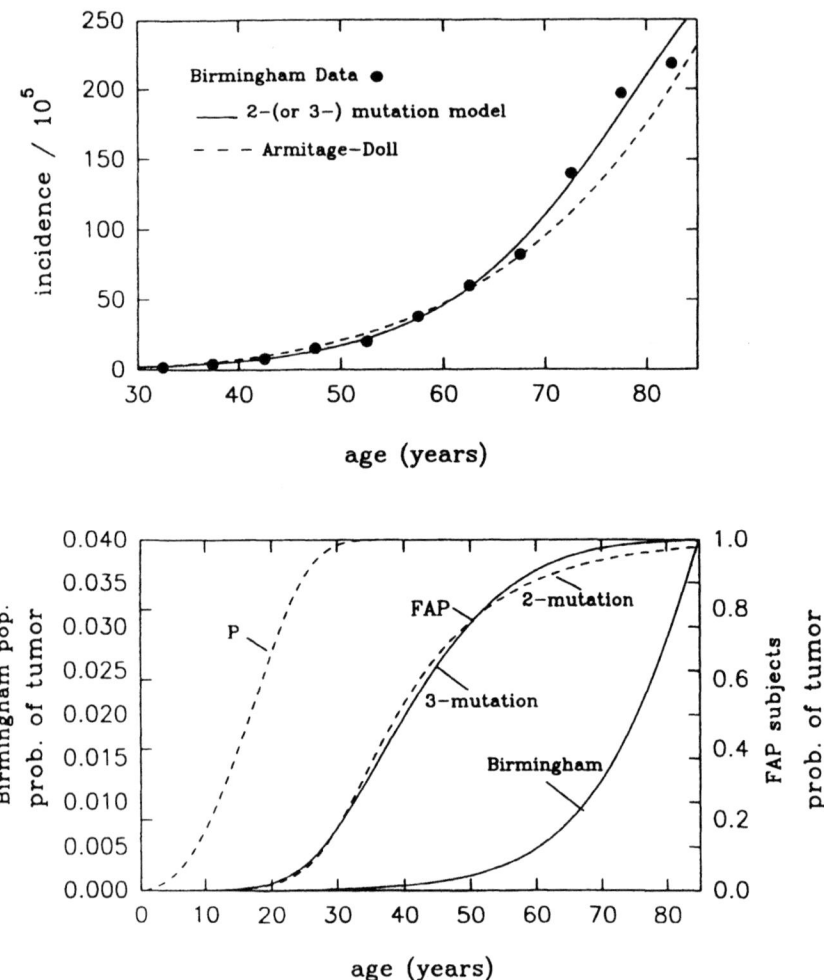

Fig. 2. Upper panel: Age-specific incidence of colon cancer among Birmingham males, and the age-specific incidence curves generated by the Armitage-Doll and the two- and three-mutation models. The curves generated by the latter two models are indistinguishable. Lower panel: Probability of tumor by a given age plotted against age for colon cancer among Birmingham males and polyposis subjects. Curves generated by the two- and three-mutation models. Note that the probability among Birmingham males, Prob(t), is related to the incidence I(t) shown in the upper panel by the relationship Prob(t)=1-exp(-h(s)ds), where h(s)=I(s)/10^5. The curve labelled P denotes the predicted probability of cancer among polyposis subjects if inheritance of the polyposis gene were equivalent to inheriting one of the necessary mutations in a three-mutation model.

I have argued in this section that cell replication kinetics play an important role in the genesis of colon cancer and in conferring the risk imposed by inheritance of the polyposis gene. There may be genes that increase the risk of cancer in the same way as promoters do: by increasing the population of critical target cells such as initiated cells.

RADON AND LUNG CANCER IN RATS

In this example I shall discuss the analysis of a large experimental data set on radon exposure and lung cancer in rats, with particular emphasis on the effect of fractionation of exposure on lung cancer risk. There are reports in the literature that for exposure to high LET radiation, fractionation of exposure increases the life-time risk of cancer. The analysis described here was undertaken to verify this finding, and to seek an explanation for it in biological terms. The experiment was performed by Dr. Fred Cross of Battelle Pacific Northwest Laboratories. Seventeen hundred and ninety-seven rats were divided

---

Table I. <u>Results of fitting the two- and three-mutation models to colon cancer incidence among polyposis subjects. The data were taken from the table in Ashley (12). The lower panel shows the observed number of colon cancer cases observed among polyposis subjects and the expected number generated by the two-mutation (A) and three-mutation (B) models.</u>

Two-Mutation Model:
    Estimated mutation rates = 4.5×10^{-9}/ cell generation
    Estimated a-b = 0.207
Three-Mutation Model:
    Estimated mutation rates = 2.6×10^{-7}/ cell generation
    Estimated $a_1 - b_1 = 0$
    Estimated $a_2 - b_2 = 0.307$

Age Group	Males Observed	Expected A	B	Females Observed	Expected A	B
0-9	0	0.4	0.1	1	0.3	0.9
11-14	1	0.9	0.6	0	0.7	0.4
20-24	6	5.9	6.8	2	1.7	1.7
15-19	0	2.4	2.4	5	3.9	4.5
25-29	14	11.4	12.3	8	6.9	7.5
30-34	17	15.2	14.5	10	6.2	9.4
35-39	8	14.2	13.4	7	9.8	9.3
40-44	14	12.1	12.2	6	8.3	8.4
45-49	6	6.2	7.1	10	6.2	7.1
50-54	3	3.6	4.7	3	2.4	3.1
55-59	2	2.4	3.5	1	1.1	1.5
60-64	1	1.6	2.5	0	0.3	0.6
65-69	1	1.0	1.9	1	0.3	0.6
70-74	2	0.7	1.3	-	-	-

into subgroups and exposed to several different regimens of radon. Within each total exposure group there were several different exposure rate regimens. For example, in the subgroup of rats exposed in total to 5000 WLM (working level months) of radon, there were two different exposure rate regimens: 500 WLM per week for 10 weeks and 50 WLM for 100 weeks. Details of the experiment and the analysis can be found in a recent paper (Moolgavkar et al., 1990).

The two-mutation model discussed in the previous section was fit to the data. There were 326 animals with histologically verified malignant lung tumors. The analysis revealed that radon exposure strongly increased the first mutation rate and also the rate of effective intermediate cell division. Further, the inverse exposure rate effect was confirmed (figure 3). The inverse exposure rate effect could be attributed to two facts. First, the first mutation rate is a sublinear function of the rate of exposure to radon; second, radon appears to have a promoting effect, i.e., it increases a-b, the net rate of intermediate cell division. Within the framework of any multistage model, promoting agents appear to have the property that exposure to them leads to an inverse dose (or exposure) rate effect.

In biological terms, why should the mutation rate be a sublinear function of high LET radiation? One plausible hypothesis is that repair is relatively more efficient at high than low doses of high LET radiation in the range of doses that do not result in cell killing. A small inhibition of cell differentiation could explain the promotional effect of high LET radiation.

DISCUSSION

The underlying theme of this paper is the importance of cell kinetics in the carcinogenic process. Incorporation of cell kinetics into models of carcinogenesis can lead to some surprising results. It has been known for some time now that incorporation of cell kinetics into models of carcinogenesis leads to a decrease in the number of stages estimated from incidence data. An unexpected result, however, is that cell kinetics may explain the inverse dose-rate effect of high LET radiation. In fact, one can make the prediction that exposure to promoting agents is characterized by the inverse dose-rate effect. Of course, as with any generalizations of this type, certain obvious qualifications are necessary. To be precise, the following statement can be made. Suppose that a-b, the net rate of intermediate cell proliferation, is a linear or sublinear function of the dose-rate of an agent, and if exposure begins early enough in life, then exposure to the agent will be characterized by an inverse dose-rate effect, i.e., for given total dose, a lower dose-rate will lead to a higher life-time probability of tumor. Although I have investigated this phenomenon only within the context of the two-mutation model, I believe that the general qualitative features remain unchanged if more than two mutations are involved in malignant transformation.

An immediate consequence of the observations I have made here is that quantitative cancer risk assessment is even trickier than has hitherto been believed. If pattern of exposure plays an important role in determining risk, then current risk assessment procedures, which consider only total exposure, are seriously flawed.

ACKNOWLEDGEMENTS

Supported by grant CA-47658 from NCI and contract DE-FG06-88GR60657 from DOE.

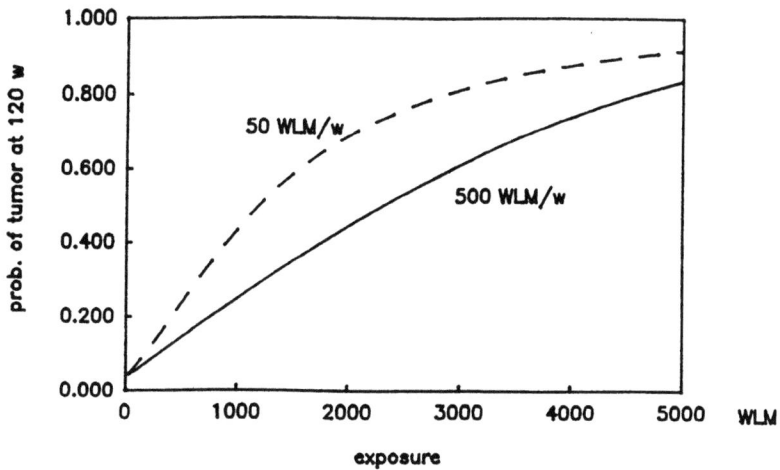

Fig. 3. Probability of malignant lung tumor plotted against total exposure to radon daughters in working level months (WLM) for two different exposure-rate regiments 50 WLM/week and 500 WLM/week. All exposures started at 14 weeks of age.

REFERENCES

Armitage, P., Doll, R. The age distribution of cancer and multi-stage theory of carcinogenesis. Br. J. Cancer, 8:1-12, 1954.
Ashley, D.J.B. Colonic Cancer arising in polyposis coli. J Med Gen 6:376-378, 1969.
Kinzler, L.W., Nilbert, M.C., Su, L.K., Vogelstein, B., et al. Identification of FAP locus genes from chromosome Sq21. Science, 253:661-664, 1991.
Knudson, A.G. Mutation and cancer: Statistical study of retinoblastoma. Proc Natl Acad Sci. USA 68:820-823, 1971.
Lipkin, M. Biomarkers of increased susceptibility to gastrointestinal cancer: New application to studies of cancer prevention in human subjects. Cancer Research 48:235-245, 1988.
Moolgavkar, S.H., Cross, F.T., Luebeck, G., Dagle, G.E. A two-mutation model for radon induced lung tumors in rats. Radiation Research 121:28-37, 1990.
Moolgavkar, S.H., Knudson, A.G. Mutation and cancer: A model for human carcinogenesis. J Natl Cancer Inst 66:1037-1052, 1981.
Moolgavkar, S.H., Luebeck, E.G. Multistage Carcinogenesis: Population-based model for colon cancer. J Natl Cancer Inst, 84: 610-618, 1992.
Moolgavkar, S.H., Luebeck, E.G. Two-event model for carcinogenesis: Biological, mathematical, and statistical considerations. Risk Analysis 10:323-341,1990.
Nishisho, I., Nakamura, Y., Miyoshi, Y., Miki, Y., et al. Mutations of chromosome Sq21 genes in FAP colorectal cancer patients. Science, 253:665-669, 1991.
Preston-Martin, S., Pike, M.C., Ross, R.K., Jones, P.A., Henderson, B.E. Increased cell division as a cause of human cancer. Cancer Res 50:7415-7421, 1990.
Veale, A.M.O. Intestinal Polyposis. Eugenics Laboratory Memoirs, No. 40, Cambridge University Press, Cambridge, 1965.

PREDICTIVE VALUE FOR CANCER RISK ASSESSMENT OF CELL- AND TISSUE- SPECIFIC FORMATION OF CARCINOGEN-DNA ADDUCTS

Ewalt Scherer

Division of Molecular Carcinogenesis, The Netherlands Cancer
Institute (Antoni van Leeuwenhoek Huis), 121 Plesmanlaan, 1066CX
Amsterdam, The Netherlands

INTRODUCTION

Carcinogenesis is a complex multistep/multistage process in which compounds of largely different working mechanisms result in a number of specific oncogenic events selected on the basis of the advantage they confer to the process of tumor development. For risk assessment of such (groups of) compounds it is important to elucidate the working mechanisms by which they contribute to the generation of the oncogenic events underlying the formation of tumor precursor stages, and their progression to tumors. Since most chemical carcinogens are also mutagens, and specific somatic mutations are frequently involved in tumor etiology, it is likely that the type and level of carcinogen-DNA adducts formed is related to the carcinogenic risk. In the present contribution I will discuss in how far this hypothesis holds.

CARCINOGEN-DNA ADDUCTS AND MUTAGENESIS[a]

Heterogeneity in DNA adduct formation and repair at both the cellular and genomic level makes it difficult, if not impossible, to deduce from overall DNA adduct formation the level of adducts in the critical target cell population at the critical gene location.

---

[a] Abbreviations: AAF=2-acetylaminofluorene; NMU=N-nitroso-N-methylurea; AFB1=Aflatoxin B_1; 7-meG=7-methylguanine; BOP=N-nitroso-N-bisoxopropylamine; O^6-meG=O^6-methylguanine; DEN=N-nitroso-N-diethylamine; DMN=N-nitroso-N-dimethylamine; NMBzA=N-nitroso-N-methylbenzylamine

Metabolic activation and DNA adducts

Although a few carcinogens can directly bind to DNA, most carcinogenic and genotoxic compounds require enzymatic metabolic activation to electrophilic species to form covalent bonds with electron-rich sites,[1,2] in DNA the nitrogen and oxygen atoms, certain unsaturated bonds of the bases and the phosphate groups of the sugar-phosphate backbone. Since each carcinogen generally leads to a multitude of DNA adducts,[2,4] it is difficult to establish which adducts are responsible for the induction of single base mutations, deletions or rearrangements of genetic material.

Table 1. Some typical point mutations induced by modified DNA bases[a]

Modified base	Mutation pattern		
O^6-methylG	GC->AT		
O^6-ethylG	GC->AT		
O^6-benzylG	GC->AT	(GC->TA)[b]	(GC->CG)
O^4-ethylT	AT->GC		
AFB_1-adducts	(GC->AT)	GC->TA	(AT->TA)

[a]modified from 9-15; [b] minor type of mutation

Site-specific mutagenesis has considerably contributed to the understanding of the mutagenic properties of individual carcinogen-DNA adducts[5]. Consistent with the earlier observation that O-alkylation and not N-alkylation correlates with the carcinogenic potency of alkylating carcinogens[6], O-alkyl-modified bases in DNA were found to be primarily responsible for alkyl-DNA adduct-induced mutations. Mispairing of modified bases during replication has since long been considered to underlie the mutation induction. The mechanisms of mispairing by O^4-meT and O^6-meG have been reviewed recently[7]. NMR studies[8] indicate that the originally proposed base pairing schemes might be incorrect. Mutation patterns typical for some carcinogen-modified bases are given in Table 1.

The preferential GC to TA transversions obtained from AFB_1-modified DNA probably depends on both the formation of aguaninic sites formed from modified guanine and the very poor recognition of a GA mispair by error-correcting processes[16]. The latter may be due to the almost undisturbed double strand conformation of DNA containing an A(*anti*)G(*anti*)duplex[17].

Heterogeneity in adduct formation and repair

Within the living organism many factors affect the tissue-specific DNA modification, and thus the likelihood of mutagenesis and tumor formation (Table 2). Adducts present in DNA reflect the net result of the interplay of many of these factors. Their cell type- and differentiation-specific expressions determine the cellular distribution of DNA modification. In view of the number of variables involved a direct relationship between carcinogen-induced DNA adduct formation in different tissues and cancer induction cannot be expected.

An ideal tool for the investigation of formation, persistence and eventual accumulation of carcinogen-DNA adducts in individual cells within their normal environment is immunocytochemistry[18,19]. For a large number of adducts conventional or monoclonal antibodies have been raised whose affinity is sufficiently high to get significant nuclear staining at experimentally relevant doses[19].

Table 2. Factors determining the tissue-specific modification of DNA, and tumor formation, by chemical carcinogens

Level of organism
  Carcinogen uptake, distribution, metabolism, excretion

....

Cellular level

....

  Amount and type of DNA adducts formed
      Cell type and differentiation/tissue location
      Carcinogen uptake/elimination
      Activating versus detoxifying metabolism
      Conjugation to GSH/reaction with non-DNA target molecules
  Repair of adducts
  Promutagenic and clastogenic relevance of adducts
  DNA synthesis/proliferation
  Number of tissue cells
  Preferential mutagenic spectrum versus type of essential molecular event
  Number of rare events/stages to reach tumor manifestation

Molecular level

  Gene-specific and sequence-specific differences in DNA modification, repair and mutation induction

---

Immunocytochemical studies revealed a pronounced heterogeneity in carcinogen metabolism and DNA adduct formation among cells of the same tissue. A typical example is the alkylation of hepatocyte DNA by N-nitrosamines. After DMN or DEN DNA of centrilobular hepatocytes becomes alkylated to a much higher extent than that of periportal hepatocytes[19,20]. Similar results were obtained for the lobular distribution of 7-meG. An unusual lobular distribution of DNA alkylation in the liver has recently been observed for another N-nitrosamine, BOP[21]: the strongest alkylation was observed in the periportal hepatocytes and adjacent Kupffer and endothelial cells. In contrast with an earlier report[20] preferential periportal localization has also been reported for livers of AAF-treated rats[22]. Other striking examples of cell-specific localization of carcinogen-metabolizing enzymes and consequently of DNA adduct formation are the pancreatic ducts of hamsters[21] and nasal epithelia and glands of several rodent species[18,23]. The immuncytochemical studies demonstrate that the level of modification as determined in DNA isolated from the entire tissue or from tissue fractions is unlikely to represent the adduct level in relevant cells. The same may apply to the loss of DNA adducts[19,24] if compared with that deduced from whole tissue DNA adduct measurements.

Heterogeneity at the molecular level of DNA adduct formation and repair

An additional aspect which complicates the assessment of the role of DNA adducts in cancer induction is the non-randomness of DNA adduct formation and repair at the level of genome organization[25,26] and DNA sequence[27-30].

NMU has been shown to preferentially modify in vitro the second dG residue in a 5'-NGGN-3' sequence.[29] The methylation efficiency of NMU toward four different guanine bases in a synthetic 17-mer which contained a known mutation hotspot of the E. coli xanthine-guanosine phosphoribosyltransferase gene was also strongly sequence dependent[30]: the most intense methylation has again been observed for the second G in a GGG neighborhood, and methylation of a G flanked by two pyrimidines (C, T) proved to be 6 times less likely.

Using a H-ras/thymidine kinase cassette in which guanine residues of codon 12 of the ras gene were replaced by the promutagenic base O^6-meG, Mitra et al.[11] observed that the transformation efficiencies of rat fibroblasts by mutated H-ras were independent of the position of the modified G. Topal and coworkers[31] analyzed the mutation pattern as obtained on in vitro replicative incorporation of O^6-medGTP into the ampicillinase gene and subsequent transfection into mutH E.coli. From the non-random mutation efficiency they deduced that the ability to tolerate the presence of a O^6-meG depends on the local sequence. The obtained consensus sequence had significant similarity with the environment of the preferentially mutated codon 12 position of the ras genes. It was proposed that O^6-meG formed by direct alkylation of guanine present within this consensus sequence may be less subject to repair. A variation of repair efficiency by at least 3- to 4-fold with position of the lesion was observed for O^6-meG-containing oligonucleotides. A 25-fold variation in repair rate of O^6-meG, depending on the flanking sequences, has been reported recently by the group of Swann.[32]

THE ROLE OF DNA ADDUCTS IN THE MULTISTEP PROCESS OF TUMOR FORMATION

DNA of tissues in which tumors develop on application of carcinogen (target tissues) generally becomes modified by the metabolically activated carcinogen. There is, however, no direct relationship between level of DNA modification and likely target cell type for initiation. In addition, non-target tissues often exhibit higher levels of DNA adducts than target tissues, and adducts may persist longer in nontarget tissues than in target tissues. Combined analysis of cell-specific DNA modification, persistence of DNA adducts and proliferative capacity supports the view that the combination of DNA adducts and DNA replication is responsible for tissue-specific tumor formation. This is in agreement with the currently held mechanism of mutation induction by mispairing of modified bases during DNA replication (see above). Persistence of adducts (reflecting also low proliferative capacity since proliferation would decrease the adduct level by dilution) may therefore even be considered as an indicator of low tumor risk. However, in cases where resting tissues containing accumulated DNA adducts can be forced to enter the replicative phase (e.g. by intoxication, wounding), mutational events and tumor formation are again expected. If such a co-carcinogenic mechanism is considered, DNA adduct formation must be regarded as a carcinogenic risk factor, irrespective of the common organ specificity of the carcinogen in question.

Initiation

Specific (point) mutations resulting from carcinogen-DNA base modifications are considered to underlie the first rare genetic event in carcinogenesis, initiation. Cells carrying such a specifically mutated gene are called initiated. The first order dose-response kinetics holding for initiation in a number of experimental model systems[33,34] indicate that a single, but dominant event (mutation?) is sufficient for initiation. If the initiating event confers proliferative advantage to the affected cells (slowly) growing foci of altered cells may develop. In several systems, like the liver and pancreas of the rat, such proliferative lesions have been shown to be the precursor cell population from which tumors develop by step-wise progression to more neoplastic phenotypes[35,36]. In this situation no promotor is thus *essential* for the generation of clones of initiated cells. If initiation does not confer proliferative advantage, like in mouse skin cells carrying a H-*ras* gene activated by point mutation[37], at least the short presence of promotor is required to allow the initiated cells to grow out to papilloma and eventually progress to carcinoma.

Promotion/progression

The observation of multiple mutated genes (oncogenes/tumor suppressor genes) in the same human or experimental tumor (see below) stresses that (point) mutations are not limited to the initiating events, but contribute also to later rare oncogenic events. The best studied example is human colon cancer[38] in which at least 4 critical mutations, among which activation of the K-*ras* oncogene by point mutation and inactivation of the *p53* tumor suppressor gene by point mutation or deletion, are involved. It is highly unlikely that all these independent mutations occurred in the same cell at the same time (initiation) as the consequence of the exposure to the initiating carcinogen. In fact, the mutational inactivation of the *p53* gene seems to be a rather late event since accumulation of mutated p53 protein is confined to the malignant lesion[39]. It is therefore indicated that induction of point mutations (and of other types of mutations) occurs throughout the neoplastic development. Under conditions of prolonged or continued carcinogen exposure - a likely situation in humans - DNA modification-induced mutations may thus play a role not only in initiation but also in later steps related to the progression phase of carcinogenesis. Pertinent experiments carried out in several laboratories showed that genotoxic carcinogens can trigger the formation of more advanced cell stages, e.g. the formation of carcinoma from skin papilloma[40] and the formation of hepatocellular carcinoma from enzyme-deficient liver cell nodules[41]. With respect to carcinogen risk assessment it is thus important to establish carcinogen metabolism and DNA modification not only for normal tissues but also for putative precancerous lesions from which tumors may develop.

Papillomas induced in rat esophagus by NMBzA have been shown by immunocytochemistry (see above) to have the same high capacity to metabolically activate NMBzA and related compounds as has the normal esophageal mucosa[42]. In contrast, nuclei of liver foci induced in the rat by DEN generally remain unstained[19] after a challenging dose of DEN sufficiently high for strong staining (indicating strong ethylation of DNA) of normal surrounding (centrilobular) nuclei. Similar results were reported for the capacity of focus cells to metabolize 2-AAF[22]. Putative precancerous nodules with strong metabolic capacity - similar to that of centrilobular hepatocytes - have been observed only occasionally (E. Scherer and M. Schwarz, unpublished observations).

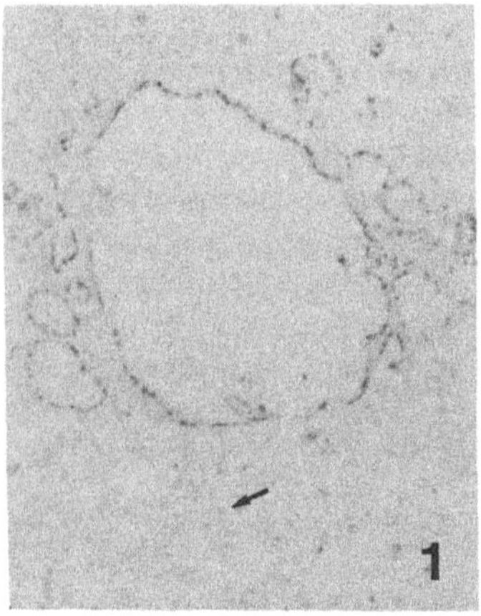

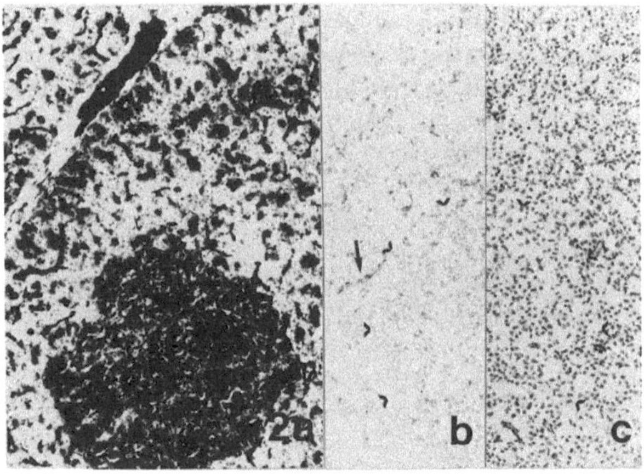

Fig. 1. Metabolic activation of BOP and formation of O^6-meG in BOP-induced ductal complexes of hamster pancreas. Three h prior to killing a challenging dose of 30 mg/kg BOP was injected. Note the pronounced O^6-meG-specific nuclear staining in one ductal lesion, and its absence from another (arrow). Immunocytochemistry.

Fig. 2. DNA damage by azaserine in acinar tissue and in a azaserine-induced focus of altered acinar cells from rat pancreas. (a): ATPase staining, strong activity in focus cells. (b,c): DNA-specific Feulgen staining with (b) and without (c) NaOH pretreatment for elution of damaged DNA; sections adjacent to (a). Note the similar loss of DNA (b) from nuclei of focal and of normal acinar tissue. Duct cells are not detectably damaged (arrow). The focus in (b,c) is outlined by arrow heads.

Ductal lesions developing in the pancreas of Syrian hamster after one or a few doses of BOP could be differentiated on the basis of their capacity to metabolize a challenging dose of BOP (Fig. 1). It is, however, still unknown which of these lesions (if any) are related to the development of pancreatic cancer in Syrian hamsters[43].

In cases where immunocytochemistry cannot be applied because either the DNA adducts are not yet characterized, or specific antibodies are not available, increased alkali instability of adducted DNA may be used for the cytochemical investigation of DNA damage in focal lesions relative to that of normal tissue. Using this method[44] we could show that azaserine-induced foci and nodules of altered acinar cells of rat pancreas become only slightly less damaged by a challenging dose of azaserine than normal acinar cells (Fig. 2). The latter indicates metabolic capacity of the focus cells that may contribute to progressive tumor development on prolonged azaserine injection.

Table 3. Mutation pattern in the K-*ras* gene of pancreatic tumors from hamster and humans

Species	Carcinogen	Mutants	Codon	% G>A	% G>T	% G>C	Ref.
Hamster	BOP	10/10	12	100			(58)
Hamster	BOP	20/20	12	90			(59)
			13	10			
	Country						
Human	Netherlands	28/30	12	32	64	4	(60)[a]
Human	Austria	49/63	12	37	31	33	(61)
Human	Japan	10/12	12	70	20	10	(62)
Human	Japan	35/38	12[b]	68	29	6	(63)
Human	North America	51/68	12[b]	49	39	10	(68)

[a] Partial reavaluation of the mutated tumor samples revealed that hte pattern is not significantly different from that of North america (J.L. Bos, Personal Communication).
[b] Including one codon 13 (G→A) mutation.

Relationship between DNA modification and organ-specific carcinogenesis

N-nitrosamines are a group of carcinogenic compounds which exhibit a pronounced organotropism of tumor induction[45] that is related to the presence of P450 isozymes necessary to bioactivate these compounds. A methylating intermediate is often formed, and methylated DNA bases are likely to be responsible for tumor initiation by most N-nitrosamines. The role of tissue-specific alkylation, repair of promutagenic lesions and cell proliferation has been reviewed.[46,47] Repair of promutagenic base modifications before fixation into a mutation during DNA replication is considered to decrease initiation, and thus to lower the tumor yield, while lack of repair and increased cell proliferation should result in more initiated cells and enhanced tumor formation. An example of a (seemingly) good correlation between tissue-specific level of methylation, persistence of formed promutagenic O^6-meG, and tumor formation is the hamster pancreas carcinoma induced by BOP. In centroacinar, pancreatic duct and ductular cells, considered to be the cell population from which the ductal adenocarcinoma develops[43], the highest level of O^6-meG and persistence for many weeks has been observed[24]. Nevertheless, there was no strict correlation between persisting O^6-meG and tumor formation since in proximal convoluted tubules of kidney high levels of persistent methylation were also found[24]. For kidney, however, only low tumor incidences were reported. Since the ducts of both organs are also only slowly proliferating one might doubt the ductal origin of pancreatic cancer[48,49].

Another example in which the alkylation level correlates with tumor induction only at first glance is the esophagus. NMBzA strongly alkylates not only the esophageal mucosa, predominantly in the basal layer[50], but also epithelia of the nontarget tissues lung and nasal cavity[23]; in the latter tissues methylated bases of DNA are even relatively stable and accumulate to a certain extent[51]. The very low level of proliferation observed in the NMBzA-metabolizing non-target tissues[51] has been put forward to explain the esophagus-specific carcinogenicity of NMBzA.

RELATIONSHIP BETWEEN CARCINOGEN AND MUTATION PATTERN IN TUMOR DNA

Mutational events in carcinogenesis

The identification of oncogenes and tumor suppressor genes as major determinants in the regulation of normal cell proliferation and differentiation has advanced tremendously our understanding of cancer as a multiple failure of critical gene functions. For a few tumor sites, most noticeable colon[38], lung[52], and breast[53], we begin to understand the sequence of oncogenic events and neoplastic stages leading to the fully malignant endpoint. As far as it is understood, aberrant gene function is the consequence of mutations in coding, regulating or other sequences (e.g. splice sites). The best studied genes with respect of point mutations are members of the *ras* family[54,55], growth factor receptor genes *(neu)*[56] and the tumor suppressor gene *p53*[57,58]. The observed dependence of the type of point mutation on either the applied carcinogen or the country from which the human tumor samples were obtained makes it likely that the specific mutations observed are the direct consequence of the carcinogenic compound(s) involved in tumor formation. Table 3 illustrates this dependence for experimental and human pancreatic cancer. The exclusive G->A transition of the second G of either codon 12 or 13 of the K-*ras* gene from hamster pancreatic cancer[59,60] is consistent with induction by BOP-induced O^6-meG. It is reasonable to speculate that the broader mutation spectrum characterizing the activating mutations of the K-*ras* gene from human pancreatic cancer reflects the exogenous and endogenous carcinogenic factors involved in its induction. Another recent example for a country-specific mutation pattern is the codon 249 mutation of the *p53* tumor suppressor gene in hepatocellular carcinoma[65,67]. Mutations of this codon were observed only for tumors obtained from countries in which the risk of dietary aflatoxin intake is high. The type of mutation (G->T, compare table 1) corresponded with that expected for aflatoxin-DNA adducts[9].

The above and the recent identification of more human tumors[57] in which sites and types of *p53* mutations differed among cancers of different tissue origin, makes it likely that mutation spectra of oncogenes and tumor suppressor genes reflect the contributing carcinogens, and thus may be useful in the identification of carcinogenic risk factors.

REFERENCES

1. Miller, J. A., and Miller, E. C., 1966, Mechanism of chemical carcinogenesis: nature of proximate carcinogens and interactions with macromolecules, Pharmacol. Rev., 18:805-838.
2. Singer, B., and Grunberger, D., 1983, Molecular Biology of Mutagens and Carcinogens, Plenum, New York, 347 pp.
3. Saffhill, R., Margison, G. P., and O'Connor, P. J., 1985, Mechanisms of carcinogenesis induced by alkylating agents, Biochem. Biophys. Acta, 823:111-145.

4. Ludlum, D. B., 1990, DNA alkylation by the haloethylnitrosoureas: Nature of modifications produced and their enzymatic repair or removal, Mutation Res., 233:117-126.
5. Singer, B., and Essigmann, J.M., 1991, Site-specific mutagenesis: retrospective and prospective, Carcinogenesis, 12:949-955.
6. Singer, B., 1976, All oxygens in nucleic acid react with carcinogenic ethylating agents, Nature, 264:333-339.
7. Swann, P., 1990, Why do O^6-alkylguanine and O^4-alkylthymine miscode? The relationship between the structure of DNA containing O^6-alkylguanine and O^4-alkylthymine and the mutagenic properties of these bases, Mutation Res., 233:81-94.
8. Loechler, E. L., 1991, Rotation about the C^6-O^6 bond in O^6-methylguanine; the syn and anti conformers can be of similar energies in duplex DNA as estimated by molecular modeling techniques, Carcinogenesis, 12, 1693-1699.
9. Foster, P, Eisenstadt, E., and Miller, J.H., 1983, Base substitution mutations induced by metabolically activated aflatoxin B_1, Proc. Natl. Acad. Sci. (USA), 80: 2695-2698.
10. Klein, J. C, Bleeker, M. J., Lutgerink, J.T., Van Dijk, W.J., Brugghe, H. E, Van Der Elst, H., Van Der Marel, G A., Van Boom, J.H., Westra, J.G., Berns, A. J.M., and Kriek, E., 1990, Use of shuttle vectors to study the molecular processing of defined carcinogen-induced DNA damage: mutagenicity of single O^4-ethylthymine adducts in HeLa cells, Nucl. Acid Res., 18:4131-4137.
11. Mitra, G, Pauly, G. T., Kumar, R., Pei, G K., Hughes, S. H., Moschel, R. C., and Barbacid, M., 1989, Molecular analysis of O^6-substituted guanine-induced mutagenesis of *ras* oncogenes, Proc. Natl. Acad. Sci. (USA) 86:8650-8654.
12. Richardson, K. K., Richardson, E. C., Crosby, R. M., Swenberg, J. A., and Skopek, T. R., 1987, DNA base changes and alkylation following in vivo exposure of *Escherichia coli* to N-methyl-N-nitrosourea or N-ethyl-N-nitrosourea, Proc. Natl Acad. Sci. (USA), 84 344-348.
13. Loechler, E. L., Green, C. L., and Essigmann, J. M., 1984, In vivo mutagenesis by O^6-methylguanine built into a unique site in a viral genome, Proc. Natl. Acad. Sci. (USA), 81:6271-6275.
14. Ellison, K. S., Dogliotti, E., Connors, T. D., Basu, A. K., and Essigmann, J.M., 1989, Site specific mutagenesis by O^6-alkylguanines located in the chromosomes of mammalian cells: influence of the mammalian O^6-alkylguanine-DNA alkyl transferase, Proc. Natl. Acad. Sci. (USA), 86:8620-8624.
15. Dosanjh, M. K., Singer, B., and Essigmann, J. M., 1991, Comparative mutagenesis of O^6-methylguanine and O^4-methylthymine in vivo, Biochemistry, 30:7027-7033.
16. Fersht, A. R., Knill-Jones, J.W., and Tsui, W.-C., 1982, Kinetic basis of spontaneous mutation, J. Mol. Biol., 156:37-51.
17. Lane, A. N., Jenkins, T. C., Brown, D. J. S., and Brown, T., 1991, NMR determination of the solution conformation and dynamics of the A.G mismatch in the d(CGCAAATTGGCG)$_2$ dodecamer, Biochem. J., 279:269-281.
18. Scherer, E., Bax, H., and Van Benthem, J., 1991, Immunocytochemical analysis of carcinogen-DNA adducts in normal and preneoplastic tissues, Prog. Histo-Cytochem., 23:77-83.
19. Den Engelse, L., Van Benthem, J., and Scherer, E., 1990, Immunocytochemical analysis of in vivo DNA modification, Mutation Res., 233:265-287.

20. Menkveld, G J. Van der Laken, C. J. Hermsen, T., Kriek, E., Scherer, E., and Den Engelse, L., 1985, Immunohistochemical localization of O^6-ethyldeoxyguanosine and deoxyguanosin-8-yl-(acetyl)aminofluorene in liver sections of rats treated with diethylnitrosamine, ethylnitrosourea or N-acetylaminofluorene, Carcinogenesis, 6:263-270.
21. Bax, J., Schippers-Gillissen, C., Woutersen, R. A., and Scherer, E., 1991, Cell specific DNA alkylation in target and non-target organs of N-nitrosobis(2-oxopropyl)amine-induced carcinogenesis in hamster and rat, Carcinogenesis, 12:583-590.
22. Huitfeldt, H. S., Hunt, J. M., Pitot, H. C., and Poirier, M. C., 1988, Lack of acetylaminofluorene-DNA adduct formation in enzyme-altered foci of rat liver, Carcinogenesis, 9:647-652.
23. Van Benthem, J., Wild, C. P, Vermeulen, E., Den Engelse, L., and Scherer, E., 1991, Immunocytochemical localization of DNA adducts induced by a single dose of N-nitroso-N-methylbenzylamine in target and non-target tissues of tumor formation in the rat, Carcinogenesis, 12:1831-1837.
24. Bax, J., Pour, P.M., Nagel, D. L., Lawson, T. A., Woutersen, R. A., and Scherer, E., 1990, Long term persistence of DNA-alkylation in hamster tissues after N-nitrosobis(2-oxopropyl)amine, J. Cancer Res. Clin. Oncol., 116:149-155.
25. Berkowitz, E. M. L., and Silk, H., 1981, Methylation of chromosomal DNA by two alkylating agents differing in carcinogenic potential, Cancer Lett., 12:311-321.
26. Durante, M., Geri, C., Bonatti, S., and Parenti, R., 1989, Non-random alkylation of DNA sequences induced in vivo by chemical mutagens, Carcinogenesis, 10:1357-1361.
27. Topal, M. D., 1988, DNA repair, oncogenes and carcinogenesis, Carcinogenesis, 9:691-696.
28. Gordon, A.J.E., Burns, P.A., and Glickman, B. W., 1990, N-Methyl-N'-nitro-N-nitrosoguanidine induced DNA sequence alteration; non-random components in alkylation mutagenesis, Mutation Res., 233:95-103.
29. Dolan, E. M., Oplinger, M., and Pegg, A. E., 1988, Sequence specificity of guanine alkylation and repair, Carcinogenesis, 9:2139-2143.
30. Richardson, E C., Boucheron, J. A., Skopek, T. R., and Swenberg, J. A., 1989, Formation of O^6-methyldeoxyguanosine at specific sites in a synthetic oligonucleotide designed to resemble a known mutagenic hotspot, J. Biol. Chem., 264:838-841.
31. Topal, M. D., Eadie, J. S., and Conrad, M., 1986, O^6-Methylguanine mutation and repair is nonuniform, J. Biol. Chem., 261:9879-9885.
32. Georgiadis, P., Smith, C. A., and Swann, P.E, 1991, Nitrosamine-induced cancer: selective repair and conformational differences between O^6-methylguanine residues in different positions in and around codon 12 of rat H-ras, Cancer Res., 51:5834--5850.
33. Emmelot, P., and Scherer, E., 1980, The first relevant cell stage in rat liver carcinogenesis. A quantitative approach, Biochim. Biophys. Acta, 605:247-304.
34. Bax, J. Schippers-Gillissen, C., Woutersen, R. A., and Scherer, E., 1990, Kinetics of induction and growth of putative acinar cell foci in azaserine-induced rat pancreas carcinogenesis, Carcinogenesis, 11:245-250.
35. Scherer, E., 1984, Neoplastic progression in experimental hepatocarcinogenesis, Biochim. Biophys. Acta, 738:219-236.
36. Scherer, E., Bax, J. and Woutersen, R. A., 1989, Pathogenic interrelationship of foci, nodules, adenomas and carcinomas in the multistage evolution of

azaserine-induced rat pancreas carcinogenesis, in: Biologically-based methods for cancer risk assessment, Travis, C.C. ed., Plenum, New York, pp. 41-54.
37. Toftgard, R., Roop, D. R., and Yuspa, S. H., 1985, Proto-oncogene expression during two-stage carcinogenesis in mouse skin, Carcinogenesis, 6:655-657.
38. Fearon, E. R., and Vogelstein, B., 1990, A genetic model for colorectal tumorigenesis, Cell, 61:759-767.
39. Bártek, J. Bártková, J. Vojtesek, B., Stasková, Z., Lucás, J. Rejthar, A., Kovarik, J. Midgley, C. A., Gannon, J. U, and Lane, D. E., 1991, Aberrant expression of the p53 oncoprotein is a common feature of a wide spectrum of human malignancies, Oncogene, 6:1699-1703.
40. Hennings, H., Shores, R., Wenk, M. L., Spangler, E. E, Tarone, R., and Yuspa, S. H., 1983, Malignant conversion of mouse skin tumors is increased by tumor initiators and unaffected by tumor promoters, Nature, 304:67-69.
41. Scherer, E., Feringa, A. W., and Emmelot, P.,1984, Initiation-promotion-initiation. Induction of neoplastic foci within islands of precancerous liver cells in the rat, IARC Scientific Publications, 56:57-66.
42. Dirsch, O. R., Koenigsmann, M., Ludeke, B. I., Scherer, E., and Kleihues, P., 1990, Bioactivation of N-nitrosomethylbenzylamine and N-nitrosomethylamylamine in oesophageal papillomas, Carcinogenesis, 11:1583-1586.
43. Pour, P. M., 1984, Histogenesis of exocrine pancreatic cancer in the hamster model, Environ. Health Perspec., 56:229-243.
44. Bax, J. Winterwerp, H. H. K., and Scherer, E., 1990, Visualization of azaserine-induced DNA damage in rat pancreas tissue sections, in: Histogenesis of pancreatic cancer induced in rats and hamsters by chemical carcinogens, Thesis J. Bax, University of Utrecht, pp. 65-70.
45. Druckrey H., Preussmann, R., Ivankovic, S., and Schmähl, D., 1967, Organotrope carcinogene Wirkungen bei 65 verschiedenen N-Nitroso-Verbindungen an BD-Ratten, Z. Krebsforsch., 69:103-201.
46. Swenberg, J. A., Richardson, E C., Boucheron, J. A., and Dyroff, M. C., 1985, Relationship between DNA adduct formation and carcinogenesis, Environ. Health Perspect., 62:177-183.
47. Hall, 1, and Montesano, R., 1990, DNA alkylation damage: consequences and relevance to tumor production, Mutation Res., 233:247-252.
48. Scarpelli, D. G, Rao, M. S., and Reddy, J. K., 1991, Are acinar cells involved in the pathogenesis of ductal adenocarcinoma of the pancreas?, Cancer Cells, 3:-275-277.
49. Sandgren, E.P., Quaife, C.J., Paulovich, A. G, Palmiter, R. D., and Brinster, R.L., 1991, Pancreatic tumor pathogenesis reflects the causative genetic lesion, Proc. Natl. Acad. Sci. (USA), 88:93-97.
50. Scherer, E., Van den Berg, T., Vermeulen, E., Winterwerp, H. H. K., and Den Engelse,L., 1989, Immunocytochemical analysis of O^6-alkylguanine shows tissue specific formation in and removal from esophageal and liver DNA in rats treated with methylbenzylnitrosamine, dimethylnitrosamine, diethylnitrosamine and ethylnitrosourea, Cancer Lett., 46:21-29.
51. Van Benthem, J. Vermeulen, E., Winterwerp, H. H. K., Wild, C.P., Scherer, E., and Den Engelse, L., 1992, Accumulation of O^6- and 7- methylguanine in DNA of N-nitroso-N-methylbenzylamine-treated rats is restricted to non-target organs for N-nitroso-N-methylbenzylamine- induced carcinogenesis, Carcinogenesis, in press.
52. Lehman, T. A., Bennett, W.P., Metcalf, R. A., Welsh, J.A., Ecker, J. Modali,

R. U, Ullrich, S., Romano, J.W., Appella, E., Testa, J.R., Gerwin, B. I., and Harris, C. C., 1991, *p53* Mutations, *ras* mutations, and *p53*-heat shock 70 protein complexes in human lung carcinoma cell lines, Cancer Res., 51:4090-4096.
53. Devilee, E.P., and Cornelisse, C.J., 1990, Genetics of human breast cancer, Cancer Surveys, 9:605-630.
54. Bos, J.L., 1989, *ras* oncogenes in human cancer: a review, Cancer Res., 49:-4682-4689.
55. Marshall, C l, 1991, How does p21ras transform cells?, Trends Genet., 7:91-95.
56. Saya, H., Ara, S., Lee, P.S.Y., Ro, J., and Hung, M.-C., 1990, Direct sequencing analysis of transmembrane region of human *neu* gene by polymerase chain reaction, Mol. Carcinogenesis, 3:198-201.
57. Hollstein, M., Sidransky, D., Vogelstein, B., and Harris, C.C., 1991, *p53* mutations in human cancers, Science, 253:49-53.
58. Levine, A. l, Momand, J.and Finlay, C. A., 1991, The *p53* tumour suppressor gene, Nature, 351:453-456.
59. Fujii, H., Egami, H., Chaney, W., Pour, P., and Pelling, J., 1990, Pancreatic ductal adenocarcinomas induced in Syrian hamsters by N-nitrosobis(2-oxopropyl)-amine contain a c-Ki-*ras* oncogene with a point-mutated codon 12, Mol. Carcinogenesis, 3:296-301.
60. Van Kranen, H. J. Vermeulen, E., Schoren, L., Bax, J. Woutersen, R.A., Van Iersel, P., and Scherer, E., 1991, Activation of c-K-*ras* is frequent in pancreatic carcinomas of Syrian hamsters, but is absent in pancreatic tumors of rats, Carcinogenesis, 12:1477-1482.
61. Smit, V. T. H. B. M., Boot, A. J.M., Smits, A. M. M., Fleuren, G J. Cornelisse, C. J. and Bos, J.L., 1988, K-*ras* codon 12 mutations occur very frequently in pancreatic adenocarcinomas, Nucl. Acid Res., 16, 7773-7782.
62. Grünewald, K., Lyons, J. Fröhlich, A., Feichtinger, H., Weger, R. A., Schwab, G, Janssen, J.W. G., and Bartram, C. R., 1989, High frequency of Ki-*ras* codon 12 mutations in pancreatic adenocarcinomas, Int. J. Cancer, 43:1037-1041.
63. Mariyama, M., Kishi, K., Nakamura, K., Obata, H., and Nishimura, S., 1989, Frequency and types of point mutation at the 12th codon of the c-Ki-*ras* gene found in pancreatic cancers from Japanese patients, Jpn. J.Cancer Res., 80:622-626.
64. Nagata, Y., Abe, M., Motoshima, K., Nakayama, E., and Shiku, H., 1990, Frequent glycine-to-aspartic acid mutations at codon 12 of c-Ki-*ras* gene in human pancreatic cancer in Japanese, Jpn. J.Cancer Res., 81:135-140.
65. Hsu, I. C., Metcalf, R. A., Sun, T., Welsh, J.A., Wang, N. J. and Harris, C. C., 1991, Mutational hotspot in the *p53* gene in human hepatocellular carcinomas, Nature, 350:427-428.
66. Bressac, B., Kew, M., Wands, J. and Ozturk,M., 1991, Selective G to T mutations of *p53* gene in hepatocellular carcinoma from southern Africa, Nature, 350:429-431.
67. Ozturk, M., and collaborators, 1991, *p53* mutation in hepatocellular carcinoma after aflatoxin exposure, Lancet, 338:1356-59.
68. Capellá, G., Cronauer-Mitra, S., Pienado, M.A., and Perucho, M. 1991. Frequency and spectrum of mutations at codons 12 and 13 of the c-K-*ras* gene in human tumors, Environ. Health Perspect., 93:125-131.

STUDIES ON THE ROLE OF PROTEIN KINASE C IN MULTISTAGE CARCINOGENESIS AND THEIR RELEVANCE TO RISK EXTRAPOLATION

Kevin R. O'Driscoll, Scott M. Kahn, Christoph M. Borner, Wei Jiang, and I. Bernard Weinstein

Comprehensive Cancer Center and Institute of Cancer Research
College of Physicians and Surgeons of Columbia University
701 West 168th Street, New York, N.Y. 10032

ABSTRACT

One of the most intensively studied non-genotoxic carcinogens is the phorbol ester tumor promoter 12-O-tetradecanoylphorbol-13-acetate (TPA). Since TPA appears to exert its biologic effects through protein kinase C (PKC), a key enzyme in signal transduction, we have studied this enzyme in considerable detail. To explore the role of PKC in carcinogenesis we developed various types of cell lines that overproduce the $PKC_{\beta I}$ isoform. To gain further insight into the relationship between PKC structure and its function in tumor promotion we expressed deletion mutants of $PKC_{\beta I}$ in a rat fibroblast cell line. These studies provide genetic evidence that PKC plays a critical role in growth control and the action of certain growth factors, tumor promoters and oncogenes. The findings are discussed within the context of multistage carcinogenesis, risk assessment, and cancer prevention.

RISK ASSESSMENT AND MULTISTAGE CARCINOGENESIS

That the assessment of the risks of developing cancers as a result of exposure to specific environmental, lifestyle and dietary substances is a daunting task is obvious. Perhaps the most difficult aspect is the frequent necessity to extrapolate to the human situation data derived largely from rodent bioassays, *in vitro* assays, or, to an increasing extent, studies done at the level of molecular biology. Although cancer prevention and public health must be the primary concern, informed and cost-effective decisions require that a variety of questions with less than obvious answers are addressed, by both quantitative and qualitative means. These questions include but are not limited to the following: What is the appropriate test system to identify and assay potentially carcinogenic compounds? What is the appropriate dose and route of administration of a particular test compound? Are there particular species or organ specificities for the test compound? What factors should be considered in extending conclusions gained in a given system or at a given dose to other systems or doses? How should low dose effects and possible thresholds be calculated, especially for agents that may not show simple non-linear dose-responses? How sensitive is a given assay, and how valid is a negative result in that assay? How can synergistic, cooperative or antagonistic interactions between multiple agents be detected, and how can cofactors that are not themselves carcinogens be identified? How can one best evaluate interspecies and interindividual variations in respect to a given agent or class of agents? How are the potential benefits of recommendations, guidelines, and regulations to be estimated? How is a decision point reached on the balance between risk and benefit for exposure of the public to a given compound? And finally, what are the societal and political barriers and consequences in the implementation of specific regulatory actions? While most of these questions are beyond the scope of this paper it seems likely that recent and current basic research studies will provide new insights and new strategies for rational approaches to carcinogen risk assessment and extrapolation.

The ultimate goal of our laboratory, and that of many others, is to elucidate basic molecular mechanisms of multistage carcinogenesis, concentrating on the molecular events of tumor promotion. We hope that this knowledge, coupled with new methods and technology, will contribute significantly to the risk assessment process, and thereby enhance primary cancer prevention, treatment and cure. One of the earliest experimental models of multistage carcinogenesis is that of mouse skin tumor formation induced by topical exposure to chemicals (reviewed by Berenblum, 1982). In this system a single application of a low dose of a genotoxic tumor initiating compound produces irreversible genetic changes to the treated area of skin but is insufficient by itself to produce a tumor. If the tumor initiation step is followed by the repeated application of a non-genotoxic tumor promoting compound, benign papillomas will eventually form at the treated site. A subset of the papillomas can eventually undergo tumor progression and will develop into invasive carcinomas of the skin. This progression stage, i.e. the conversion of benign papillomas to carcinomas, is markedly enhanced by treatment of the papillomas with certain genotoxic agents (reviewed by Nowell, 1986). Thus the mouse skin carcinogenesis model defines at least three distinct stages in tumor development that are termed initiation, promotion and progression.

This model has provided a paradigm for other examples of multistage carcinogenesis, revealed from the diverse fields of epidemiology, genetics, clinical pathology, and the molecular genetics of cancer. Taken together these findings indicate that the emergence of a malignant tumor requires multiple genetic and epigenetic events that involve multiple factors that act at discrete stages. An excellent example is that of human colon cancer, whose genesis is thought to involve the sequential acquisition of a combination of mutations in particular genes. Some of these mutations may be either inherited through the germline, for example the FAP gene in familial polyposis coli, or may alternatively arise in the affected tissue later in life (Vogelstein et al.,1988; Fearon et al.,1990; Kinzler et al.,1991; Joslyn et al.,1991; Nishisho et al.,1991). The effects of such genetic lesions may be compounded by the potential tumor promoting effects of chemicals generated in the colon such as bile acids (Fitzer et al.,1987; Nair et al. 1988) and diacylglycerol (Friedman *et al.*,1989; Moritomi *et al.*, 1990), which can activate PKC. Indeed we have demonstrated that colon carcinomas relative to adjacent normal colonic mucosa possess changes in the expression of PKC, and of other genes that are regulated through PKC (Giullem *et al.*, 1987; *ibid.*, 1990, Levy *et al.*, manuscript in preparation), thus suggesting that alterations in PKC pathways may contribute to carcinogenesis in the colon.

Recent studies on the mutational changes in tumor cells have revealed a remarkable number of changes in the DNA and chromosomes of tumors, both in experimental animals and in humans. It is now clear that the multistage carcinogenic process leads to the activation of cellular proto-oncogenes, and the deletion or inactivation of tumor suppressor genes that normally function to inhibit growth (reviewed by Weinberg, 1989; Bishop, 1991). It has also become apparent that most of the known viral oncogenes and cellular proto-oncogenes code for proteins that function in biochemical signalling pathways to regulate cellular growth and gene expression (i.e. growth factors, cell surface receptors, GTP-binding proteins, protein kinases, transcription factors, etc.) (reviewed by Cantley *et al.*, 1991). Therefore, a common characteristic of many of the multiple genes involved in carcinogenesis is that they encode components of the signal transduction systems of cells. The term "signal transduction" in this context refers to biochemical mechanisms that permit complex changes in the cytoplasm and in gene expression in the nucleus to be controlled by extracellular ligands that act through receptors, second messenger molecules and protein kinases. The subsequent phosphorylation of proteins at several subcellular localizations leads in yet to be determined ways, to altered gene expression (both positive and negative), and thus to epigenetic control of cellular growth and/or differentiation.

Unlike tumor initiating carcinogens which are often alkylating agents, for example certain nitrosamines, or which form bulky covalent adducts, for example the polycyclic aromatic hydrocarbons or the N-substituted aromatic compounds (reviewed by Jeffreys, 1987), tumor promoters appear to act through epigenetic events, i.e. non-mutational alterations in gene expression (reviewed by Diamond, 1987). A large body of evidence supports the view that these epigenetic events are also mediated by cellular signal transduction systems (reviewed by Weinstein, 1988a). Thus, although genotoxic and non-genotoxic carcinogens have quite different initial cellular targets and mechanisms of action, both types of agents can ultimately influence signal transduction pathways and gene expression. The major distinction is that a genotoxic carcinogen induces an irreversible and stable alteration by mutation, whereas a non-genotoxic carcinogen (since it not mutagenic) induces changes which during the early phases of its action are reversible, and which therefore require multiple applications for effective tumor promotion. At a later phase of the process tumor cells become independent of the tumor promoting agent, by unknown mechanisms.

A specific example of this is provided by the role of the PKC signalling pathway. In the case of TPA-induced tumor promotion, the tumor promoter binds directly to PKC and thereby activates this enzyme (Castagna *et al.*, 1982). On the other hand in transformed fibroblasts, bearing a mutationally activated *ras* oncogene, alterations in the PKC signal transduction pathway occur indirectly, apparently because *ras*-transformed fibroblasts produce increased levels of the second messenger diacylglycerol, which activates PKC (Fleischmann *et al.*, 1986; Preiss *et al.*, 1986; Lacal *et al.*, 1987a; *ibid.*,

1987b; Wolfman & Macara,1987; Huang *et al.*,1988a; Price *et al.*,1989). Thus both the phorbol ester tumor promoter TPA which activates PKC allosterically, and the polycyclic aromatic hydrocarbon tumor initiator dimethyl benzanthracene which produces activating c-Ha-*ras* mutations (Balmain & Pragnell, 1983; Balmain *et al.*, 1984; Quintanilla *et al.*, 1986), can lead to alterations in the PKC signalling pathway. Therefore, although various types of carcinogens have different initial mechanisms of action they can produce somewhat similar disturbances in signal transduction. Further evidence for interactions between TPA, PKC and an activated *ras* oncogene are described below. The implications for cancer risk assessment are that a deeper understanding of the target molecules of various carcinogens (both genotoxic and non-genotoxic), and their molecular mechanisms of action, will lead to better, and perhaps simplified, methods of assessing the risk of human exposure to those agents.

EPIGENETICS, TUMOR PROMOTION AND PROTEIN KINASE C

As discussed above, genetic changes may constitute only part of the carcinogenic process. There are several types of evidence that non-genotoxic agents, acting on cellular targets other than DNA and through epigenetic mechanisms, also play major roles in the development and maintenance of a tumor (Weinstein, 1988a). 1) The first type of evidence comes from detailed studies on the phenomenon of tumor promotion, particularly studies with the phorbol ester tumor promoters. 2) There is increasing evidence that a large number of other types of compounds that induce tumors in rodents do not have genotoxic effects (Ashby & Tennant, 1988). Although it is possible that further studies will indicate that some of these compounds do damage DNA, perhaps indirectly through oxidative damage, it is likely that many of them act through epigenetic mechanisms. 3) The incidence of the major human tumors (other than lung cancer) in the United States, i.e. breast, prostate and colon cancer, appears to be related to hormonal factors or dietary lipid, which are non-genotoxic agents. 4) There are numerous examples in which the growth and differentiation of malignant tumor cells can be modulated by non-genotoxic agents, despite the presence of activated oncogenes and/or chromosomal rearrangements. 5) The last type of evidence is more speculative and is based on the following logic. There is considerable evidence that the growth, development and differentiation of normal mammalian cells both in the embryo and the adult does not involve mutational changes in DNA (except in the immune system). Therefore, epigenetic events play a profound role in determining the phenotypes of normal cells and can be quite stable at the somatic cell level, even though the mechanisms are largely unknown (reviewed by Holliday, 1990). Since cancer cells often display abnormalities in differentiation, it seems reasonable to assume that carcinogenesis includes aberrations in these epigenetic mechanisms.

The most detailed information on tumor promotion comes from studies on the potent phorbol ester tumor promoter 12-*O*-tetradecanoylphorbol-13-acetate (TPA). A number of years ago our laboratory and others obtained evidence that TPA is a potent modulator of growth, differentiation and gene expression, and that its earliest effects are exerted at the level of the plasma membrane (reviewed by Diamond, 1987). A major advance in our understanding of mechanisms of tumor promotion came when it was demonstrated that TPA directly activates PKC, a calcium and phospholipid-dependent serine/threonine protein kinase involved in signal transduction (Castagna *et al.*, 1982). Thus, PKC is the major cellular receptor for TPA. The normal physiological ligand that activates PKC is the second messenger diacylglycerol, which is commonly produced by receptor-mediated hydrolysis of the inositol phospholipids, but which can also be generated from phosphatidyl choline degradation and from receptor- or voltage-mediated Ca^{2+} gate opening (reviewed by Farago & Nishizuka, 1990). At least eight distinct subspecies of PKC, encoded by seven distinct genes, have since been identified, which are designated α, βI, βII, γ, δ, ϵ, η and ζ (Knopf et al., 1986; Ono et al., 1986; Parker et al., 1986; Housey et al., 1987; Ohno et al., 1988; Ono et al., 1988; Bacher et al., 1990; Osada et al., 1990). The δ, ϵ, η and ζ isoforms constitute a distinct subfamily of PKC isoforms that lack a putative calcium-binding protein domain present in the α, β and γ isoforms.

The existence of a PKC gene family whose members are related by various degrees of amino acid sequence homology implies that individual PKC isoforms contribute in specialized ways to cellular signalling (reviewed by Nishizuka, 1988). This heterogeneity may be due to upstream agonist-dependent specificities (Kiley et al., 1990; Strulovici et al., 1991a), activating ligand specificities (Huang et al. 1988b, Burns et al., 1990; Pfeffer et al., 1991), as well as to substrate specificities (Schaap et al., 1989; Marais et al., 1990), and to differences in intracellular localization (Hyatt et al., 1990; C. Borner and I.B. Weinstein, unpublished studies). In addition, the differential expression of these isoforms in specific tissues (Housey et al., 1987; Bacher et al., 1990; Osada et al. 1990), as well as tissue specific alternative mRNA splicing variants (Coussens et al., 1987; Schaap et al., 1990), suggests that individual PKC isoforms may also differ with respect to their roles in tissue specific cellular signal transduction (Nishizuka, 1988 and references therein).

In addition to the capacity for phorbol esters to bind and to activate PKC, it is well known that in a variety of cell types treatment with TPA leads to a rapid translocation of PKC from a cytosolic to a membrane-bound localization (Kraft & Anderson, 1983; Anderson et al., 1985; Ballester & Rosen, 1985; Ashendel & Minor, 1986; Woodgett & Hunter, 1987; Kiley et al., 1990; Strulovici et al., 1991a). The prolonged treatment of a variety of cultured cells with TPA eventually results in a dose-dependent down-regulation of the enzyme, i.e. a marked loss of activity from both the cytosol and the membrane fractions (Rodriguez-Pena & Rozengurt, 1984; Ballester & Rosen, 1985; Fournier & Murray, 1987; Woodgett & Hunter, 1987). This TPA-induced down-regulation of PKC occurs subsequent to translocation of PKC to the membrane fraction, and is mediated by increased post-translational degradation (Ballester & Rosen, 1985; Woogett & Hunter, 1987; Young et al., 1987; Borner et al., 1988). Limited proteolysis of purified PKC *in vitro* can produce a protein fragment with unregulated kinase activity (Kishimoto et al., 1983; ibid., 1989). A similar phenomenon may occur in the cell due to the action of calpain on the PKC V3 protein domain (Melloni et al., 1986; Kishimoto et al., 1989; Schaap et al., 1990) or to stable expression of proteins corresponding to the catalytic domain of PKC in carcinoma-derived cell lines (Strulovici, 1991b). Although a physiological role for the unregulated protein kinase remains to be demonstrated, several studies have suggested that truncation or deletion mutants of PKC from which regulatory domain sequences have been removed possess constitutive biological activities, i.e. protein kinase activity in the absence of the usually required cofactors phospholipid, DAG or TPA (Hata et al., 1989; Muramatsu et al., 1989; Kaibuchi et al., 1989; O'Driscoll et al., manuscript in preparation).

The allosteric activation of kinase activity by phorbol esters and phospholipids probably involves a disruption of the interaction of the pseudosubstrate domain (located near the amino terminus of the protein) with the catalytic site of the protein kinase domain. Evidence for this mechanism of PKC activation comes from studies in which a peptide corresponding to the pseudosubstrate sequence was shown to be a potent and highly-specific PKC inhibitor (House & Kemp, 1987); and also from evidence that an antibody directed against the pseudosubstrate domain activates PKC *in vitro* (Makowske & Rosen, 1989). Additional studies supporting the pseudosubstrate hypothesis have used site-directed mutagenesis of the pseudosubstrate protein coding sequence of PKCα which was found both to activate protein kinase activity constitutively, and to activate TPA-responsive promoters in the absence of phorbol esters or other agonists (Kaibuchi *et al.*, 1989; Pears *et al.*, 1990). There is direct evidence that the cysteine-rich C1 domain of PKC, which contains two putative zinc finger-like regions, contains the phorbol ester binding site(s) of the intact enzyme (Kaibuchi *et al.*, 1989; Ono *et al.*, 1989; Burns & Bell, 1991). While these studies have investigated the enzymology, expression and regulation of various PKC isoforms, relatively little is understood of the role of individual PKC isoforms, or of the structure/function correlates of PKC, in tumor promotion and multistage carcinogenesis.

RECENT STUDIES ON PKC IN MULTISTAGE CARCINOGENESIS AND SIGNAL TRANSDUCTION

To gain further insight into the role of PKC in signal transduction and multistage carcinogenesis, and to define better the biochemical properties and physiologic effects of a specific isoform of PKC, our strategy has been to construct a series of cell lines that constitutively overproduce high levels of the $PKC_{\beta I}$ isoform. We have utilized the retroviral expression vector pMV7 (Kirschmeier *et al.*, 1988) to transduce the $PKC_{\beta I}$ DNA into a variety of cell types, while the pMV7 vector lacking a cDNA insert was used to generate matched control cell lines. For instance an R6 rat embryo fibroblast cell line (designated R6-PKC3) infected with the pMV7-$PKC_{\beta I}$ virus, expresses about 40-fold increased levels of PKC enzyme activity and shows several disturbances in growth control relative to control cell lines (Housey *et al.*, 1988). These findings, as well as results obtained with derivatives of both the murine fibroblast cell line C3H-10T1/2 (Krauss *et al.*, 1989) and rat liver epithelial cell lines (Hsieh *et al.*, 1989), that also overproduce $PKC_{\beta I}$, provide direct evidence that this isoform of PKC can alter cell morphology and disturb growth control, particularly in the presence of TPA.

Although the R6-PKC3 cells are not fully transformed, they display a marked increase in susceptibility to transformation by an activated c-H-*ras* oncogene (Hsiao *et al.*, 1989). This finding provides genetic evidence for a cooperative interaction between PKC and the c-Ha-*ras* oncogene in the process of cell transformation. Strangely, TPA treatment strongly reduced the frequency of *ras* oncogene transformation of the R6-PKC3 cells but did not affect the frequency of transformation of control cells, and TPA also inhibited the colony forming ability of c-H-*ras* transformed R6-PKC3 clones (Hsiao *et al.*, 1989). The studies with R6-PKC3 cells indicate therefore that an appropriate balance between the levels of PKC activation and of c-H-*ras* oncogene expression is required for cell transformation. These results can be rationalized in terms of the cooperative and intersecting contributions of a c-Ha-*ras* oncogene and the overproduced

$PKC_{\beta I}$ isoform (Hsiao et al., 1989), and provide a model system in which multiple disturbances in a signal transduction pathway act together to regulate a balance between normal cell growth, cell transformation and cell death (Krauss et al., manuscript submitted). The susceptibility of R6-PKC3 cells to transformation is not confined to *ras* oncogenes, since these cells also display increased susceptibility to transformation by both the v-*myc* and v-*fos* oncogenes (W.-L.W. Hsiao and I.B. Weinstein, unpublished studies). These results indicate that enhanced expression of $PKC_{\beta I}$ in R6 fibroblasts can also cooperate in cell transformation with oncogenes whose encoded proteins are localized to the nucleus and which function as transcription factors. Thus, one can argue that PKC is a shared effector for enhanced cellular transformation by rather diverse oncogenes. The level of PKC activity in a given tissue may, therefore, be a rate-limiting determinant during multistage carcinogenesis.

It is of interest that when R6 fibroblasts become transformed by a c-H-*ras* oncogene they show a marked increase in endogenous PKC_{α} and a marked decrease in endogenous PKC (Borner et al., 1990). Since this modulation occurs both in the presence and absence of the ectopically overproduced $PKC_{\beta I}$ isoform, and is unaffected by a specific inhibitor of PKC or by TPA-induced downregulation of PKC, it appears that a PKC-independent signalling pathway mediates PKC isoform switching during oncogene-induced transformation. A similar involvement of PKC in oncogene-mediated cell transformation has been reported in which PKC_{δ} expression is reduced while that of PKC_{ζ} is increased following N-*myc*-mediated transformation of neuroblastoma cells (Bernards, 1991). Thus, in addition to the well established cooperativity between *ras* oncogenes and the PKC signalling pathway mediated by enhanced diacylglycerol production in transformed cells, PKC may also be linked to the action of *ras*, and other oncogenes through subtle modulations in the pattern of expression of PKC isoforms.

Other studies from our laboratory have examined the role of PKC in the regulation of growth of human tumors and tumor-derived tissues. The human colon carcinoma - derived cell line HT29 has been used as a recipient to generate a series of cell lines that overproduce $PKC_{\beta I}$. This procedure yielded derivatives that display an 11-15 fold increase in PKC activity (Choi et al., 1990). These cells grow normally but, in contrast to control HT29 cells, when treated with TPA they display dramatic changes in morphology and their growth in both monolayer culture and in agar suspension is markedly inhibited. The HT29 cells that overproduce $PKC_{\beta I}$ display a marked reduction in tumorigenicity in nude mice, thus suggesting that in HT29 cells $PKC_{\beta I}$ acts as a tumor suppressor. It is curious that overexpression of $PKC_{\beta I}$ in HT29 colon cancer cells produces effects on growth which are exactly the opposite of those produced by overexpression of $PKC_{\beta I}$ in rodent fibroblast and rodent epithelial cells (see above). It is apparent, therefore, that the biologic effects of a specific isoform of PKC depend on the cell in which it is expressed. The same principle can apply to certain oncogenes since although an activated H-*ras* oncogene transforms rodent fibroblasts, when introduced into PC12 pheochromocytoma cells the same oncogene inhibits proliferation and induces differentiation (Bar-Sagi & Feramisco, 1985; Noda et al., 1985). The reciprocal effects of PKC activation in different cell types is obviously relevant to risk extrapolation, especially in terms of the tissue specificities of agents that act through the PKC pathway.

The human promyelocytic leukemia cell line HL-60 is a model system used extensively to investigate the balance between cellular differentiation and malignant growth, since these cells differentiate *in vitro* when treated with various inducers. Alteration in the expression of specific isoforms of PKC have been seen in HL-60 cells when they are induced to differentiate into granulocytes in response to retinoic acid, dimethylsulfoxide (Makowske *et al.*, 1988; Hashimoto *et al.*, 1989), or to differentiate into monocytes in response to $1\alpha,25$-dihydroxyvitamin D_3 (Solomon *et al.*, 1991). Since TPA itself is a potent inducer of the monocytic differentiation program in this cell line (Rovera *et al.*, 1979), these findings suggest a link between PKC and the differentiation of this leukemic cell line into cells having characteristics of either monocytes or granulocytes. It remains to be determined whether these changes in expression of specific isoforms of PKC reflect different functional roles of these isoforms with respect to the control of growth and differentiation. Studies are in progress to investigate the roles of specific isoforms of PKC in the differentiation of HL60 and PC12 cell lines by ectopic expression of the PKC_α and PKC_β isoforms (K. O'Driscoll and I.B. Weinstein, unpublished data).

The exposure of certain cells to TPA leads to rapid induction in the expression of several mitogen-responsive genes, including c-*fos*, c-*myc*, c-*jun*, ornithine decarboxylase (ODC), and a gene we identified as "phorbin" which encodes the metalloprotease inhibitor TIMP-1 (Johnson *et al.*, 1987). The precise mechanism(s) by which this occurs and the role of specific isoforms of PKC in this process are not known. It was of interest, therefore, to examine the effects of overexpression of $PKC_{\beta I}$ on TPA-induced expression of some of these genes. We observed an exaggerated induction by TPA in the levels of phorbin and ODC mRNAs (Hsieh *et al.*, 1989), and the levels of c-*myc* RNA were constitutively higher, in the $PKC_{\beta I}$ overexpressor cells when compared to control cells (Hsieh *et al.*, 1989; Krauss *et al.*, 1989). More recently, we have examined TPA-induced expression of the AP1/c-*jun* transcription factor in the C3H-10T1/2-PKC4 and R6-PKC3 overexpressor cell lines, and the respective control cell lines (K. O'Driscoll and I.B. Weinstein, unpublished studies). This factor is of particular interest since in cooperation with the c-*fos* proto-oncogene product it mediates the positive regulation of several TPA-responsive genes (Angel *et al.*, 1987; Bohman *et al.*, 1987; Lee *et al.*, 1987; Chiu *et al.*, 1988; Rauscher *et al.*, 1988; Sassone-Corsi *et al.*, 1988). Immunoprecipitation of extracts of [^{35}S]methionine-labelled cells, using a polyclonal antibody to c-*jun* encoded proteins (Chiu *et al.*, 1988), revealed exaggerated induction of a 39 kD c-*jun* protein in the two cell lines that overexpressed $PKC_{\beta I}$ when compared to their respective control cell lines. The immunoprecipitates of the R6-PKC3 cells also displayed increased induction by TPA of immunologically-related or co-immunoprecipitated 34, 35, 39 and 46 kD proteins. These results provide direct evidence that $PKC_{\beta I}$ can mediate the expression of diverse genes including phorbin, ODC, c-*myc* and the transcription factor AP1/c-*jun*, which itself modulates the expression of several genes.

The above studies from our laboratory have shown that the stable overproduction of $PKC_{\beta I}$ in rodent fibroblast cell lines is a useful model system to study the role of PKC in multistage carcinogenesis (Housey *et al.*, 1988; Hsiao *et al.*, 1989; Krauss *et al.*, 1989). To delineate protein domains within the primary sequence of $PKC_{\beta I}$ that may contribute to the action of this enzyme in altering cellular growth properties, a series of deletion mutants were expressed stably in R6 fibroblasts. Precisely defined deletion mutants within the regulatory domain of the $PKC_{\beta I}$ cDNA coding sequence were generated using a polymerase chain reaction-based technique termed thermal cycled fusion (Kahn *et al.*, 1990). Mutant cDNA's were constructed that lack: *i.*) one of the two zinc finger-like, cysteine-rich protein regions (ΔCys2); *ii.*) the C1 region, including both of the cysteine-rich repeats (ΔC1); and *iii.*) the entire regulatory domain, including the V1-3 and C1-2 protein regions (ΔReg). These mutant cDNA constructs were introduced into R6 cells following subcloning into the pMV7 retroviral vector and helper-free retroviral packaging in the ψ2 system (O'Driscoll *et al.*, manuscript in preparation), Clonal cell lines were expanded and screened by Northern and Southern blot analyses to detect clones that functionally express each of the mutant forms of $PKC_{\beta I}$.

Characterization by immunoblotting with PKC-specific monoclonal antibodies revealed that the mutant $PKC_{\beta I}$ proteins containing deletions of either both (ΔC1) or only the second of two cysteine-rich repeats in the C1 protein domain (ΔCys2) have distinct migrations on sodium dodecyl sulfate-polyacrylamide gels. Extracts of cytosolic proteins from R6 cells expressing the ΔC1 and ΔCys2 mutants have no increase in phorbol ester binding capacity, in contrast to the dramatic increase observed in control cell lines that overproduce the non-mutant form of $PKC_{\beta I}$. Protein kinase assays of partially-purified PKC enzyme preparations revealed that the mutant lacking one of the two cysteine-rich repeats (Δcys2) had an altered enzymatic activity since it eluted from DEAE-Sephacel columns at a higher salt concentration than the non-mutant form, and since it exhibited significant constitutive kinase activity in the absence of allosteric activators of PKC. While the tumor promoter TPA had a dramatic effect on the morphology of R6 cells overproducing the non-mutant $PKC_{\beta I}$ isoform, it had little effect on the morphology of R6 cells that expressed mutant forms of $PKC_{\beta I}$. These findings indicate that deletion of sequences encoding the cysteine-rich repeats in the C1 protein domain of $PKC_{\beta I}$, dramatically reduces the ability of the mutant protein, when expressed in R6 fibroblasts to bind phorbol esters and to mediate a morphologic response to TPA. The fact that one of the C1 domain-specific deletion mutants (ΔCys2) encoded a constitutively active form of $PKC_{\beta I}$, and yet this enzyme failed to cause the recipient R6 cells to become morphologically transformed (even in the presence of TPA), suggests that activation of the kinase activity of $PKC_{\beta I}$ is insufficient to induce cell transformation. Further experiments are in progress to determine which of the other structural aspects of $PKC_{\beta I}$ contribute to its function in controlling cell growth and transformation.

DISCUSSION

In view of the pleiotropic effects of overexpression of a specific isoform of PKC seen in the above studies, it is difficult to rationalize these diverse effects in terms of simple linear pathways of signal transduction. Studies investigating the mechanisms of action of oncogenes, including those utilizing the transgenic mouse lead to similar conclusions (see other papers in this symposium). It is apparent, therefore, that the signal transduction pathways of mammalian cells constitute a dynamic and highly interactive network, that includes considerable "cross-talk" and both positive and negative feedback control mechanisms. The complexity in signal transduction probably explains why multiple types of exogenous agents, multiple mechanisms and multiple genes are involved in the carcinogenic process. This complexity also helps to explain why the phenotypes of tumors can be highly heterogeneous, even when they are induced by the same agent and in the same tissue. It is for these reasons that we have previously defined multistage carcinogenesis as a "progressive disorder in signal transduction" (Weinstein *et al.*, 1990).

Our studies might be relevant, in several ways, to human carcinogenesis and to the problem of assessing the risks of exposure to various environmental agents. The first is that we have provided direct genetic evidence that PKC plays an important role in growth control and in the action of TPA in tumor promotion. We have also shown that the effects of a specific isoform of PKC on growth depends on the specific cell type in which it is expressed. Further studies are needed to see if PKC signalling pathways are involved directly or indirectly in tumor promotion in specific organ systems (i.e. colon, breast, liver etc.) and in the action of various types of tumor promoters, such as halogenated organic compounds (i.e.. polychlorinated biphenyls) and dioxin, as well as other environmental pollutants. The systems discussed in this paper could provide useful models for assaying potential tumor promoters in various environmental sources, and for studying their mechanism(s) of action. Development of similar cell culture systems from other species and tissues, and of assay systems for other types of tumor promoters (i.e. those that do not activate PKC directly), may provide further insights into the mechanism(s) of action of various non-genotoxic carcinogens, and into the basis of the species and organ specificities of these compounds. Such studies might also lead to the development of rapid and inexpensive screening tests for non-genotoxic chemicals that might enhance carcinogenesis, and for identifying inhibitors of tumor promotion. Risk assessment protocols might in the future be modified to include *in vitro* assays for tumor promotion, thus reducing costs, and providing better data for informed decisions related to the assessment of cancer hazards for humans. Further studies on the mechanism of action of tumor promoters might also provide biomarkers that can be used in molecular epidemiology studies in human populations (Weinstein, 1991b). This is of particular importance because most of the current biomarkers (DNA adducts, etc.) relate to genotoxic agents. We are hopeful, therefore, that studies on the cellular and molecular mechanisms of action of tumor promoters and on the PKC signal transduction pathways will provide new insights into risk assessment and extrapolation, thus facilitating the goal of primary cancer prevention.

The fields of oncogenes, and of tumor suppressor genes are burgeoning areas of research (see reviews by Klein, 1987; Weinberg, 1989; Bishop, 1991; Hunter, 1991). Their fundamental contributions to understanding carcinogenesis are enormous, yet at the same time they tend to lead to a reductionist mode of thinking that may oversimplify the multistage carcinogenic process. It seems likely that a multitude of genetic elements are involved. As stressed above epigenetic events induced by hormones or exogenous compounds such as tumor promoters or other cocarcinogens, can also produce dramatic effects on cellular growth and differentiation. The argument has recently been made that mitogenesis per se can be responsible for cancer induction in certain rodent bioassays (Ames & Gold, 1990). We have argued, however, that while mitogenesis is undoubtedly one contributory factor in the carcinogenic process, it is unlikely to be solely responsible for the accumulation of the multiple genetic and epigenetic events that culminate in the emergence of a tumor (Weinstein, 1991a). The latter argument concludes that, although continued basic research on cancer causation is essential to our better understanding of mechanisms, and to better mechanism-based methods for detecting potential human carcinogens, at the present time rodent bioassays remain the mainstay of potential carcinogen testing protocols.

While complexities such as those alluded to at the beginning of this paper underscore the difficulty of risk assessment, especially taking into account the multifactorial nature of cancer, it is likely that rapid progress in the field of molecular epidemiology (Perera, 1987) and molecular biomarkers (Santella, 1988) will greatly improve present methods for assessing cancer risks in human populations and in specific individuals. Indeed there is considerable evidence to suggest that cancer is a preventable disease since most cancers are due to exogenous exposures (Weinstein, 1991b). It follows that improved risk assessment will lead to increased cancer prevention. In cases where the causative agent cannot be removed from the environment, or where individuals may have already had significant exposures or have already suffered a cancer, or where a known genetic predisposition to cancer exists, the simple approach of avoiding exposure to causative agents is obviously not sufficient. To combat these latter situations it may become necessary to actively intervene in the initiation, promotion and progression stages of cancer development through early detection and chemoprevention using newly emerging methods (Weinstein, 1988b; *ibid.*, 1991b). Although at the present time our understanding of signal transduction mechanisms in cancer cells is limited, it is sufficient nevertheless to suggest new approaches to inhibiting the multistage carcinogenesis process (Weinstein, 1988b). Hopefully the increased identification of potential human carcinogens and the development of better methods of cancer risk extrapolation, coupled with advances in our understanding of the molecular basis of multistage carcinogenesis, will lead to reductions in both the incidence and mortality of human cancer within the next few decades.

ACKNOWLEDGEMENTS

The authors gratefully acknowledge our coworkers Drs. Polly Gazetas, Wendy Hsiao, Rob Krauss, Sarah Nichols-Guadagno, and Marius Ueffing for their comments, expertise and collaborative spirit. The authors acknowledge support from NIH/NCI Grants CA021111 and CA26056 (to I.B.W.), and additional support from the American Cancer Society, the Markey Charitable Trust, the National Foundation for Cancer Research (to I.B.W.), and also a National Research Service Award CA09529-05 Postdoctoral Fellowship (to K.O'D.). K.O.'D. is exceedingly grateful to Helen Choy O'Driscoll for allowing the presentation of this paper during their honeymoon.

REFERENCES

Ames, B.N. and Gold, L.S., 1990, Too many rodent carcinogens: mitogenesis increases mutagenesis, Science, 249:970-971.
Anderson, W.B., Estival, A., Tapiovaara, H., and Gopoalakrishna, R., 1985, Altered subcellular distribution of protein kinase C (a phorbol ester receptor): Possible role in tumor promotion and the regulation of cell growth, Adv. Cycl. Nucl. Res., 19: 287-306.
Angel, P., Imagawa, M., Chiu, R., Stein, B., Imbra, R.J., Rahmsdorf, H.J., Jonat, C., Herrlich, P., and Karin, M., 1987, Phorbol ester-inducible genes contain a common

cis element recognized by a TPA-modulated *trans*-acting factor, Cell, 49:729-731.

Ashby, J., and Tennant, R.W., 1988, Summary of tumorigenicity data for 222 chemicals evaluated by the United States National Toxicology Program: Correlations of chemical structure, mutagenicity to salmonella and tumor sites, Mutation Res., 204:17-115.

Ashendel, C.L., and Minor, P.L., 1986, Mechanism of phorbol ester activation of calcium activated phospholipid-dependent protein kinase. Carcinogenesis, 7:517-521.

Bacher, N., Zisman, Y., Berant, E., Livneh, E., 1991, Isolation and characterization of PKC-L, a new member of the protein kinase C-related gene family specifically expressed in lung, skin and heart, Molec. and Cell. Biol., 11:126-133.

Ballester, R. and Rosen, O.M., 1985, Fate of immunoprecipitable protein kinase C in GH_3 cells treated with phorbol 12-myristate 13-acetate, J. Biol. Chem., 260: 15194-15199.

Balmain, A. and Pragnell, I.B., 1983, Mouse skin carcinomas induced *in vivo* by chemical carcinogens have a transforming Harvey-*ras* oncogene, Nature, 303:72-74.

Balmain, A., Ramsden, M., Bowden, G.T., and Smith, J., 1984, Activation of the mouse cellular Harvey-*ras* gene in chemically induced benign skin papillomas, Nature, 307:658-660.

Bar-Sagi, D. and Feramisco, J.R., 1988, Microinjection of the *ras* oncogene protein into PC12 cells induces morphological differentiation, Cell, 42:841-848.

Berenblum, I., 1982, Sequential aspects of chemical carcinogenesis: skin. *In:* "Cancer: A Comprehensive Treatise," Vol. 1, Ed. 2., F.F. Becker, ed., Plenum Publishing Corp, New York.

Bernards, R., 1991, N-*myc* disrupts protein kinase C-mediated signal transduction in neuroblastoma, EMBO J., 10:1119-1125.

Bishop, J.M., 1991, Molecular themes in oncogenesis, Cell, 64:235-248.

Bohmann, D., Bos, T.J., Adman, A., Nishimura, T., Vogt, P.K., and Tjian, R., 1987, Human proto-oncogene c-*jun* encodes a DNA binding protein with structural and functional properties of transcription factor AP-1, Science, 238:1386-1391.

Borner, C., Eppenberger, U., Wyss, R., and Fabbro, D., 1988, Continuous synthesis of two protein kinase C-related proteins after down-regulation by phorbol esters, Proc. Natl. Acad. Sci. USA., 85:2110-2114.

Borner, C., Nichols-Guadagno, S., Hseih, L.L., Hsiao, W.-L.W., and Weinstein I.B., 1990, Transformation by a *ras* oncogene causes increased expression of PKCα and decreased expression of PKCϵ, Cell Growth and Differ., 1:653-660.

Burns, D.J., Bloomenthal, J., Lee, M.H., and Bell, R.M., 1990, Expression of the α, βII, and γ protein kinase C isozymes in the baculovirus-insect cell expression system, J. Biol. Chem., 265:12044-12051.

Burns, D.J. and Bell, R.M., 1991, Protein kinase C contains two phorbol ester binding domains, J. Biol. Chem., 266:18330-18338.

Castagna, M., Takai, Y., Kaibuchi, K., Sano, K., Kikkawa, U., Nishizuka, Y., 1982, Direct activation of calcium-activated phospholipid-dependent protein kinase by tumor-promoting phorbol esters, J. Biol. Chem., 257:7847-7854.

Cantley, L.C., Auger, K.R., Carpenter, C., Duckworth, B., Graziani, A., Kapeller, R., and Soltoff, S., 1991, Oncogenes and signal transduction. Cell, 64:281-302.

Chiu, R., Boyle, W.J., Meek, J., Smeal, T., Hunter, T., and Karin, M., 1988, The c-*fos* protein interacts with c-*jun*/AP-1 to stimulate transcription of AP-1 responsive genes, Cell, 54:541-547

Choi, P.M., Tchou-Wong, K.-M., and Weinstein, I.B., 1990, Overexpression of protein kinase C in HT29 colon cancer cells causes growth inhibition and tumor suppression, Molec. and Cell. Biol., 10:4650-4657.

Coussens, L., Rhee, L., Parker, P.J., and Ullrich, A., 1987, Alternative splicing increases the diversity of the human protein kinase C family, DNA, 6:389-394.

Diamond, L., 1987, Tumor promoters and cell transformation, *In:* "Mechanisms of Cellular Transformation by Carcinogenic Agents" (pp 73-133), D. Grunberger and S.P. Goff, eds., Pergamon Press, New York.

Farago, A., and Nishizuka, Y., 1990, Protein kinase C in transmembrane signalling, FEBS Lett., 268:350-354.

Fearon, E.R., Cho, K.R., Nigro, J.M., Kern, S.E., Simons, J.W., Ruppert, J.M., Hamilton, S.R., Preisinger, A.C., Thomas, G., Kinzler, K.W., and Vogelstein, B., 1990, Identification of a chromosome 18q gene that is altered in colorectal cancers, Science, 247:49-56.

Fitzer, C.J., O'Brian, C.A., Guillem, J.G., and Weinstein, I.B., 1987, The regulation of protein kinase C by chenodeoxycholate, deoxycholate and several related bile acids, Carcinogenesis, 8:217-220.

Fleischmann, L.F., Cuahwala, S.B., and Cantley, L., 1986, *ras*-transformed cells: altered levels of phosphatidylinositol 4,5 bisphosphate and catabolites, Science, 231: 407-410.

Fournier, A., and Murray, A.W., 1987, Application of phorbol ester to mouse skin causes a rapid and sustained loss of protein kinase C, Nature, 330:767-769.

Friedman, E., Isaksson, P., Rafter, J., Marian, B., Winawer, S., and Newmark, H., 1989, Fecal diglycerides as selective endogenous mitogens for premalignant and malignant human colonic epithelial cells, Cancer Research, 49:544-548.

Guillem, J.G., Levy, M.F., Hsieh, L.L., Johnson, M.D., LoGerfo, P., Forde, K.A. and Weinstein, I.B., 1990, Increased levels of phorbin, c-*myc* and ornithine decarboxylase in human colon cancer, Molec. Carcinogenesis, 3:68-74.

Guillem, J.G., O'Brian, C.A., Fitzer, C.J., Forde, K.A., LoGerfo, P., Treat, M., and Weinstein, I.B., 1987, Altered levels of protein kinase C and Ca^{2+}-dependent protein kinases in human colon carcinomas, Cancer Research, 47:2036-2039.

Hashimoto, K., Kishimoto, A., Hiroaki, A., Yasuda, I., Mikawa, K., and Nishizuka, Y., 1990, Protein kinase C during differentiation of human promyelocytic leukemia cell line HL-60, FEBS Lett. 263:31-34.

Hata, A., Akita, Y., Konno, Y., Suzuki, K., and Ohno, S., 1989, Direct evidence that the kinase activity of protein kinase C is involved in transcriptional activation through a TPA-responsive element, FEBS Lett., 252:144-146.

Holliday, R., 1990, DNA methylation and epigenetic inheritance, Phil. Trans. Roy. Soc. Lond., 326:329-338.

House, C., and Kemp, B.E. Protein kinase C contains a pseudosubstrate prototope in its regulatory domain. Science 238:1726-1728 (1987).

Housey, G.M., O'Brian, C.A., Johnson, M.D., Kirschmeier, P.T., and Weinstein, I.B., 1987, Isolation of cDNA clones encoding protein kinase C: Evidence for a novel protein kinase C-related gene family, Proc. Natl. Acad. Sci. USA, 84:1065-1069.

Housey, G.M., Johnson, M.D., Hsiao, W.-L., O'Brian, C.A., Murphy, J.P., Kirschmeier, P. and Weinstein, I.B., 1988, Overproduction of protein kinase C causes disordered growth control in rat fibroblasts, Cell, 52:343-354.

Hsiao, W.-L.W., Housey, G.M., Johnson, M.D. and Weinstein, I.B., 1989, Cells that overproduce protein kinase C are hypersensitive to transformation by an activated

H-*ras* oncogene, Molec. and Cell. Biol., 9:2641-2647.

Hsieh, L.L., Hoshina, G. and Weinstein, I.B., 1989, Phenotypic effects of overexpression of PKC$_{\beta I}$ in rat liver epithelial cells, J. Cell. Biochem., 41:179-188.

Huang, M., Chida, K., Kamata, N., Nose, K., Kato, M., Homma, M.Y., Takenawa, T., and Kuroki, T., 1988a, Enhancement of inositol phospholipid metabolism and activation of protein kinase C in *ras*-transformed rat fibroblasts, J. Biol. Chem., 263:17975-17980.

Huang, K.P., Huang, F.L., Nakabayashi, H., and Yoshida, Y., 1988b, Biochemical characterization of rat brain protein kinase C isozymes, J. Biol. Chem., 263:14839-14845.

Hunter, T., 1991, Cooperation between oncogenes, Cell, 64:249-270.

Hyatt, S.L., Klauck, T., and Jaken, S., 1990, Protein kinase C is localized in focal contacts of normal but not transformed fibroblasts, Molec. Carcinogenesis, 3:45-53.

Jeffreys, A.M., 1987, DNA modification by chemical carcinogens. *In:* "Mechanisms of Cellular Transformation by Carcinogenic Agents," (pp 33-71), D. Grunberger and S.P. Goff, eds., Pergamon Press, New York.

Johnson, M.D., Housey, G.M., Kirschmeier, P.T., and Weinstein, I.B., 1987, Molecular cloning of gene sequences regulated by tumor promoters and mitogens through protein kinase C, Molec. and Cell. Biol., 7:2821- 2829.

Joslyn, G., Carlson, M., Thliveris, A., Albertsoen, H., Gelbert, L., Samowitz, W., Groden, J., Stevens, J., Spirio, L., Robertson, M., Sargeant, L., Krapcho, K., Wolff, E., Burt, R., Hughes, J.P., Warrington, J., McPherson, J., Wasmuth, J., LePaslier, D., Aderrahim, H., Cohen, D., Leppert, M., and White, R., 1991, Identification of deletion mutations and three new genes at the familial polyposis locus, Cell, 66:601-613.

Kahn, S.M., Jiang, W., Borner, C., O'Driscoll, K., and Weinstein, I.B., 1990, Construction of defined deletion mutants by thermal cycled fusion: applications to protein kinase C, Technique, 2:27-30.

Kaibuchi, K., Fukumoto, Y., Oku, N., Takai, Y., Arai, K., and Muramatsu, M., 1989, Molecular genetic analysis of the regulatory and catalytic domains of protein kinase C, J. Biol. Chem., 264:13489-13496.

Kishimoto, A., Kajikawa, N., Shiota, M., and Nishizuka, Y., 1983, Proteolytic activation of calcium-activated, phospholipid-dependent protein kinase by calcium-dependent neutral protease. J. Biol. Chem., 258:1156-1164.

Kishimoto, A., Mikawa, K., Hashimoto, K., Yasuda, I., Tanaka, S., Tominaga, M., Kuroda, T., and Nishizuka, Y., 1989, Limited proteolysis of protein kinase C subspecies by calcium-dependent neutral protease (Calpain), J. Biol. Chem., 264:4088-4092.

Kiley, A., Schaap, D., Parker, P., Hsieh, L.L., and Jaken, S., 1990, Protein kinase C heterogeneity in GH$_4$C1 rat pituitary cells, J. Biol. Chem., 265:15704-15712.

Kinzler, K.W., Nilbert, M.C., Vogelstein, B., Bryan, T.M., Levy, D.B., Smith, K.J., Preisinger, A.C., Hamilton, S.R., Hedge, P., Markham, A., Carlson, M., Joslyn, G., Groden, J., White, R., Miki, Y., Miyoshi, Y., Nishisho, I., and Nakamura, Y., 1991, Identification of a gene located at chromosome 5q21 that is mutated in colorectal cancers, Science, 251:1366-1370.

Kirschmeier, P.T., Housey, G.M., Johnson, M.D., Perkins, A.S. and Weinstein, I.B., 1988, Construction and characterization of a retroviral vector demonstrating efficient expression of cloned cDNA sequences, DNA, 7:219-225.

Klein, G., 1987, The approaching era of the tumor suppressor genes, Science, 238:1539-1546.

Knopf, J., Lee, M.-H., Sultzman, L.A., Kriz, R.W., Loomis, C.R., Hewick, R.M., and Bell, R.M., 1986, Cloning and expression of multiple protein kinase C cDNAs, Cell, 46:491-502.

Kraft, A.S. and Anderson, W.B., 1983, Phorbol esters increase the amount of Ca^{2+}, phospholipid-dependent protein kinase associated with plasma membrane, Nature, 301:621-623.

Krauss, R.S., Housey, G.M., Johnson, M.D., and Weinstein, I.B., 1989. Disturbances in growth control and gene expression in a C3H-10T1/2 cell line that stably overproduces protein kinase C, Oncogene, 4:991-998.

Krauss, R.S., Guadagno, S.N., Weinstein, I.B. Novel revertants of H-*ras* oncogene-transformed R6-PKC3 cells, manuscript submitted.

Lacal, J.C., Moscat, J., and Aaronson, S.A., 1987a, Novel source of 1,2-diacylglycerol in cells transformed by a Ha-*ras* oncogene, Nature, 330:259-272.

Lacal, J.C., Flemming, T.P., Warren, B.S., Blumberg, P.M. and S.A. Aaronson, S.A., 1987b, Involvement of functional protein kinase C in the mitogenic response to the H-*ras* oncogene product, Molec. and Cell. Biol., 7:4146-4151.

Lee, W., Mitchell, P. and Tjian, R., 1987, Purified transcription factor AP-1 interacts with TPA-inducible enhancer elements, Cell, 49:741-747.

Levy, M.F., Pocsidio, J., Guillem, J.G., Forde, K., LoGerfo, P., and Weinstein, I.B., Decreased levels of protein kinase C enzyme activity and PKC mRNA in primary human colon tumors, manuscript in preparation.

Makowske, M., Ballester, R., Cayre, Y. and Rosen, O.M., 1988, Immunochemical evidence that three protein kinase C isozymes increase in abundance during HL-60 differentiation induced by dimethyl sulfoxide and retinoic acid, J. Biol. Chem., 263:3402-3410.

Makowske, M. and Rosen O.M., 1989, Complete activation of protein kinase C by an antipeptide antibody directed against the pseudosubstrate prototope, J. Biol. Chem., 264:16155-16159.

Marais, R.M., Nguyen, O., Woodgett, J.R., and Parker, P.J., 1990, Studies on the primary sequence requirements for PKC-α,-βI and -γ peptide substrates, FEBS Lett., 277:151-155.

Melloni, E., S. Pontremeli, M. Michetti, O. Sacco, B. Sparatore, and Horecker, B.L., 1986, The involvement of calpain in the activation of protein kinase C in neutrophils stimulated by phorbol myristic acid, J. Biol. Chem., 261: 4101-4104.

Morotomi, M., Guillem, J.G., LoGerfo, P., and Weinstein, I.B., 1990, Production of diacylglycerol, an activator of protein kinase C, by human intestinal microflora, Cancer Research, 50:3595-3599.

Muramatsu, M.A., Kaibuchi, K., and Arai, K.I., 1989, A protein kinase C cDNA without the regulatory domain is active after transfection *in vivo* in the absence of phorbol ester, Molec. and Cell. Biol. 9:831-836.

Nair, P.P., 1988, Role of bile acids and neutral steroids in carcinogenesis, Am. J. Clin. Nutr., 48:768-774.

Nishisho, I., Nakamura, Y., Miyoshi, Y., Miki, Y., Ando, H., Horii, A., Koyama, K., Utsunomiya, Baba, S., Hedge, P., Markham, A., Krush, A.J., Peterson, G., Hamilton, S.R., Nilbert, M.C., Levy, D.B., Bryan, T.M., Preisinger, A.C., Smith, K.J., Su, L.-K., Kinzler, K.W., and Vogelstein, B., 1991, Mutations of chromosome 5q21 genes in FAP and colorectal cancer patients, Science, 253:665-669.

Nishizuka, Y., 1988, The molecular heterogeneity of protein kinase C and its implications for cellular regulation, Nature, 334:661-665.

Noda, M., Ko, M., Ogura, A., Liu, D., Amano, T., Takano, T., and Ikawa, Y., (1985), Sarcoma viruses carrying *ras* oncogenes induce differentiation-associated properties in a neuronal cell line, Nature, 318:73-75.

Nowell, P.C., 1986, Mechanisms of tumor progression, Cancer Research, 46:2203-2207.O'Driscoll, K.R., Kahn, S.M., Borner, C.B., Jiang, W., Fabrick, K.C.,and Weinstein, I.B. Stable expression and phenotypic analyses of regulatory domain mutants of protein kinase C βI in rat fibroblasts, manuscript in preparation.

Ohno, S., Akita, Y., Konno, Y., Imajoh, S., and Suzuki, K., 1988, A novel phorbol ester receptor/protein kinase, nPKC distantly related to the protein kinase C family, Cell, 53:731-737.

Ono, Y., Kurokawa, T., Fujii, T., Kawahara, K., Igarashi, K., Kikkawa, U., Ogita, K., and Nishizuka, Y., 1986, Two types of complementary DNAs of rat brain protein kinase C, FEBS Lett., 206:347-352.

Ono, Y., Fujii, T., Ogita, K., Kikkawa, U., Igarashi, K., and Nishizuka, Y., 1988, The structure, expression, and properties of additional members of the protein kinase C family, J. Biol. Chem., 263:6927-6932.

Ono, Y., Fujii, T., Igarashi, K., Kuno, T., Tanaka, C., Kikkawa, U., and Nishizuka, Y., 1989, Phorbol ester binding to protein kinase C requires a cysteine-rich, zinc-finger-like sequence, Proc. Natl. Acad. Sci. USA., 86:4868-4871.

Osada, S., Mizuno, K., Saido, T.C., Akita, Y., Suzuki, K., Kuroki, T., and Ohno, S., 1990, A phorbol ester receptor/protein kinase, nPKCη, a new member of the protein kinase C family predominantly expressed in lung and skin, J. Biol. chem., 265:22434-22440.

Parker, P.J., Coussens, L., Totty, N., Rhee, L., Young, S., Chen, E., Stabel, S., Waterfield, M.D., and Ullrich A., 1986, The complete primary sequence of protein kinase C- The major phorbol ester receptor, Science, 233:853-859.

Pears, C.J., Kour, G., House, C., Kemp, B.E., and Parker, P.J., 1990, Mutagenesis of the pseudosubstrate site of protein kinase C leads to activation, Eur. J. Biochem, 194:89-94.

Perera, F.P., 1987, Molecular cancer epidemiology: a new tool in cancer prevention, J. Natl. Cancer Int., 78:887-898.

Pfeffer, L.M., Strulovici, B., and Saltiel, A.R., 1990, Interferon-α selectively activates the β isoform of protein kinase C through phosphatidyl choline hydrolysis, Proc. Natl. Acad. Sci. USA., 87:6537-6541.

Price, B.D., Morris, J.D.H., Marshall, C.J. and Hall, A., 1989, Stimulation of phosphatidyl choline hydrolysis, diacylglycerol release, and arachidonic acid production by oncogenic *ras* is a consequence of protein kinase C activation, J. Biol. Chem., 264: 16638-16643.

Priess, J., Loumis, C.R., Bishop, W.R., Stein, R., Niedel, J.E., and Bell, R.M., 1986, Quantitative measurement of *sn*-1,2-diacylglycerols present in platelets, hepatocytes, and *ras*- and *sis*-transformed normal rat kidney cells, J. Biol. Chem., 261: 8597-8600.

Quintanilla, M., Brown, K., Ramsden, M., and Balmain, A., 1983, Carcinogen-specific mutation and amplification of Ha-*ras* during mouse skin carcinogenesis, Nature, 322:78-80.

Rauscher, F.J., Cohen, D.R., Curran, T., Bos, T.J., Vogt, P.K., Bohmann, D., Tjian, R., and B.R. Franza, 1988, *Fos*-associated protein p39 is the product of the *jun* proto-oncogene, Science, 240:1010-1017.

Rodriguez-Pena, A. and Rozengurt, E., 1984, Disappearance of Ca^{2+}-sensitive, phospholipid-dependent protein kinase activity in phorbol ester-treated 3T3 cells, Biochem. Biophys. Res. Comm., 120:1053-1059.

Rovera, G., Santoli, D., and Damsky, C., 1979, Human promyelocytic leukemia cells in culture differentiate into macrophage-like cells when treated with a phorbol diester, Proc. Natl. Acad. Sci. USA., 76:2779-2783.

Santella, R.M., 1988, Application of new techniques for the detection of carcinogen adducts to human population monitoring, Mutation Research, 202:271-282.

Sassone-Corsi, P., Lamph, W.W., Kamps, M., and Verma, I.M., 1988, Fos-associated cellular p39 is related to nuclear transcription factor AP-1, Cell, 54:553-560.

Schapp, D., Parker, P.J., Bristol, A., Kriz, R., and Knopf, J., 1989, Unique substrate specificity and regulatory properties of PKCε: A rationale for diversity, FEBS Lett., 243:351-357.

Schaap, D., Hsuan, J., Totty, N., and Parker, P.J., 1990, Proteolytic activation of protein kinase C-ε, Eur. J. Biochem., 191:431-435.

Solomon, D.H., O'Driscoll, K.R., Sosne, G., Weinstein, I.B., and Cayre, Y., 1991, 1α,25-Dihydroxyvitamin D_3-induced regulation of protein kinase C gene expression during HL-60 cell differentiation, Cell Growth and Differ., 2:187-194.

Strulovici, B., S. Daniel-Issakani, G. Baxter, J. Knoph, L. Sultzman, H. Cherwinski, J. Nestor Jr., D.R. Webb, and Ransom, J., 1991a, Distinct mechanisms of regulation of protein kinase Cε by hormones and phorbol diesters, J. Biol. Chem., 256:168-173.

Strulovici, B., Oto, E., Issakani, I., and Baxter G., 1991b, Characterization of PKC isoforms in an established SCLC cell line: regulation of PKCε and its catalytic fragment, J. Cell. Biochem. Suppl., 15B:166.

Vogelstein, B., Fearon, E.R., Hamilton, S.R., Kern, S.E., Preisinger, A.C., Leppert, M., Nakamura, Y., White, R., Smits, A.M.M., Bos, J.L., 1988, Genetic alterations during colorectal-tumor development. New Engl. J. Med., 319:525-531 (1988).

Weinberg, R.A., 1989, Positive and negative controls on cell growth, Biochemistry, 28:8263-8269.

Weinstein, I.B., 1988a, The origins of human cancer: Molecular mechanisms of carcinogenesis and their implications for cancer prevention and treatment. Twenty-seventh G.H.A. Clowes Memorial Award Lecture, Cancer Research, 38:4135-4143.

Weinstein, I.B., 1988b, Strategies for inhibiting multistage carcinogenesis based on signal transduction pathways, Mutation Research, 202:413-420.

Weinstein, I.B., Borner, C.M., Krauss, R.S., O'Driscoll, K.R., Choi, P.M., Morotomi, M., Hoshina, S., Hsieh, L.L., Tchou-Wong, K.M., Guadagno, S.N., Ueffing, M., and Guillem, J.G., 1990, Pleiotropic effects of protein kinase C and the concept of carcinogenesis as a progressive disorder in signal transduction, In: "The Origins of Human Cancer, Cold Spring Harbor Laboratory, Cold Spring Harbor, New York.

Weinstein, I.B., 1991a, Mitogenesis is only one factor in carcinogenesis, Science, 251:387-388.

Weinstein, I.B., 1991b, Cancer prevention: recent progress and future opportunities, Cancer Research (Suppl.), 51:5080s-5085s.

Wolfman, A., and Macara, I.G., 1987, Elevated levels of diacyglycerol and decreased phorbol ester sensitivity in ras-transformed fibroblasts, Nature, 325:359-362.

Woodgett, J.R. and Hunter, T., 1987, Immunological evidence for two physiological forms of protein kinase C, Molec. and Cell. Biol., 7: 85-96.

Young, S., Parker, P.J., Ullrich, A., and Stabel, S., 1987, Down-regulation of protein kinase C is due to an increased rate of degradation, Biochem. J., 244:775-779.

ONCOGENE ACTIVATION AND HUMAN CANCER

Demetrios A. Spandidos[1,2] and Margaret L.M. Anderson[3]

[1]Inst. Biolog. Res. Biotech., NHR, Athens, Greece
[2]Medical School, University of Crete, Heraklion, Greece
[3]Dept. Biochem., University of Glasgow, Scotland

MECHANISMS OF ONCOGENE ACTIVATION

Cellular oncogenes are mutated derivatives of normal cell genes "proto-oncogenes" which play a role in the control of cell growth and differentiation. Two main approaches have lead to the identification of oncogenes (reviewed in ref. 1). The first came from the studies of acutely transforming retroviruses. These viruses transform susceptible cells and their transforming genes e.g. *ras*, *myc*, *myb*, etc., have nucleotide sequence homology to cellular sequences (2). Approximately 25 such genes have been cloned and characterized (2). The second approach was based on the observation that chromosomes or DNA from malignant cells could transform cells in culture (3,4). Subsequently, "normal" NIH 3T3 fibroblasts were transfected with DNA from tumor-derived cell lines or fresh tumors and the transforming sequences were found to be homologues to oncogenes already identified in retroviruses e.g. H-*ras* and K-*ras*. However, other genes had not previously identified e.g. *neu* and N-*ras* (5). Approximately 25 more oncogenes have been characterized by this gene transfer approach.

The activation of proto-oncogenes to oncogenes can occur by a variety of ways. These include:

Transduction

This involves recombination between retroviral and host DNA which leads to incorporation of host sequences within the viral DNA (6). Although, observed in cats it is not thought to play a role in the generation of human tumors.

Insertional mutagenesis

Slowly transforming retroviruses may integrate next to a cellular proto-oncogene and cause its transcriptional activation through its regulatory sequences in the long terminal repeat (LTR) of the viral DNA (6). Unscheduled expression of the proto-oncogene has been observed in mice (mouse mammary tumor virus) and chicken (avian leukosis virus). So far have been no reports of insertional mutagenesis in humans.

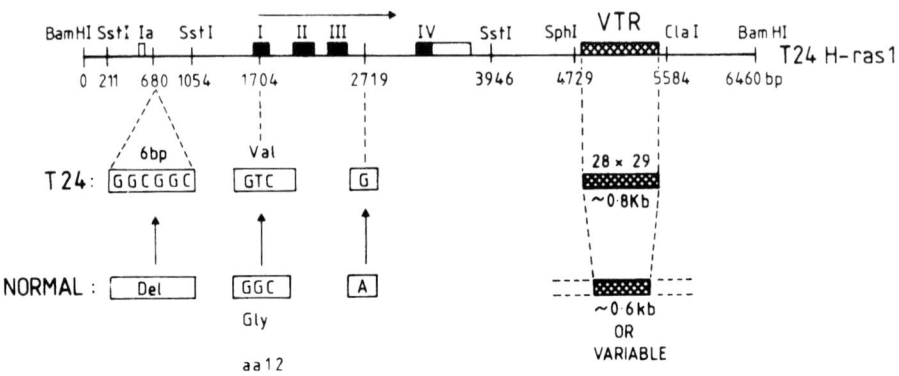

Fig. 1. The structure of the human H-*ras*1 proto-oncogene and a transforming oncogene derivative, T24. The nucleotide sequence is from Capon et *al.* (12). The coding sequences are designated by black boxes and the non-coding exon by open boxes. The variable tandem repeat (VTR) is indicated by a hatched box and the transcriptional orientation by an arrow. There are four differences between the normal H-*ras*1 protooncogene and the T24 H-*ras*1 oncogene. T24 H-*ras*1 has an extra six base pairs (GGCGGC) in the first intron, a G to T base conversion in the first coding exon (at aa 12), an A to G base conversion in the fourth intron and a longer (0.8 kb instead of 0.6 kb) VTR region.

Chromosome translocation

A high proportion of patients with certain cancers e.g follicular lymphoma or chronic myelocytic leukemia carry specific chromosomal translocations. It has been found that proto-oncogenes adjacent to chromosomal breakpoints are activated such that either a hybrid protein is produced or an unscheduled synthesis occurs (7,8).

Mutations

An analysis of human tumors has shown that certain proto-oncogenes are often mutated in certain cancers. For example, specific mutations are found in N-*ras* in a relatively high proportion of patients with leukemias, K-*ras* in lung and colon carcinomas and H-*ras* in a small proportion of breast and bladder carcinomas. Because these specific mutations cause transformation in vitro they are thought to contribute to malignant transformation *in vivo* (9).

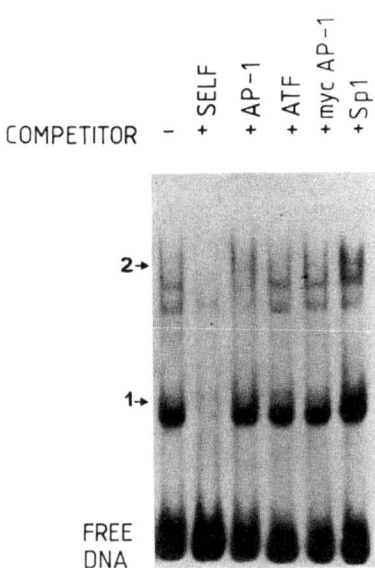

Fig. 2. Gel retardation assay. A 30 mer oligonucleotide (carrying the 6 bp present in the first non-coding exon of the T24 H-ras1 gene was incubated with nuclear preparations of HeLa cells. Competitor 30mer oligonucleotides (17) were present at 100x molar excess as indicated at the top of the gel. The arrows indicate DNA-protein complexes, that were competed out by the homologous oligonucleotides.

Amplification

The N-*myc* proto-oncogene is amplified in a significant proportion of patients with neuroblastoma or small cell lung carcinoma. As amplification is associated with a more aggressive stage of the disease and with poor prognosis, high levels of expression of this gene may be involved in the pathogenesis of these cancers (10).

ras ONCOGENES

ras genes are members of a multigene family (11) . They are highly conserved in evolution from yeast to man and play a role in cell growth and differentiation (11). Their protein products share sequence homology with those regions of G proteins involved in GDP and GTP binding (11) . Both G and *ras* proteins are located in the inner site of the cytoplasmic membrane and both have GTPase activity (11). Since G proteins are known to be signal transducers, *ras* proteins are thought to act in a similar manner.

The structure and organization of the human H-*ras*1 proto-oncogene and a derivative oncogenic mutant T24 are shown in Fig. 1. Apart from the mutation in amino acid 12 other differences involve a 6 bp addition in the T24 oncogene in the first intron (12), a conversion of A to G in the fourth intron (13) and a longer variable tandem repeat (VTR) of 0.8 kb instead of 0.6 kb (12) at the 3' end of the gene. The mutation in the first coding exon which leads to the conversion of glycine to valine results in the activation of the proto-oncogene (11). The other changes are located in regulatory sequences and all may act to increase transcription of the gene (12-14) . At least *in vitro*, increased expression of the H-*ras*1 contributes to the development of cancer (13,15). Therefore, it is possible that alterations in the above regulatory sequences could also participate in the development of cancer.

Many cis-acting regulatory sequences bind proteins which mediate their effect (16). To determine if the regulatory sequence in the first intron of the H-*ras*1 gene binds proteins a gel retardation assay is performed. Results are shown in Fig. 2. Two retarded bands disappeared when excess homologous cold competitor oligonucleotide was added but not when competitors with consensus binding sites for AP-1 and Sp1 proteins were added. This suggests that the T24 gene does indeed contain specific protein binding site(s). The fact that the two bands do not disappear probably means that there is non--specific protein binding even though poly dI:poly dC was added to reduce it. These experiments provides preliminary evidence for protein binding to the mutant T24 sequence and should provide the basis for further studies comparing the normal and mutant H-*ras*1 sequences.

The transmission of a proliferation signal from the cell surface to DNA in the nucleus leads to transcription of specific genes such as *fos* and AP-1/*jun* both of which are proto-oncogenes. Elevated levels of T24 H-*ras*1 lead to increased levels of both *fos* RNA and protein (18,19) . Since *fos* and AP-1/*jun* form a heterodimer which binds to DNA leading to transcription of specific genes it was of interest to determine whether the T24 H-*ras*1 gene would also increase the levels of AP-1/ *jun* .

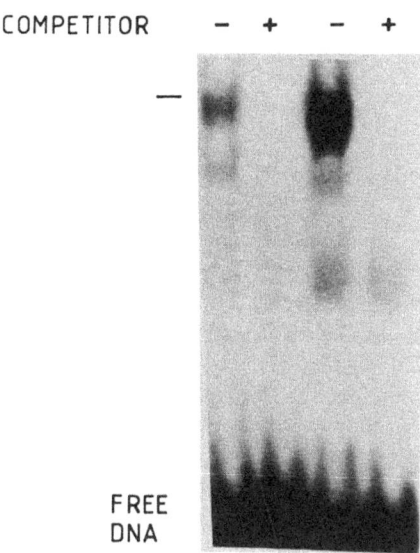

Fig. 3. Gel retardation assay of the adenovirus E3 promoter AP1 site oligonucleotide (AP-1) by the nuclear preparations of rat 208F and FRFAGT1-1 cells (which were derived after transfection of 208F with plasmid pAGT1 carrying the human T24 H-*ras*1 gene) . The arrows show AP-1/oligonucleotide complexes.

It has previously shown that the RFAGT1-1 cells derived by transfection of rat 208F fibroblasts by a plasmid carrying the T24 H-*ras*1 gene express elevated levels of the (mutant) *ras* p21 protein as compared with the non-transfected parent (18). As shown in Fig. 3 nuclear extracts from 208F cells bound much less AP-1 than RFAGT1-1 cells. This suggests that over expression of the mutant ras gene does indeed lead to an increase in the level of AP-1/ *jun*. It would be of interest to determine if increased expression of the normal gene also causes this effect.

The tumor promoter TPA (a phorbol ester) acts epigenetically (through protein kinase C?) to increase transcription from both the normal and the mutant T24 H-*ras*1 promoters in functional *in vitro* assays (20, 21) . It may be true of tumor promoters in general that they increase expression of H-*ras*1 genes.

The promoter of H-*ras*1 which is located 5' of the noncoding exon contains four consensus elements for AP-1/*jun* binding (20, 21) . They may be required for H-*ras*1 gene expression but this has not been shown. It is tempting to speculate that elevated levels of H-*ras*1 may lead to elevated levels of *fos* and *jun* and elevated levels of the latter may lead to elevated levels of the H-*ras*1. Thus, there may be a circuit between the expressions of these nuclear and cytoplasmic oncoproteins. It is of interest that AP-1/ *jun* (22) and H-*ras*1 (23, 24) are both elevated in a high proportion of breast tumors which is consisted with the above hypothesis. Future studies will focus on this relationship.

REFERENCES

1. Spandidos D.A. and Anderson M.L,M, J. Pathol. 157, 1-10, 1989.
2. Bishop J.M. Science 235, 305-311, 1987.
3. Spandidos D.A. and Siminovitch L. Cell 12, 675-682, 1977.
4. Spandidos D.A. and Siminovitch L. Nature 271, 259-261, 1978.
5. Weinberg R.A. Science 230, 770-776, 1985.
6. Varmus H. Ann. Rev. Gen, 18, 553-612, 1984.
7. Haluska F.G., Tsijimoto Y. and Croce C. M. Ann. Rev. Gen. 21, 321-345, 1987.
8. Cory S. Adv. Cancer Res. 47, 18~-239, 1986.
9. Bos J.L. Cancer Res. 49, 4682-4689, 1989.
10. Brodeur D . M., Seeger R . C ., Schwab M. *et al* . Science 224, 1121-1124, 1984.
11. Barbacid M. Ann. Rev. Biochem. 56, 779-827, 1987.
12. Capon D.J., Chen E.Y., Levinson A.D., Seeburg P.H. and Goeddel D.V. Nature 302, 33-37, 1983.
13. Cohen J.B. and Levinson A.D. Nature 334, 119-123, 1988.
14. Spandidos D.A. and Holmes L.   FEBS Lett. 218, 41-46,1987.
15. Spandidos D.A. and Pintzas A. FEBS Lett. 232, 269-274, 1978.
16. Jones, N. Sem. Cancer Biol. 1, 5-17, 1990.
17. Whitelaw, B., Wilkie, N.M., Jones, K.A., Kadonga, J.T., Ti jan, R . and Lang, J . C . In Growth Factors, Tumor Promoters and Cancer Genes, Publ. Alan Liss Inc. pp 337-351, 1988 .
18. Wyllie A.H., Rose K.A., Morris R.G., Steel C.M., Foster E. and Spandidos D.A.. Br. J. Cancer 56, 251-259, 1987.
19. Spandidos D.A. and Lang J.C. CRC Crit. Rev. Oncogen. 1, 195-209, 1989.
20. Spandidos D.A. and Riggio M. FEBS Lett. 203, 169-174, 1986.
21. Spandidos D.A., Nichols R.A.B., Wilkie N.M. and Pintzas A. FEBS Lett. 240, 191-195, 1988 .
22. Linardopoulos S., Malliri A., Pintzas A, Vassilaros S., Tsikinis A. and Spandidos, D.A. Anticancer Res. In Press.
23. Spandidos D.A. and Agnantis N. Anticancer Res. 4, 269- 272, 1984.
24. Spandidos D.A. Anticancer Res. 7, 991-996, 1987.

GENETICS OF HEPATOCARCINOGENESIS IN MOUSE AND MAN

Tommaso A. Dragani, Giacomo Manenti,
Manuela Gariboldi, F. Stefania Falvella,
Marco A. Pierotti, and Giuseppe Della Porta

Division of Experimental Oncology A
Istituto Nazionale Tumori
Via G. Venezian 1, 20133 Milan, Italy

ABSTRACT

Mice genetically susceptible to hepatocarcinogenesis carry a germ-line genetic alteration, which has liver-specific effects. It should be considered as an initial step in the multistage process of hepatocarcinogenesis. Any other step could add to the genetic alteration, resulting in a high susceptibility to spontaneous and carcinogen induced liver tumorigenesis. A quantitative evaluation of carcinogen-induced liver tumors showed that the genetic alteration controls liver tumor growth, but not liver tumor frequency. Our recent observations indicate that Ha-*ras* gene mutations at codon 61 are frequently involved in the pathogenesis of liver tumors in mice genetically susceptible to hepatocarcinogenesis, but not in genetically resistant mice. Some rare human genetic syndromes are characterized by a high predisposition to hepatocellular carcinoma development, through the prior induction of cirrhosis. However, these syndromes do not represent the human counterparts of the genetic alterations responsible for their high susceptibility to hepatocarcinogenesis of the C3H and of the CBA mouse strains.

INTRODUCTION

Tumor development can be modulated by different factors, i.e., age, sex, diet, chemical carcinogens, and genetic constitution. The role of genetic constitution in tumorigenesis has become more clear in the few last years, concomitantly with the identification of the genes and of the chromosomal regions linked with various types of tumors (1-3). Experimental evidence has also shown that artificially introduced germ-line gene alterations behave as genetic traits conferring a high susceptibility to carcinogenesis in transgenic mice (4).

Taking into account the multistep nature of carcinogenesis, we can clearly agree that multiple genetic alterations are needed for neoplastic transformation, tumor promotion and progression (5,6). If any of the different genetic alterations involved in tumorigenesis are already present in the germ-line of a given individual or animal, one is genetically prone to a high level of cancer susceptibility.

MURINE MODELS

The genetics of liver tumor susceptibility has been extensively studied in the past. The genetically susceptible C3H and CBA inbred strains spontaneously develop hepatocellular tumors at a high incidence whereas other strains display a low incidence of spontaneous liver tumors (Table 1). Strains C3H and CBA are closely related, since both arose in 1920 from a single cross of the Bagg and DBA lines (14, 15). Phylogenetic analysis based on genetic variation at 97 loci confirmed the close relationship between the CBA and the C3H mice (16). Therefore, we can hypothesize that the same genetic alteration(s) are responsible for the high susceptibility to hepatocarcinogenesis for both strains.

Table 1. Murine inbred strains with different susceptibility to the development of spontaneous hepatocellular tumors[1]

Strain/ hybrid	% Liver tumor incidence (range)	Duration[2] of the experiment (range)	References
129/J	0	LS	7
A/He	0	12 mo	8
BALB/c	0-9	1 yr-140 wks	9,10
C3H	14-67	1 yr-LS	7, 9, 11-13
C57BL	0-3	1 yr-LS	9, 11, 7
CBA	15-23	1 yr-LS	9, 7
DBA/2J	1	LS	7
IF	0	1 yr	9
LP/J	6	LS	7
SWR	0	100 wks	11

[1] Male mice only. [2] LS, lifespan.

There is a correlation between the degree of spontaneous liver tumor incidence and the susceptibility to the development of liver tumors after treatment with chemical carcinogens (9,11,17-20).

Genetically susceptible mice develop liver tumors at a higher incidence and in a shorter period of time with respect to their control groups when they are treated with some non-genotoxic chemicals, whereas genetically resistant mice do not develop liver tumors in the same experimental conditions (Table 2). Among the chemicals causing such effect, phenobarbital (PB) induces liver hyperplasia, increases the levels of different liver enzymes involved in drug metabolism and it is a tumor promoter of liver carcinogenesis in rats and mice (24). Tetrachlorodibenzo-p-dioxin (TCDD) is also an enzyme inducer and a promoter of rat liver carcinogenesis (25). Rifampicin (RFP) is an antibiotic whose liver tumor effects were observed only in the females of the C3Hf strain, but not in the C3Hf males, nor in both sexes of BALB/c mice. A RFP-induced hormonal imbalance was probably responsible for the RFP liver carcinogenesis effects in C3Hf female mice (12).

The liver carcinogenesis effects observed with non-genotoxic chemicals in genetically susceptible, but not in resistant mice, suggest that the susceptible mice carry in their germ-line an alteration at a genetic locus involved in the pathogenesis of liver tumors. In comparison with the transgenic mice, other somatically acquired gene alterations (either spontaneously, or after treatment with chemical carcinogens or with non-genotoxic liver tumor promoters) are needed to confer the full tumor phenotype to hepatocytes. The need of multiple genetic alterations in the same hepatocyte explains why few liver tumors develop in the susceptible mice, although all their hepatocytes already carry a genetic alteration.

Susceptibility genes control nodule growth

Quantitative estimations of the degree of difference in susceptibility to hepatocarcinogenesis among strains showed wide variations, depending on the examined parameters,

Table 2. Different liver carcinogenesis response among murine inbred strains and F1 hybrids to non genotoxic chemicals administered alone (i.e., without previous initiating carcinogen treatment)

Chemical	Strain/Hybrid	Liver tumor incidence (%)		References
		Control	Treated	
PB	C3Hf	41	94	13
	C3H/HeN	61	100	21
	C3H/HeN	0	40	18
	B6C3	29	100	21
	C57BL/6N	0	0	21
	C57BL/6N	0	0	18
	DBA/2	0	0	18
	A/He	0	0	8
	BALB/c	0	0	22
RFP	C3Hf	23	80	12
	BALB/c	3	0	12
TCDD	B6C3	20	58	23
	B6C	3	0	23

e.g., tumor incidence, tumor multiplicity, and tumor size (11,18-21,26). Quantitative parameters should allow estimation of tumor frequency and size, because the genetically susceptible phenotype may result from a high susceptibility either to initiation (high tumor number) or to promotion/progression (high tumor size). We used stereological methods that have been shown to be very useful for the quantitative analysis of hepatocarcinogenesis and, as quantitative indexes of susceptibility, we considered the number of nodules/cm^3 of liver (N/cm^3) at different cutoffs (an estimate of tumor frequency) and the percentage of liver volume occupied by nodules (V%) (an estimate of both tumor frequency and size, i.e., it represents the volume fraction of liver occupied by tumors).

Recently, we have carried out a liver carcinogenesis experiment with mice of the A/J, BALB/c, C3H/He strains and of the AC3 and CC3 F1 hybrids treated with a single dose (300 mg/kg, s.c. at 1 wk of age) of the carcinogen urethan and then kept without further treatment until 30 and 40 wks (males), or 30 and 65 wks (females). The degree of difference in susceptibility to hepatocarcinogenesis between the susceptible C3H/He and AC3 and the resistant A/J and BALB/c mice, in the different age and sex groups, was 4-12-fold as indicated by the analysis of the N/cm^3, and was more than 400-fold as indicated by the V% (Table 3 and ref. 27).

Table 3. Liver tumor susceptibility in A/J, BALB/c, C3H/He, AC3 and CC3 male mice treated once with urethan (300 mg/kg s.c.) at 1 wk of age and killed at 40 wks of age [a]

Strain/Hybrid	No. of mice	Incidence(%)	Liver tumors[b] N/cm^3	V%
A/J	19	74	11.8±3.3	0.06±0.02
BALB/c	20	40	3.7±1.6	0.10±0.06
C3H/He	17	100	44.1±4.8	47.40±5.07
AC3	18	100	45.0±7.9	34.55±5.17
CC3	13	100	43.8±5.2	2.82±0.74

[a] Values represent mean±SE. [b] N/cm^3 and V% evaluated at cutoff for radii=120 μm

The latter results confirmed our previous findings in diethylnitrosamine (NDEA)-treated mice and the observation of another group (20,26), indicating that the frequency of carcinogen-induced liver lesions was about the same in susceptible and resistant strains, but tumors tended to be much larger in susceptible than in resistant mice. Therefore, the gene(s) responsible for the genetic susceptibility to hepatocarcinogenesis control liver tumor growth (tumor progression).

Ha-*ras* gene mutations

As a first approach to studying molecular alterations that could lead to a higher growth rate of liver tumors in the genetically susceptible rather than in resistant strains, we have analyzed the frequency of Ha-*ras* gene mutations in liver tumors developed in mice genetically susceptible and resistant to liver tumorigenesis.

Mutations of the Ha-*ras* gene, mostly localized at codon 61, have been previously reported in a high percentage of mouse liver tumors, either spontaneously developed or induced by chemical carcinogens in the genetically susceptible B6C3 hybrid (28-30). We found it interesting to investigate whether or not the mutations at codon 61 of Ha-*ras* gene are present with the same frequency in hepatocellular tumors developed in murine strains and hybrids genetically susceptible (C3Hf, B6C3) and resistant (BALB/c, B6C) to hepatocarcinogenesis, since no previous comparative analysis was available.

The incidence of mutations at codon 61 of the Ha-*ras* gene was analyzed by selective oligonucleotide hybridization of polymerase chain reaction amplified tumor DNAs. Of 27 hepatocellular tumors from untreated C3Hf and B6C3 male mice, 16 (59%) showed mutations at codon 61 of Ha-*ras* gene. By contrast, only one out of 15 (7%) BALB/c liver tumors from untreated mice had a mutation (31). In urethan-induced liver tumors, only 1 out of 11 tumors developed in B6C male mice contained a mutation. In contrast, among a total of 15 urethan-induced liver tumors analyzed in the B6C3 male mice, 9 tumors contained mutations at codon 61 of the Ha-*ras* gene (31).

Our results indicate that liver tumors developed in genetically resistant mice (BALB/c, B6C) display a low frequency of the Ha-*ras* mutations compared to the frequency found in the corresponding tumors of genetically susceptible mice (C3Hf, B6C3). Recent results of Buchmann et al. (32) are in agreement with our findings.

The higher mutation frequency at codon 61 of Ha-ras gene in liver tumors seen in genetically susceptible mice compared with genetically resistant mice could be the result of different mechanisms. Obviously, the Ha-*ras* gene does not represent the gene responsible for genetic susceptibility to hepatocarcinogenesis, because the Ha-*ras* gene is ubiquitously expressed (33) and is not liver-specific, and because mutations at codon 61 are somatically acquired in liver tumors and are not present in the germline. However, the specificity of liver tumors developed in the genetically susceptible mice, with regard to the Ha-*ras* mutations, should be taken into account in the carcinogenicity risk assessment of chemicals able to induce liver tumors only in B6C3 mice, but not in other murine strains nor in rats (34,35).

HUMAN MODELS

The hepatocellular carcinoma (HCC) is a relatively rare cancer in developed countries, with an incidence of 1-4 cases per 100,000 males per year, but it is very common in high risk areas, with an incidence of up to 150 cases per 100,000 males per year (36). A close association exists between HCC and hepatitis B virus (HBV) infection, that increases the relative risk of developing HCC up to 200-fold (37). It has been proposed that other factors, such as aflatoxin exposure, alcohol consumption, etc., are involved in HCC development. However, a small percentage of HCCs could be attributed to genetic factors. In fact, the following series of rare genetic syndromes is associated with an increased risk of developing HCC (Table 4).

Alpha$_1$-antitrypsin deficiency

Alpha$_1$-antitrypsin deficiency (α_1AT) is an autosomal recessive disease resulting from mutations in the gene α_1AT (38) at the locus PI located on chromosome 14q32.1 (39). α_1AT is a glycoprotein synthesized at high levels in the liver, and it is the main protease inhibitor in human serum. The biochemical function of α_1AT is to inhibit neutrophil elastase (38).

The PI locus is pleiomorphic, with approximately 75 alleles so far identified (38). These alleles can be classified as "normal" (alleles coding for α_1AT proteins can be seen in normal amount and with normal functions) and "at risk" (alleles coding for either α_1AT proteins can be seen in serum at levels lower than normal, or no α_1AT proteins in serum, or α_1AT proteins with altered functions) (38). The two parental alleles are both expressed, so that inheritance of a normal allele protects against the risk conferred by a second at risk allele, and α_1AT deficiency develops only when the individuals bear two at risk alleles.

Table 4. Genetic syndromes associated with the development of hepatocellular carcinoma in humans

Syndrome	Locus Name	Chromosome	Mode of Inheritance[1]	Excess Risk
Tyrosinemia I	FAH	15q23-q25	AR	very high
α1-antitrypsin deficiency	PI	14q32.1	AR	20
Hemochromatosis	HFE	6p21.3	AR	200
Acute intermittent porphyria	PBGD	11q23.2-qter	AD	60
Porphyria cutanea tarda	UROD	1p34	AD	100-200
Glycogen storage disease type I	–	?	AR	high

[1] AR, autosomal recessive; AD, autosomal dominant.

Patients with α_1AT deficiency can develop pulmonary emphysema and liver disease. The mechanism of disease is probably related to the altered protease/anti-protease balance. α_1AT deficiency is associated with an increased risk of liver cirrhosis and HCC, especially in males (40). The overall relative risk of HCC in α_1AT deficiency is 20, when compared with the control population (40).

Hereditary tyrosinemia type I

Hereditary tyrosinemia type I is an autosomal recessive disorder caused by a deficiency of fumaryl acetoacetate hydrolase, the last enzyme in the catabolic pathway of tyrosine (41,42). The accumulation of intermediates of tyrosine catabolism, due to the enzymatic deficit, is believed to be the cause of the disease which results either in acute hepatic failure in infancy or in a chronic liver disease associated with cirrhosis and HCC development (43). The risk of the occurrence of HCC is very high. About 40% of patients surviving beyond 2 years of age develop HCC in their childhood (43,44). The locus of fumaryl acetoacetate hydrolase (FAH) has been mapped on chromosome 15q23-q25 (39).

Hemochromatosis

This disease is characterized by an excessive gastrointestinal iron absorption and by a consequent increased iron storage occurring in all organs, except for the brain and nervous tissue (45, 46). The biochemical defect responsible for the increased iron absorption is unknown. Fibrous tissue reaction develops where the iron is deposited; limited liver dysfunction are present in early stages, but ultimately hepatic cirrhosis develops (46). Primary HCC develops in about 14% of cases, reflecting a 200-fold excess risk over the normal population (47, 48).

Genetic hemochromatosis (GH) is inherited as an autosomal recessive trait. Heterozygotes comprise 6-8% and homozygotes about 0.3-0.6% of the Caucasian population. However, clinical expression of the disease does not occur in many homozygotes (49). The hemochromatosis locus (HFE) has been mapped on chromosome 6p21.3, in close proximity to the HLA-A (39), but the product coded for by that gene is unknown.

Porphyrias

Porphyrias are a group of inherited diseases resulting from defects in the heme biosynthesis pathway. The acute intermittent hepatic porphyria (AIP) is clinically characterized by occasional acute attacks of abdominal pain and neuropsychiatric symptoms. Cirrhosis is not frequent in patients with AIP, but morphologic abnormalities of hepatocytes at liver biopsy and altered liver biochemical functions have been reported (50). The disease is inherited as an autosomal dominant trait, and it is related to a deficiency in porphobilinogen deaminase (PBG deaminase) (51). The locus of AIP (PBGD) has been mapped on chromosome 11q23.2-qter (39). Compared with the total population, the risk of HCC is increased about 60-fold in AIP (52).

Porphyria cutanea tarda (PCT) is transmitted as an autosomal dominant trait. The disease is due to a deficiency in uroporphyrinogen decarboxylase gene (51) localized on chromosome 1p34 (locus UROD) (39). PCT is associated with subacute hepatitis and liver cirrhosis (53). Several studies have indicated a 100 to 200-fold relative risk of HCC in PCT patients (54-56).

Glycogen storage disease type I

Glycogen storage disease type I (Von Gierke's disease) is caused by the impairment of glucose-6-phosphatase (G6Pase) activity, with a consequent excess glycogen storage in the liver. Two subtypes of this disease are recognized: type Ia, in which there is a complete absence of G6Pase, and type Ib, in which there is a deficiency of the glucose-6-phosphate translocase at the endoplasmic reticulum membrane (57, 58). The clinical features are generally similar in both subtypes: hepatomegaly, fasting hypoglycemia, lactic acidosis, hyperlipidemia, hyperuricemia, and growth retardation (59, 60). Liver functions are usually normal or show only minor deviations. Cirrhosis does not develop. Hepatocellular adenomas develop in most patients with type Ia glycogen storage disease by their second or third decade of life (59-61). About 10% of type Ia glycogenesis patients with liver adenoma later develop an HCC (60-62). The disease is inherited as an autosomal recessive trait.

Role of cirrhosis

Cirrhosis usually results as a reaction to past or ongoing liver necrosis. Hepatocellular necrosis leads to regeneration, cell turnover and proliferation of hepatocytes. Inflammation accompanying the hepatocellular necrosis stimulates fibroplasia. These processes produce cirrhosis which is accompanied by further reactive proliferation of hepatocytes.

A close association exists between HCC and cirrhosis. Patients known to have cirrhosis manifest an increased risk of HCC development, and liver cirrhosis is frequently found in patients with HCC. The type and etiology of the cirrhosis are important factors in determining the risk of HCC, which is greater with macronodular cirrhosis (with 15 to 55% of patients who develop HCC) and lower with micronodular cirrhosis (3 to 10% of cases developing HCC). The macronodular type is most often caused by chronic HBV infection, whereas theMicronodular one results from alcohol abuse (63-65). The prevalence of cirrhosis in non-neoplastic liver of patients with HCC varies among reports and geographical regions, ranging from 60 to 90% (63, 65, 66). On the whole, cirrhosis appears to constitute a major risk factor in the development of HCC (63-66).

It is worth noting that the human genetic disorders associated with an increased risk of the development of HCC are also associated with the development of cirrhosis, except for the glycogen storage disease type I. Therefore, the increased risk of the development of HCC may not be directly linked to the genetic syndromes, but it may be a consequence of the cirrhosis caused by metabolic disorders. So far, the genes whose alterations are involved with the liver metabolic disorders constitute an unique example of germ-line genetic alterations affecting indirectly the pathogenesis of a neoplasm, through mechanisms mediated by chronic damage of the target organ (cirrhosis).

COMPARATIVE ASPECTS

The existence of human genetic syndromes associated with an increased risk of HCC indicates that genetic susceptibility to hepatocarcinogenesis is not limited to the murine models. However, in comparing the human and the murine models of genetic susceptibility to hepatocarcinogenesis we must consider that there are important differences in the pathogenesis of liver tumors between mouse and man, although their morphologic, biochemical and biological characteristics are similar in both species (Table 5) (67,68). In humans, cirrhosis is an important component in the pathogenesis of HCC and, in fact, the human genetic disorder associated with an increased risk of HCC are also associated with the development of cirrhosis or of liver dysfunctions prior to HCC. In mice, cirrhosis develops very rarely, even after treatment with chemical carcinogens that rapidly induce hundreds of neoplastic lesions in the liver. Murine strains genetically susceptible to hepatocarcinogenesis do not manifest any apparent liver disfunction, nor any sign of liver cirrhosis that can be related to liver tumor development.

Table 5. Comparative aspects in hepatocellular carcinoma development

Alteration	Human	Mouse (B6C3)
Morphological		
-trabecular pattern	+	+
Biochemical		
-AFP	+	+
Biological		
-vein invasion	+	+
-metastasis	+	+
-transplantation	+	+
Cirrhosis	+	-
ras gene mutations	-	+
p53 mutations	+	?
Loss of heterozygosity	+	?

With regard to the mutation frequency of *ras* genes in liver tumors, humans appear similar to genetically resistant mice, since they both show a low mutation frequency (69, 70). However, we cannot exclude the possibility that a subset of human HCCs, deriving from genetically susceptible individuals, display a high mutation frequency of *ras* genes. Interestingly, it has been recently reported that mutations of the K-*ras* gene identify a subset of human lung adenocarcinomas, characterized by a poor prognosis, compared to *ras* mutation-negative counterparts (71).

Other genetic alterations found in human HCCs, i.e., p53 gene mutations, loss of heterozygosity at specific loci (72, 73) have not yet been studied in mouse liver tumors. Future studies can define similarities and differences between mouse and human hepatocellular tumors for these findings. The results of these studies could provide important insights for the interpretation of long-term carcinogenesis bioassays and for the risk assessment of the potential carcinogenicity of chemicals to humans.

ACKNOWLEDGEMENTS

This investigation was supported by grants from the Consiglio Nazionale delle Ricerche (Finalized Project "Oncology"), Rome, Italy, and Associazione Italiana Ricerca sul Cancro. The authors wish to thank Miss Sally Clegg and Mrs. Anna Grassi for assistance in the manuscript preparation.

REFERENCES

1. Li, F.P. Cancer families: Human models of susceptibility to neoplasia-The Richard and Hinda Rosental Foundation Award Lecture. Cancer Res., 48:5381-5386, 1988.
2. Knudson, A.G. Hereditary cancers: Clues to mechanisms of carcinogenesis. Br. J. Cancer, 59: 661-666, 1989.
3. Ponder, B.A.J. Inherited predisposition to cancer. Trends in Genetics, 6: 213-218, 1990.
4. Hanahan, D. Dissecting multistep tumorigenesis in transgenic mice. Ann. Rev. Genet., 22: 479-519, 1988.
5. Weinstein, I.B. The origins of human cancer: Molecular mechanisms of carcinogenesis and their implications for cancer prevention and treatment-Twenty-seventh G.H.A. Clowes Memorial Award Lecture. Cancer Res., 48: 4135-4143, 1988.
6. Weinberg, R.A. Oncogenes, antioncogenes, and the molecular bases of multistep carcinogenesis. Cancer Res., 49: 3713-3721, 1989.
7. Smith, G.S., Walford, R.L., and Mickey, M.R. Lifespan and incidence of cancer and other diseases in selected long-lived inbred mice and their F1 hybrids. J. Natl. Cancer Inst., 50: 1195-1213, 1973.
8. Naito, M., Naito, Y., and Ito, A. Effect of phenobarbital on the development of tumors in mice treated neonatally with N-ethyl-N-nitrosourea. Gann, 73: 111-114, 1982.
9. Flaks, A. The susceptibility of various strains of neonatal mice to the carcinogenic effects of 9,10-dimethyl-1,2-benzanthracene. Eur. J. Cancer, 4: 579-585, 1968.
10. Dragani, T.A. Analysis of tumor incidence in BALB/c mice used as controls in carcinogenicity experiments. Tumori, 65: 665-675, 1979.
11. Della Porta, G., Capitano, J., Parmi, L., and Colnaghi, M.I. Urethan carcinogenesis in newborn, suckling, and adult mice of C57BL, C3H, BC3F1, C3Hf and SWR strains. Tumori, 53: 81-102, 1967.
12. Della Porta, G., Cabral, J.R., and Rossi, L. Carcinogenicity study of rifampicin in mice and rats. Toxic. Appl. Pharmacol., 43: 293-302, 1978.
13. Peraino, C., Fry, R.J.M., and Staffeldt, E. Enhancement of spontaneous hepatic tumorigenesis in C3H mice by dietary phenobarbital. J. Natl. Cancer Inst., 51: 1349-1350, 1973.
14. Strong, L.C. The origin of some inbred mice. Cancer Res., 2: 531-539, 1942.
15. Morse, H.C. The laboratory mouse - a historical perspective. In: H.L. Foster, J.D. Small and J.G. Fox (eds.), The mouse in biomedical research, pp. 1-16, New York: Academic Press Inc. 1981.
16. Fitch, W.M. and Atchley, W.R. Evolution in inbred strains of mice appears rapid. Science, 228: 1169-1175, 1985.
17. Dragani, T.A., Sozzi, G., and Della Porta, G. Spontaneous and urethan-induced tumor incidence in B6C3F1 versus B6CF1 mice. Tumori, 70: 485-490, 1984.
18. Diwan, B.A., Rice, J.M., Ohshima, M., and Ward, J.M. Interstrain differences

in susceptibility to liver carcinogenesis initiated by N-nitrosodiethylamine and its promotion by phenobarbital in C57BL/6NCr, C3H/HeNCr^{MTV-} and DBA/2NCr mice. Carcinogenesis, 7: 215-220, 1986.

19. Drinkwater, N.R., and Ginsler, J. Genetic control of hepatocarcinogenesis in C57-BL/6J and C3H/HeJ. Carcinogenesis, 7: 1701-1707, 1986.
20. Dragani, T.A., Manenti, G., and Della Porta, G. Genetic susceptibility to murine hepatocarcinogenesis is associated with high growth rate of NDEA-initiated hepatocytes. J. Cancer Res. Clin. Oncol., 113: 223-229, 1987.
21. Becker, F.F. Morphological classification of mouse liver tumors based on biological characteristics. Cancer Res., 42: 3918-3923, 1982.
22. Cavaliere, A., Bufalari, A., and Vitali, R. Carcinogenicity and cocarcinogenicity test of phenobarbital sodium in adult BALB/c mice. Tumori, 72: 125-128, 1986.
23. Della Porta, G., Dragani, T.A., and Sozzi, G. Carcinogenic effects of infantile and long-term 2,3,7,8-tetrachlorodibenzo-p-dioxin treatment in the mouse. Tumori, 73: 99-107, 1987.
24. McClain, R.M. Mouse liver tumors and microsomal enzyme inducing drugs: experimental and clinical perspectives with phenobarbital. Prog. Clin. Biol. Res., 331: 345-365, 1990.
25. Pitot, H.C., Goldsworthy, T., Campbell, H.A., and Poland, A. Quantitative evaluation of the promotion by 2,3,7,8-tetrachlorodibenzo-p-dioxin of hepatocarcinogenesis from diethylnitrosamine. Cancer Res., 40: 3616-3620, 1980.
26. Hanigan, M.H., Kemp, C.J., Ginsler, J.J., and Drinkwater, N.R. Rapid growth of preneoplastic lesions in hepatocarcinogen-sensitive C3H/HeJ male mice relative to C57BL/6J male mice. Carcinogenesis, 9: 885-890, 1988.
27. Dragani, T.A., Manenti, G., and Della Porta, G. Quantitative analysis of genetic susceptibility to liver and lung carcinogenesis in mice. Cancer Res., 51: 6299-6303, 1991.
28. Reynolds, S.H., Stowers, S.J., Patterson, R.M., Maronpot, R.R., Aaronson, S.A., and Anderson, M.W. Activated oncogenes in B6C3F1 mouse liver tumors: implications for risk assessment. Science, 237: 1309-1316, 1987.
29. Stowers, S.J., Wiseman, R.W., Ward, J.M., Miller, E.C., Miller, J.A., Anderson, M.W., and Eva, A. Detection of activated proto-oncogenes in N-nitrosodiethylamine induced liver tumors: a comparison between B6C3F1 mice and Fisher 344 rats. Carcinogenesis, 9: 271-276, 1988.
30. Wiseman, R.W., Stowers, S.J., Miller, E.C., Anderson, M.W., and Miller, J.A. Activating mutations of the c-Ha-*ras* protooncogene in chemically induced hepatomas of the male B6C3 F1 mouse. Proc. Natl. Acad. Sci. USA, 83: 5825-5829, 1986.
31. Dragani, T.A., Manenti, G., Colombo, B.M., Falvella, F.S., Gariboldi, M., Pierotti, M., and Della Porta, G. Incidence of mutations at codon 61 of the Ha-ras gene in liver tumors of mice genetically susceptible and resistant to hepatocarcinogenesis. Oncogene, 6: 333-338, 1991.
32. Buchmann, A., Bauer-Hofmann, R., Mahr, J., Drinkwater, N.R., Luz, A., and Schwarz, M. Mutational activation of the c-Ha-ras gene in liver tumors of different rodent strains: Correlation with susceptibility to hepatocarcinogenesis. Proc. Natl. Acad. Sci. USA, 88: 911-915, 1991.
33. Leon, J., Guerrero, I., and Pellicer, A. Differential expression of the *ras* gene family in mice. Mol. Cell. Biol., 7: 1535-1540, 1987.
34. Maronpot, R.R., Haseman, J.K., Boorman, G.A., Eustis, S.E., Rao, G.N. and

Huff, J.E. Liver lesions in B6C3F1 mice: the National Toxicology Program, experience and position. Arch. Toxicol. Suppl. 10: 10-26, 1987.
35. Della Porta, G., and Dragani, T.A. Long-term assays for carcinogenicity. Terat. Carcinog. Mutag., 10: 137-145, 1990.
36. WHO. Cancer Incidence in Five Continents, Vol. V, Muir C., Waterhouse J., Mach T., Powell J., and Whelan S. (eds). IARC Sci. Publ. No. 88, IARC, Lyon, 1987.
37. Beasley, R.P., Hwang, L.Y., Lin, C.C., and Chien, C.S. Hepatocellular carcinoma and hepatitis B virus. A prospective study of 22707 men in Taiwan. Lancet, 2: 1129-1133, 1981.
38. Crystal, R.G. The α1-antitrypsin gene and its deficiency states. Trends in Genetics, 5: 411-417, 1989.
39. Harper, P.S., Frezal, J., Ferguson-Smith, M.A., and Schinzel, A. Report of the committee on clinical disorders and chromosomal deletion syndromes. In: Human Gene Mapping 10: Tenth International Workshop on Human Gene Mapping. Cytogenet Cell Genet, 51: 563-661, 1989.
40. Eriksson, S., Carlson, J., and Velez, R. Risk of cirrhosis and primary liver cancer in alpha$_1$ antitrypsin deficiency. N. Engl. J. Med., 314: 736-739, 1986.
41. Lindblad, B., Lindstedt, S., and Steen, G. On the enzymatic defects in hereditary tyrosinemia. Proc. Natl. Acad. Sci. USA, 74: 4641-4645, 1977.
42. Furukawa, N., Hayano, T., Sato, N., Inoue, F., Machida, Y., Kinugasa, A., Imashuku, S., Kusunoki, T., and Takamatisu, T. The enzyme defects in hereditary tyrosinemia type I. J. Inherited Metab. Dis., 7 (suppl 2): 137-138, 1984.
43. Dehner, L.P., Snover, D.C., Sharp, H.L., Ascher, N., Nakhleh, R., and Day, D.L. Hereditary tyrosinemia type I (chronic form): Pathologic findings in the liver. Hum. Pathol., 20: 149-158, 1989.
44. Weinberg, A.G., Mize, C.E., and Worthen, H.G. The occurrence of hepatoma in the chronic form of hereditary tyrosinemia. J. Pediatr., 88: 434-438, 1976.
45. Jacobs, A. The pathology of iron overload. In: Jacobs A, Worwood M (eds.) "Iron in biochemistry and medicine. II". London and New York: Academic Press, pp.428-461, 1980.
46. Grace, N.D., and Powell, L.W. Iron storage disorders of the liver. Gastroenterology, 64: 1257-1283, 1974.
47. Bradbear, R.A., Bain, C., Siskind, V., Schofiel, F.D., Webb, S., Axelsen, E.M., Halliday, J.W., Bassett, M.L., and Powell, L.W. Cohort study of internal malignancy in genetic hemochromatosis and other chronic nonalcoholic liver diseases. J. Natl. Cancer Inst., 75: 81-84, 1985.
48. Niederau, C., Fisher, R., Sonnenberg, A., Stremmel, W., Trampisch, H.J., and Strohmeyer, G. Survival and cause of death in cirrhotic and in noncirrhotic patients with primary hemochromatosis. N. Engl. J. Med., 313: 1256-1262, 1985.
49. Edwards, C.Q., Griffen, L.M., Goldgar, D., Drummond, C., Skolnick, M.H., and Kushner, J.P. Prevalence of hemochromatosis among 11,065 presumably healthy blood donors. N. Engl. J. Med., 318: 1355-1362, 1988.
50. Ostrowski, J., Kostrzewska, E., Michalak, T., Zawirska, B., Medrzejewski, W., and Gregor, A. Abnormalities in liver function and morphology and impaired aminopyrine metabolism in hereditary hepatic porphyrias. Gastroenterology, 85: 1131-1137, 1983.
51. Straka, J.G., Rank, J.M., and Bloomer, J.R. Porphyria and porphyrin metabolism. Ann. Rev. Med., 41: 457-469, 1990.

52. Kauppinen, R., and Mustajoki, P. Acute hepatic porphyria and hepatocellular carcinoma. Br. J. Cancer, 57: 117-120, 1988.
53. Cortes, J.M., Oliva, H., Paradinas, F.J., and Hernandez-Guio, C. The pathology of the liver in porphyria cutanea tarda. Histopathology, 4: 471-485, 1980.
54. Kordac, V. Frequency of occurrence of hepatocellular carcinoma in patients with porphyria cutanea tarda in long-term follow up. Neoplasma, 19: 135-139, 1972.
55. Solis, J.A., Betancor, P., Campos, R., Enriquez de Salamanca, R.E., Rojo, P., Marin, I., and Schuller, A. Association of porphyria cutanea tarda and primary liver cancer. Report of ten cases. J. Dermatolog., 9: 131-137, 1982.
56. Salata, H., Cortes, J.M., Enriquez de Salamanca, R. , Oliva, H., Castro, A., Kusak, E., Carreno, V., and Hernandez Guio, C. Porphyria cutanea tarda and hepatocellular carcinoma. Frequency of occurrence and related factors. J. Hepatol., 1: 477-487, 1985.
57. Lange, A.J., Arion, W.J., and Beaudet, A.L. Type 1b glycogen storage disease is caused by a defect in the glucose-6-phosphate translocase of the microsomal glucose-6-phosphatase system. J. Biol. Chem., 255: 8381-8384, 1980.
58. Burchell, A. Molecular pathology of glucose-6-phosphatase. FASEB J., 4: 2978--2988, 1990.
59. Howell, R.R., and Williams, J.C. The glycogen storage disease. In: The Metabolic Basis of Inherited Disease, ed 5. Stanbury, J.B., Wyngaarden, J.B., Fredrickson, D.S. et al. (eds). New York, McGraw-Hill, pp 141-166, 1983.
60. Coire, C.I., Qizilbash, A.H., and Castelli, M.F. Hepatic adenomata in type Ia glycogen storage disease. Arch. Pathol. Lab. Med., 111: 166-169, 1987.
61. Howell, R.R., Stevenson, R.E., Ben-Menachem, Y., Phyliky, R.L., and Berry, D.H. Hepatic adenomata with type 1 glycogen storage disease. J. Am. Med. Assoc., 236: 1481-1484, 1976.
62. Limmer, J., Fleig, W.E., Leupold, D., Bittner, R., Ditschuneit, H., and Beger, H.-G. Hepatocellular carcinoma in type I glycogen storage disease. Hepatology, 8: 531-537, 1988.
63. Kew, M.C., and Popper, H. Relationship between hepatocellular carcinoma and cirrhosis. Semin. Liver Dis., 4: 136-146, 1984.
64. Zaman, S.N., Melia, W.M., Johnson, R.D., Portmann, B.C., Johnson, P.J., and Williams, R. Risk factors in development of hepatocellular carcinoma in cirrhosis: prospective study of 613 patients. Lancet, ii: 1357-1360, 1985.
65. Johnson, P.J., and Williams, R. Cirrhosis and the aetiology of hepatocellular carcinoma. J. Hepatol., 4: 140-147, 1987.
66. Nakashima, T., and Kojiro, M. Hepatocellular carcinoma and liver cirrhosis. In: Hepatocellular Carcinoma, An Atlas of Its Pathology. Springer-Verlag, Tokyo--Berlin, pp 185-204, 1987.
67. Ward, J.M. Morphology of potential preneoplastic hepatocyte lesions and liver tumors in mice and a comparison with other species. In: Mouse Liver Neoplasia: Current Perspectives. (Ed. JA Popp). Hemisphere Publ. Co., Washington, pp.1-38, 1984.
68. Farber, E. Pathogenesis of experimental liver cancer: comparison with humans. Arch Toxicol Suppl 10: 281-288, 1987.
69. Tsuda, H., Hirohashi, S., Shimosato, Y., Ino, Y., Yoshida, T., and Terada, M. Low incidence of point mutation of c-Ki-ras and N-ras oncogenes in human hepatocellular carcinoma. Jpn. J. Cancer Res., 80: 196-199, 1989.
70. Tada, M., Omata, M., and Ohto, M. Analysis of *ras* gene mutations in human

hepatic malignant tumors by polymerase chain reaction and direct sequencing. Cancer Res., 50: 1121-1124, 1990.
71. Slebos, R.J.C., Kibbelaar, R.E., Dalesio, O., Kooistra, A., Stam, J., Meijer, C.J.L.M., Wagenaar, S.S., Vanderschueren, R.G.J.R.A., van Zandwijk, N., Mooi, W.J., Bos, J.L., and Rodenhuis, S. K-*ras* oncogene activation as a prognostic marker in adenocarcinoma of the lung. N. Engl. J. Med., 323: 561-565, 1990.
72. Hsu, I.C., Metcalf, R.A., Sun, T., Welsh, J.A., Wang, R.J., and Harris, C.C. Mutational hot shot in the *p53* gene in human hepatocellular carcinomas. Nature, 350: 427-428, 1991.
73. Fujimori, M., Tokino, T., Hino, O., Kitagawa, T., Imamura, T., Okamoto, E., Mitsunobu, M., Ishikawa, T., Nagagama, H., Harada, H., Yagura, M., Matsubara, K., and Nakamura, Y. Allelotype study of primary hepatocellular carcinoma. Cancer Res., 51: 89-93, 1991.

PROTEIN KINASE C IN CARCINOGENESIS: COMPARATIVE EXPRESSION OF ONCOGENES AND PROTEIN KINASE C IN RHABDOMYOSARCOMA MODELS

Monique Castagna[1], Jacqueline Robert Lezenes[2],
Xupei Huang[1] and Nicole Hanania[1]

[1]Groupe de laboratoires de l'IRSC, 7, rue Guy Moquet, 94802 Villejuif, France
[2]ICIG, 16, rue Paul Vaillant Couturier, 94801 Villejuif, France

INTRODUCTION

In the past several decades, evidence has accumulated that cancer is a multistep, multifactorial process. Two discrete stages, initiation and promotion, were first described in the model of chemical carcinogenesis developed on mouse skin (1). Initiation resulted from the action of potent mutagens likely responsible for activation of protooncogenes or inactivation of growth suppressor genes, whereas promotion was associated with epigenetic alterations. Cooperation between events occurring during these two stages was found to be required for conversion of normal cells to malignant. It has been well documented that transfection of the activated Ha-*ras* oncogene, which serves as an initiator of mouse skin carcinogenesis, did not lead to tumor formation unless tumor promoters were repeatedly applied to the skin (2,3). Similarly, the expression of oncogenes in transgenic mice carrying specific oncogenes evoked hyperplasia, with tumors arising as a rare clonal outgrowth, presumably as a result of promoting events. Ha-*ras* transgene also replaced the initiation step in mouse skin, and sensitized transgenic mice to the action of tumor promoters (4). Oncogene products, the functions of which are known, are proteins from growth signalling pathways acting as surface receptors, growth factors, G proteins, transcription factors etc., (as reviewed in ref. 5). Tumor-promoting phorbol esters interact with protein kinase C (PKC), an enzyme which plays a pivotal role in signalling pathways, exerting both positive and negative controls in cell functions and growth (as reviewed in ref. 6). In the present report, we emphasize the role of PKC in carcinogenesis, via tumor promoter-mediated activation and changes in the levels of expression.

PROTEIN KINASE C ACTIVATION AND CARCINOGENESIS

Tumor promotion occurs in response to a number of structurally unrelated chemicals and to chronic injury. It was initially shown that tumor-promoting phorbol esters bind to PKC and presumably exert their carcinogenic effects through this interaction (9). More recently, several other promoters have been shown to be activators of PKC (as reviewed in refs. 7,8). However, we have shown that tumor promoter-medi-

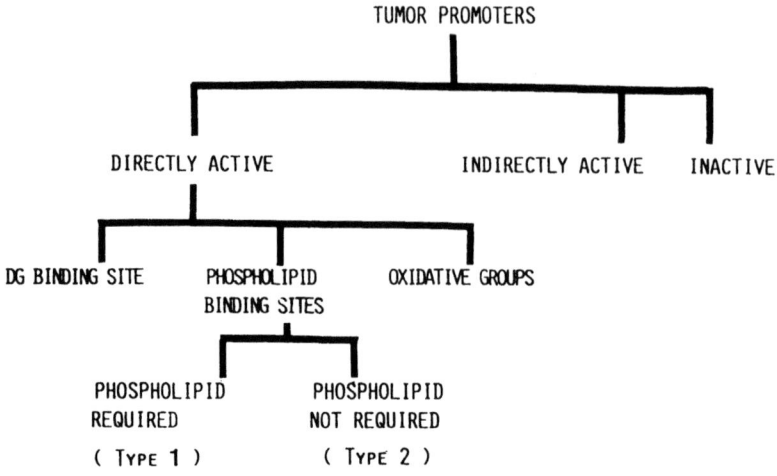

Fig. 1. Classification of tumor promoters based on their effects on protein kinase C

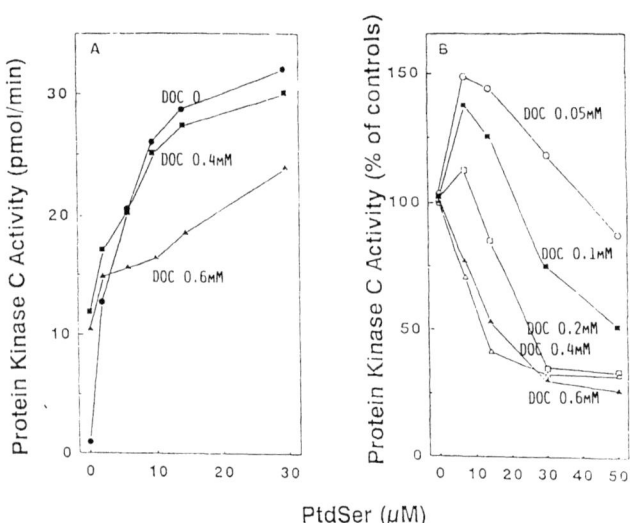

Fig. 2. Effects of phosphatidyl serine (PtdSer) on sodium deoxycholate (DOC)-supported PKC activation in 0.5 mM $CaCl_2$ (A) and 0.5 mM EGTA (B). PKC activity was assayed at various concentrations of PtdSer in the presence of DOC, as indicated. Results represent the mean values of 3 determinations performed in duplicate.

ated enzyme activation exhibits different characteristics and proceeds through various mechanisms. The classification of tumor promoters is displayed in Fig. 1.

Promoters from the series of indole alkaloids and polyacetates isolated by Sugimura and coworkers, such as teleocidin, lyngbyatoxin and aplysiatoxin, are potent

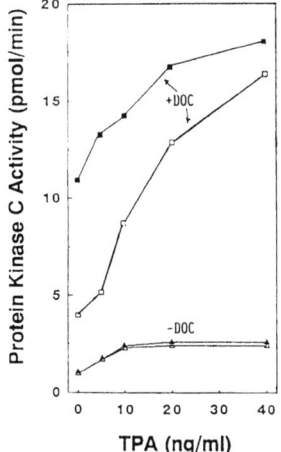

Fig. 3. DOC supported TPA-mediated PKC activation. The reaction was carried out in 0.5 mM CaCl$_2$ (■,▼) or in 0.5 mM EGTA (□,▽) in the presence or absence of 0.4 mM DOC (▼,▽) at various concentrations of TPA. Each point represents the average determinations of 3 separate experiments in duplicate.

activators of PKC in platelets which compete with phorbol esters on the diacylglycerol (DG) binding site (10). DG, as the second messenger of extracellular ligands including growth factors, is the physiological effector of PKC.

Within the group of tumor promoters which do not compete with phorbol esters, two types can be distinguished:

- Type 1 includes several organic solvents, such as chloroform, carbon tetrachloride, methylene chloride, benzene and toluene which activate PKC in intact platelets. This activation, as shown in *in vitro* studies, is phospholipid-dependent and Ca^{2+}-activated (11,12).
- Type 2 includes bile salts which, at variance with type 1, compete with phospholipid according to rather complex kinetics, suggesting that these detergents interfere with some of the phospholipid-binding sites of the enzyme, as illustrated in fig. 2. Phorbol ester or DG-mediated PKC activation, as shown in fig. 3, is supported by sodium deoxycholate (DOC). This is in agreement with the finding that DOC acts as a phosphatidyl serine-like cofactor, in providing the hydrophobic environment required for enzyme activation. Epidemiological and experimental evidence has established that bile salts play a promoting role in colorectal cancer. We have shown that DOC is able to activate PKC in intact colon cells (HT29 cell line, clone 19A) at physiological concentrations. Interestingly, DOC activated PKC in platelets only

at doses 10-fold higher (0.5 mM), indicating tissue specificity in the response to DOC (data not shown).

Unsaturated fatty acids are PKC activators (13) which also exhibit characteristics of type 2 tumor promoters (manuscript in preparation). However, these fatty acids, essentially arachidonic acid and linolenic acid, as well as the related methyl esters, may, in addition, indirectly activate PKC in platelets (14). Arachidonic acid- and methyl arachidonate-mediated PKC activation was evidenced by increased phosphorylation of

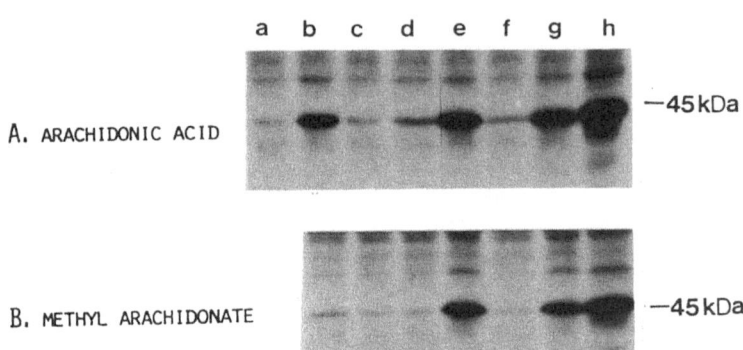

Fig.4. Autoradiogram of ^{32}P-labelled 43-kD protein in platelets. Platelets were preincubated for 2 min with ETYA, aspirin, or 20 mM Tris-HCl buffer (pH 7.4) before exposure for 2 min to the protein kinase C activators, arachidonic acid (A) and methyl arachidonate (B). Cells were: unstimulated (lane a), stimulated with 5 μM of each activator in the absence (lane b) or presence of 5 μM ETYA (lane c) or 0.2 mM aspirin (lane d), or stimulated with 50 μM of each activator in the absence (lane e) or presence of 50 μM ETYA (lane f) or 0.2 mM aspirin (lane 9). Cells were maximally stimulated with 100 ng/ml TPA (lane h). Reprinted from ref. 14.

the major substrate of this enzyme (a 43 kD protein) in platelets. However, this reaction was suppressed by inhibitors of cyclooxygenase and lipoxygenase, suggesting that PKC activation resulted mainly from the action of arachidonic acid metabolites (fig. 4). The group of promoters not competing with phorbol esters includes several which have oxidative functions, such as *m*-periodate, which, at low doses reversibly activates PKC by impairing the regulatory domain whereas at high doses, it inactivates the enzyme via alterations in the catalytic domain (15).

Not all promoters activate PKC. Interestingly, such promoters are often described as impairing growth signalling pathways by interacting with sites other than PKC, such as the EGF receptor. These promoters include palytoxin (16), chrysarobin (17), thapsigargin (18), phenobarbital (19), tetrachlorodibenzodioxine (20) and ultraviolet B radiation (21). Okadaic acid, a newly identified tumor promoter (22), is an inhibitor of serine threonine-specific phosphatases 1 and 2A which subsequently contributes to control of the activity of critical phosphoproteins in signal transduction pathways. Therefore, alterations in PKC activation and in signalling in general, seem to be associated with tumor promotion. However, some potent activators of PKC, such as bryostatin 1 and mezerein, are not tumor promoters, suggesting that activation of growth signalling pathways is not enough to promote cancer.

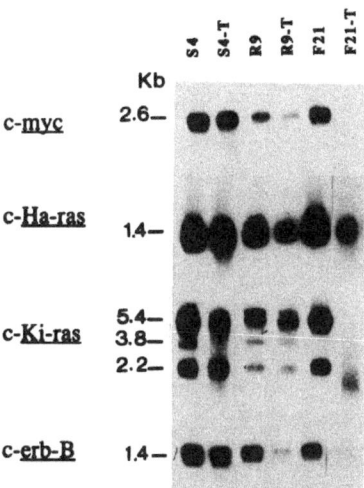

Fig. 5. Northern blot analysis of c-*onc* related transcripts in S4-t, R9-T, F21-T tumors and in S4, R9, and F21 cultured cells from which the tumors were derived. 5µg samples of poly(A+)RNA were analyzed by agarose gel electrophoresis and hybridized to nick-translated ^{32}P-labeled probes (about 1-2x10^6 cpm/ml for 18h). The same blot was hybridized with each probe. The length of RNA (kb) were estimated relative to 285 and 185 bp rRNA markers. Reprinted from Ref. 26.

Molecular cloning analysis has provided evidence that PKC consists of a family of multiple isoforms. The expression of an individual enzyme is tissue-specific, and recent reports have shown that these isoforms are functionally heterogenous (23,24). In order to understand the role of PKC in tumor promotion, we explored the hypothesis that promoters impair the expression of PKC isoforms present in a given tissue, thereby generating a cellular phenotype which prevents cells from returning to the quiescent state.

COMPARATIVE EXPRESSION OF PKC ISOFORMS AND PROTOONCOGENES IN RHABDOMYOSARCOMA-DERIVED CELL LINES

We examined a number of cell lines derived from Ni-induced rat rhabdomyosarcoma selected for their metastatic potential and degree of tumorigenicity, as previously described (25,26). Table 1 shows the characteristics of these cell lines. RMS 9-4/0, F21, S4 sublines grow in syngeneic rats and exhibit differing metastatic potential. C8 and R9 are devoid of tumorigenicity in syngeneic rats, but grow in nude mice. The former sublines will be referred to as Syn$^+$ and the latter as Syn$^-$.

Hybridization analysis of poly(A+) RNA extracted from these various cells was performed using several protooncogene- and PKC isoform-related probes. In a

Table 1. Characteristics of Rhabdomyosarcoma-Derived Cells

Cells	Tumorigenicity	Metastatic Potential
RMS 9-4/0	+	+
S4	+	+++
F21	+	-
C8	-	
R9	-	

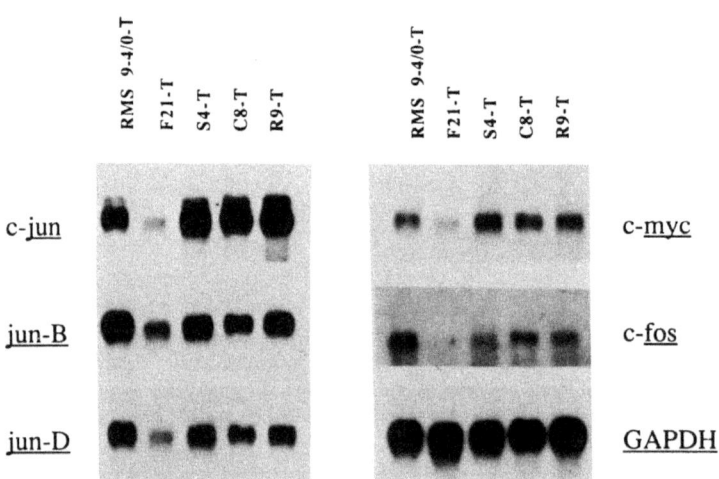

Fig. 6 Northern blot analysis. c-*jun*, *jun*-B, *jun*-D, c-*myc*, and c-*fos* related poly (A+) RNA (5μg) from the weakly metastatic RMS 9-4 0-T, the non-metastatic F21-T, and from the non-tumorigenic R9-T and C8-T sublines. Hybridization to random-primed ^{32}P-labeled DNA probes (3-7x10^8 cpm/μg) was performed for 18h. Exposure times for autoradiograms were 3-5h for c-*jun* and *jun*-D, 18-24h for c-*myc* and *jun*-B, and 3d for c-*fos*. The amount of RNA was determined by hybridization to a GAPDH-specific probe. Reprinted from Ref. 26

preliminary study, we compared the c-*myc*, c-Ha-*ras*, c-Ki-*ras* and c-*erb*B related poly (A+) RNA of S4-T, R9-T and F21-T tumors with those of cultured cells from which these tumors were derived (S4, R9 and F21). Cell lines R9 and F21 expressed higher levels of c-*onc* transcripts than the related tumors, as shown in fig. 5 (26). In contrast, no difference was observed between cell line S4 and its related tumor S4-T.

Northern blot analysis of c-*myc*, c-*fos*, c-*jun*, *jun*-B and *jun*-D-related transcripts in various tumors failed to show any significant correlation between the levels of expression of c-*onc*-related RNA and either the metastatic potential or the degree of tumorigenicity (fig. 6). In contrast, analysis of *sis*-related transcripts (c-*sis* gene encodes for the B chain of growth factor PDGF) showed strong overexpression of these transcripts in Syn$^-$ tumors which was not detected in Syn$^+$ tumors (fig. 7).

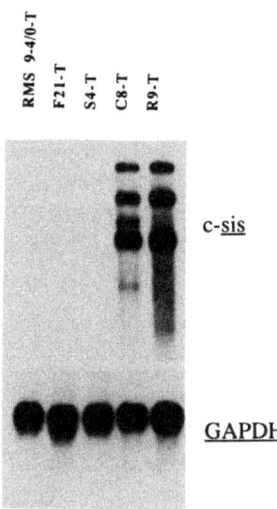

Fig. 7. Northern blot analysis of c-*sis*-related transcripts in the RM5 Y-4/0-T, F21-T, S4-T cells (Syn^{+}), C8-T, and R9-T (Syn^{-}). Hybridization was performed on the same blot as in Fig. 6. Exposure time was 5h. Reprinted from ref. 26.

Hybridization analysis of poly(A+) RNA extracted from Syn^{+} (RMS 9-4/0-T, F21-T and S4-T) and Syn^{-} (C8-T and R9-T) tumors using a PKCα-related probe showed

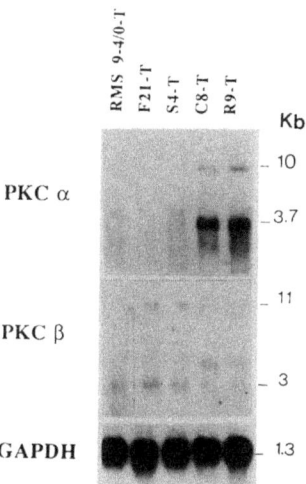

Fig. 8. Expression of PKCα- and PKCβ-related transcripts in Syn^{+} and Syn^{-} tumors. Poly (A^{+})s RNA were analyzed as described in legend to Fig. 5. Exposure times of autoradiograms was 16h and 55h for PKCα and PKCβ probes, respectively.

that the expression of this isoform in the various materials is similar to that of c-*sis* gene (fig. 8). Both genes are overexpressed in cells which do not grow in syngeneic rats, (Syn^{-}) and underexpressed in cells which grow in syngeneic rat (Syn^{+}).

At variance, PKCβ-related transcripts were more strongly expressed in Syn$^+$ than in Syn$^-$ tumors. The weak 3.7 kb band detected by the PKCβ probe in Syn$^-$ cells was likely due to non-specific hybridization of this probe to the PKCα-related 3.7 kb transcript overexpressed in these cells. PKCγ was not expressed in any sublines.

In Syn$^-$ cells (and not in Syn$^+$ cells), southern blot analysis of *Eco* RI-digested DNA from these materials showed increased hybridization using PKCα probe, whereas the PKCβ probe was evenly hybridized. The rates of hybridization were quantitated by densitometer scanning and the ratio PKC α/PKCβ indicated that the PKCα gene is 3-fold amplified in Syn$^-$ cells (data not shown).

To summarize, this study on rhabdomyosarcoma-derived cell lines has provided evidence that expression of the PKCα isoform gene and the c-*sis* gene is correlated in these cells and related to malignancy. Indeed both genes are repressed in malignant Syn$^+$ and overexpressed in Syn$^-$ cells which do not grow in syngeneic rats. Experiments are in progress to determine whether overexpression of the *sis* gene is associated with increased PDGF release and PKC-mediated autocrine proliferation in Syn$^-$ cells. In addition, this report has emphasized that experiments on oncogene expression should be performed on solid tissues rather than on cultured cells to yield conclusive results.

In conclusion, changes in the expression of individual isoforms of PKC appear to play a major role in determining tumor cell phenotype. It is hypothesized that alterations in the expression of PKC isoforms result from an adaptation process in response to chronic exposure of the cells to agents which damage the activity of this key enzyme. Although further investigation is required, it is suggested that the relative levels of expression of PKC isoforms may allow to assess the carcinogenicity of some chemicals with potential promoting activity.

ACKNOWLEDGMENTS

The authors would like to thank H. Bouzinba for technical assistance. This research was supported by ARC and la Fondation pour la Recherche Medicale.

REFERENCES

1. Berenblum, I. (1941). The cocarcinogenic action of croton resin. Cancer Res. 1:44-48.
2. Brown, K., Quintanilla, M., Ramsoen, M., Kerr, I.B., Young, S and Balmain, A. (1986). ras genes from Harvey and BALBc mouse sarcoma viruses can act as initiators of two stage mouse skin carcinogenesis. Cell 46:447-456.
3. Quintanilla, M., Brown, K., Ramsoen, M. and Balmain, A. (1986). Carcinogen-specific mutation and amplification of Ha-ras during mouse skin carcinogenesis. Nature (London). 322:78-80.
4. Leder, A., Kuo, A., Cardiff, R.D., Sinn, E. and Leder, P.(1990). Ha-ras transgene also gates the initiation step in mouse skin tumorigenesis: effects of phorbol esters and retinoic acid. 87: 9178-9182.

5. Weinstein, I.B. (1988). The origins of human cancer: molecular mechanisms of carcinogenesis and their implications for cancer prevention and treatment. Cancer Res. 48:4135-4143.
6. Cantley, L.C. , Auger, K.R., Carpenter, C., Duckworth, B., Oraziani, A., Kapeller, R. and Soltoff, S. (1991). Oncogenes and signal transduction. Cell. 64:281-302.
7. Martelly, I. and Castagna, M. (1989). Protein kinase C and tumor promoters. Current Opinion in Cell Biology. 1: 206-210.
8. Castagna, M. and Martelly, I. (1990) Role de la proteine kinase C dans les regulations cellulaires et la promotion tumorale. Bull. Cancer. 77: 489-494.
9. Castagna, M., Takai, Y., Kaibuchi, K., Sano, K., Kikkawa, U. and Nishizuka, Y. (1982). Direct activation of calcium activated, phospholipid-dependent protein kinase by tumor promoting phorbol esters. J. Biol. Chem. 257:7841-7851.
10. Fugiki, H., Suganuma, M., Suguri, H., Yoshizawa, S., Hirota, M., Takagi, K. and Sugimura, T. (1989). Diversity in the chemical nature and mechanism of response to tumor promoters. Skin carcinogenesis. 281-291.
11. Roghani, M., Da Silva, C. and Castagna, M. (1987). Tumor promoter chloroform is a potent protein kinase C activator. Biochem. Biophy. Res. Commun. 142:738-744.
12. Roghani, M., Da Silva, C., Guveli, D. and Castagna, M. (1987). Benzene and toluene activate protein kinase C. Carcinogenesis. 8: 1105-1107.
13. McPhail, L.C., Clayton, C.C. and Snyderman, P.S. (1984) A potential second messenger role for unsaturated fatty acids: activation of Ca^{2+}-dependent protein kinase. Science. 224:622-624.
14. Fan, X.T., Huang, X.P., Da Silva, C. and Castagna, M. (1990). Arachidonic acid and related methyl ester mediated protein kinase C activation in intact platelets through the arachidonate metabolism pathways. Biochem. Biophys. Res. Commun. 169:933-940.
15. Gopalakrishna, R. and Anderson, W.B. (1991). Reversible oxidative activation and inactivation of protein kinase C by the mitogen/tumor promoter periodate. Arch. Biochem. Biophys. 285:382-387.
16. Wattenberg, E.V., Fujiki, H. and Rosner, M.R. (1987). Heterologous regulation of the epidermal growth factor receptor by palytoxin, a non 12-O-tetradecanoyl phorbol--13-acetate-type tumor promoter. Cancer Res. 47:4618-4622.
17. Imamoto, A., Beltran, L.M. and Digiovanni, J. (1990). Differential mechanism for the inhibition of epidermal growth factor binding to its receptor on mouse keratinocytes by anthrones and phorbol esters. Carcinogenesis 11: 15431549.
18. Friedman, B.A., Van Amstradam, J., Fujiki, H. and Rosner, M.R. (1989). Phosphorylation at threonine-654 is not required for negative regulation of the epidermal growth factor receptor by non-phorbol tumor promoters. Proc. Natl. Acad. Sci. USA. 86: 812-816.
19. Digiovanni, J., Decina, P.C., Prichett, W.P., Cantor, J., Aalfs, K.K., and Coombs, M.M. (1985). Mechanism of mouse skin tumor promotion by chrysarobin. Cancer Res. 45: 25842589.
20. Madhukar, B.V., Brewster, D.W. and Matsumura, F. (1984). Effects of in vivo-administrated 2,3,7,8-tetrachlorodibenzodioxin on receptor binding of epidermal growth factor in the hepatic plasma membrane of rat, guinea pig, mouse and hamster. Proc. Natl. Acad. Sci. USA. 81:7407-7411.
21. Matsui, Ml.S., Laufer, L., Schede, S. and Deleo (1989). Ultraviolet B (290-320

22. nm)-irradiation inhibits epidermal growth factor binding to mammalian cells. J. Invest. Dermatol. 92:617-622.
22. Suganuma, M., Sullajit, M., Suguri, H., Ojika, M., Yamada, K. and Fujiki, H. (1989). Specific binding of okadaic acid, a new tumor promoter in mouse skin. FEBS Lett. 2: 615-618.
23. Cooper, D.R., Watson, J.E., Acevedo-Duncan, M., Pollet, R.J., Standaert, M.L. and Farese, R.V. (1989). Retection of specific protein kinase C isoenzymes following chronic phorbol ester treatment in BC3H-1 myocytes. Biochem. Biophys. Res. Commun. 161:327-334.
24. Ken-Ichi, K., Yasuhiro, K., Hisashi, F., Masatoshi, H., Hiroyoshi, H., Yasuo, F. and Yoshimi, T. (1989). Two types of protein kinase C with different functions in cultured aorta smooth muscle cells. Biochem. Biophys. Res. Commun. 161: 1020-1027.
25. Sweeney, F.L., Pot-Deprun, J., Poupon, M.F. and Chouroulinkov, I. (1982). Heterogeneity of the growth and metastatic behavior of cloned cell lines derived from a primary rhabdomyosarcoma. Cancer Res. 42: 3776-3782.
26. Hanania N., Boyano, M.D., Mangin, C. and Poupon, M.F. (1991), Oncogene and MDR1 gene expression in rat rhabdomyosarcoma sublines of different metastatic potential. Anticancer Res. 11:473-480.

ONCOGENES AND HUMAN CANCERS

J. S. Rhim, J. H. Yang, I. H. Lee, M. S. Lee and J. B. Park

National Cancer Institute
Building 37, Room 1E24
Bethesda, MD 20892

INTRODUCTION

Recent molecular genetic studies have shown that most human cancers display a series of multiple genetic alterations. These alterations presumably underlie the multifactorial etiology and multistage progression of the tumors, although in most cases the precise sequence of genetic events remains unclear. The mechanisms of carcinogenesis are probably the most important single question in cancer research, because when we understand the process of malignant transformation, we are likely to be able to devise means of cancer prevention.[1-3] So, many distinct genes and gene products have been implicated in cancer and a few basic concepts have recently emerged.

Two major classes of genes, the oncogene and the tumor-suppressor genes, have been implicated in cancer pathogenesis. The cellular oncogenes are widely believed to arise as activated versions of normal cellular growth-promoting genes or proto-oncogenes. The tumor suppressor genes are thought to function in normal cells to halt proliferation; the consequence of their inactivation in a cellular genome is loss of the ability of the cell to inhibit its own growth. Thus, cancer can be thought to arise as a result of the hyperactivity of growth-promoting genes and the inactivation of growth-inhibiting genes (the Yin and Yang concept of cancer). Several gene changes during a multistep process appear to be necessary to cause most common cancers. Not only do one or more growth-stimulating oncogenes have to be activated, but the growth inhibitory genes that would otherwise suppress tumor formation have to be inactivated.

In the present decade remarkable progress has been made in the knowledge of the neoplastic process, not least in research on oncogenes and so-called tumor suppressor or anti-oncogenes.[4-6] The properties of retroviral oncogenes, oncogenes of DNA tumor viruses, tumor suppressor genes, and the properties of their products will be summarized. Some of mechanisms by which cellular genes can become activated or inactivated during tumor development will be discussed. Recent *in vitro* human cell

Table 1. Proto-oncogenes

Protooncogene	Mode of initial identification[a]	Biochemical properties[b]	Human chromosomal location[c]
c-src-1	chicken RSV	cytoplasmic tyr k	20q12-q13
c-src-2		cytoplasmic tyr k	1q36-p34
c-abl	mouse Ab MuLV	cytoplasmic tyr k	9q34
c-fgr	cat GR-FeSV	cytoplasmic tyr k	1p36.1-36.2
c-yes	chicken Y73	cytoplasmic tyr k	1821.
c-fes	cat ST-FeSV GaFeSv H21-FeSv	cytoplasmic tyr k	15q25 q26
c-sea	chicken AEV-s13	cytoplasmic tyr k	
tck	rv insertion	cytoplasmic tyr k	
trk	NIH/3T3 transfection	cytoplasmic tyr k	5q34
c-fms	cat SM-FeSV H25-FeSV	CSF-1 receptor tyr k	7q13-q11.2
c-erbB-1	chicken AEV-H	EGF receptor tyr k	17
c-erbB-2	NIH/3T3 transfection	receptor-like tyr k	4q11-q21
c-kit	cat H24-FeSV	receptor-like tyr k	6q22
c-ros	chicken UR2	receptor-like tyr k	7p11-4
c-met	NIH/3T3 transfection	receptor-like tyr k	
met	NIH/3T3 transfection	receptor-like tyr k	
c-raf1	chicken MH2/mouse 3611-MSV	cytoplasmic ser/thr k	3p25
c-raf2			
c-m os	mouse M-MSV	cytoplasmic ser/thr k	8q22
pim-1	rv insertion	cytoplasmic ser/thr k	
c-sis	monkey SSV	PDGF ,β chain	22q12-q13
int-2	rv insertion	growth factor/like	17

Gene	Method	Function	Location
hst	NIH/3T3 transfection	growth factor/like	
N-ras	NIH/3T3 transfection	membrane bound GTPase	1cen-1p21
c-Ha-ras1	rat Ha-MSV	membrane bound GTPase	11p15
c-Ki-ras2	rat Ki-MSV	membrane bound GTPase	12p12-pter
c-myc	chicken MC29 MH2	nuclear proteins	8q24
N-myc	X-hybridization	nuclear proteins	2p23-pter
L-myc	X-hybridization	nuclear proteins	1p32
c-ski	chicken SKV-rASV	nuclear proteins	1q22-qter
c-ets-1	chicken E26	nuclear proteins	11q23-q24
c-ets-2		nuclear proteins	21q22.3
c-myb	chicken AMV	nuclear proteins	6q22-q24
c-fos	mouse FBJ-MSV	nuclear proteins	14q21-q31
c-rel	turkey REV-T	nuclear proteins	2
p53	association with SV40 T antigen	transcription factor	17
c-jun	chicken ASV-17		
c-erbAl	chicken AEV-H	thyroid hormone receptor	17
gli	gene amplification	unknown	12q13q-q14.3
bcl-1	translocation breakpoint	unknown	11q13
bcl-2	translocation breakpoint	unknown	18q21
tcl-1	translocation breakpoint	unknown	14q32
fis-1	rv insertion	unknown	
fim-1	rv insertion	unknown	
dbl	NIH/3T3 transfection	unkncown	

[a]NIH transfection, NIH/3T3 gene transfer, rv insertion, preferred integraton site recognized by retrovirus insertion; X-hydribization, cross-hybridization of amplified sequences to oncogene probe. [b]tyr k, tyrosine kinase. [c]In some cases there is more than one human gene homologous to a retrovirus oncogene. Revised from Glover and Hames[55]

Table 2. Oncogenes of DNA tumor viruses

Virus	Gene	Protein	Location	Function
Polyomavirus	A	large T	nucleus	binds DNA
	hr-t	middle T	PM	complexes with c-*src* protein
		small t	cytoplasm	?
SV40	A	large T	nucleus/PM	binds DNA, p53, and RB gene product
		small t	cytoplasm	?
Bovine papillomavirus	E5	44 a.a	?	transformation
	E6	15kD	?	?
Human papilloma	E6/E7	17kD(E7)	?	binds DNA, p53, and RB gene, immortalization
Adenoviruses	E1A	289 a.a., 243 a.a.	nucleus	transcriptional regulators; bind RB gene, immortalization
Epstein-Barr virus	E1B	53-58kD, 15kD	nucleus (?)	?
	EBNA1-4	several	nucleus	immortalization (EBNA-2)
	LMP	44 kD	PM	transformation

(EBNA) Epstein-Barr nuclear antigen; (LMP) latent membrane protein; (PM) plasma membrane; (RB) retinoblastoma.

experimental evidence will be presented in support of the concept that carcinogenesis is, indeed, a multistep process, as has been demonstrated *in vitro* in rodent cells and suggested in clinically observed human cancers.[7,8] The usefulness of *in vitro* human cell models for studying the mechanisms of multistep carcinogenesis will be shown.[9,10]

ONCOGENES AND THEIR FUNCTIONS

Proto-oncogenes are evolutionarily conserved genes that are expressed and code for proteins believed to play essential but poorly understood roles in the growth and differentiation of normal mammalian cells.[1,2] Proto-oncogenes may become mutated and/or rearranged and subsequently transduced by retroviruses, leading to the formation of viral oncogenes. Oncogenes have been isolated from acute transforming retroviruses and by transfection of DNA from transformed cell lines and tumors into NIH/3T3 cells[2,3] and have been shown to be activated forms of normal cellular proto-oncogenes.

The task of identification and characterization of oncogenes is now well into its second decade and at least two facts are clear: a relatively large number of these genes

Table 3. Oncogenes Detected by Gene Transfer

Oncogene	Cancer	Activation mechanism
H-*ras*	human and rodent	point mutation
K-*ras*	carcinoma, sarcomas	
N-*ras*	neuroblastomas, leukemia & lymphomas	
neu	rat neuroblastomas and glioblastomas	point mutation
met	chemically transformed human osteosarcoma cell line	recombinant fusion protein
trk	human colon carcinoma	recombinant fusion protein

exist and their activation, primarily through point mutations, amplification, or rearrangement, may result in positive signalling for cell proliferation at virtually any point in the complex biochemical and molecular pathways that affect this process. More than three dozen oncogenes, identified to date, constitute a heterogenous group of genes that are remarkably conserved among highly divergent species (Table 1). Functions of the oncogene proteins are being identified: they include unique protein kinase activity, growth factors or their receptor-like properties, regulatory protein activity in signal transduction and nuclear DNA-binding polypeptides. One thing that has become clear is that no generalizations can be made about the manner in which the oncogenic potential of the genes is realized. In some cases, this maybe the result of increased synthesis of the protein product of the genes; in others, an altered gene may produce a defective

Table 4. Oncogene Activation and Human Cancers

Oncogene	Cancer	Activation mechanism
c-*myc*	Burkitt's lymphoma	translocation
	leukemia, breast, stomach, lung, and colon cancers	amplification
	neuroblastomas and gliomas	
N-myc	neuroblastoma & retinoblastomas	amplification
	lung cancers (SCLC)	amplification
L-myc	lung cancers	amplification
abl	chronic myelogenous leukemia	translocation
	acute lymphocytic leukemia	
bcl-2	follicular B-cell lymphoma	translocation
*erb*B	glioblastomas, squamous cell cancers	amplification
*erb*B-2	breast, salivary gland and ovarian cancers	amplification
K-*ras*	colon and pancreatic cancers	point mutation
	lung, ovarian & bladder cancers	amplification
H-*ras*	bladder cancers	amplification
N-*ras*	acute myeloid and lymphoid leukemia	point mutation
	breast carcinoma cell line	amplification
Int-2	breast, bladder, stomach and squamous cell carcinomas	amplification
hst	breast, bladder and stomach cancer	amplification
myb	colon cancers, leukemia	amplification
ets-1	lymphomas	

product. Also, more than one oncogene may have to be activated to transform a cell to malignancy.

ONCOGENES OF DNA TUMOR VIRUSES

Whereas retroviral oncogenes are derived from the host genome, the oncogenes of DNA viruses represent information unique to the virus and required for its replication and survival. Most of the retroviral oncogenes are altered components of mitotic signaling pathways and act by disrupting positive growth regulation. DNA viral oncogenes interact with cellular tumor suppressor genes, thereby disrupting negative growth regulation (Table 2). In recent years, refined molecular dissection of the region of DNA tumor virus genomes implicated in transformation has contributed to some current concepts in oncogenesis. Research in DNA tumor viruses has demonstrated: (1) cooperation between two oncogenes during transformation (*e.g.*, the middle and large T antigen genes of polyomavirus[13] or the EIA and EIB genes of adenoviruses);[14] (2) multifunctional properties of proteins encoded by a single oncogene (*e.g*, the large T antigen of SV40);[15] and (3) interactions between proteins encoded by viral and cellular

Table 5. Loss of Tumor Suppressor Genes in Human Tumors

Tumor suppressor gene or chromosome	Tumor
Rb (chromosome 13)	retinoblastoma, osteosarcoma breast, bladder, and small cell lung carcinoma (SCLC)
p53 (chromosome 17)	colorectal SCLC, and liver cancer
DCC, the deleted colon carcinoma (chromosome 18)	colorectal cancer
WT-1 (chromosome 11)	Wilms tumor, hepatoblastoma, adrenal carcinoma, rhabdomyosarcoma, bladder and breast cancer
NF-1 (chromosome 17)	neurofibromatosis
FAP (chromosome 5)	Familial adenomatous polyposis: colorectal cancer
chromosome 1	neuroblastoma
chromosome 3	lung and renal cell carcinoma
chromosome 22	acoustic neuroma, meningioma

genes (*e.g*, complexes formed between SV40 T antigen and p53,[16] between polyoma middle T antigen and the c-*src* protein,[17] and between several DNA tumor virus proteins and the product of the Rb gene.[18]

HUMAN ONCOGENES DETECTED BY GENE TRANSFER

The most exciting initial development in cancer research has been the recognition and investigation of individual genes important in tumor development. Such genes were first identified in mammalian cells through homology with retroviral oncogenes and through the transfection assay, a sensitive method of gene transfer. Human oncogenes were detected by gene transfer/DNA transfection analysis by the groups of Weinberg,[19] Cooper,[20] Wigler,[21] and Barbacid.[22] In 1981, DNA extracted from the human bladder carcinoma cell line EJ was found to induce the transformation of NIH/3T3 cells with high efficiency in a gene transfer assay. This DNA therefore provided the first example of a biologically active human tumor oncogene. Continuing studies have now identified active oncogenes in a large number of human (and rodent) tumors of a wide variety of cell types, including sarcomas, carcinomas of many different tissues, neuroblastomas, leukemias, and lymphomas.[11,12,19-27] Overall, about 20% of individual tumor DNAs induce transformation (Table 3). The majority of the identified human transforming genes are related to the *ras* family. These include not only homologues of v-H- and K-*ras* genes, but also an additional gene, N-*ras*, which has been cloned from both a neuroblastoma cell line and the HL-60 acute promyelocytic leukemia cell line and has relatively weak homology to the H- and Ki-*ras* genes. Other transforming genes in human tumors have been identified (e.g., *myc* in Burkitt's lymphoma). It is possible

Table 6. Biological Properties of Human Keratinocytes exposed to Ki-MSV, Ad12-SV-40, Chemicals and/or X-ray

Cells	Passage	Tumorigenicity	Oncogenes
Primary human keratinocytes	< 3	-	
+ Ki-MSV	< 3	-	
+Ad12-SV40	>50	negative	
+Ad12-SV40+Ki-MSV	>50	positive	
+chemical carcinogen (MNNG or 4NQO)	< 3	-	
+Ad12-SV40 + MNNG	>50	positive	non-*ras*
+AD12-SV40 + 4NQO	>50	positive	non-*ras*
x-ray	< 3	-	
+Ad12-SV40 + X-ray	>50	positive	non-*ras*

that identification of mostly *ras* transforming genes in human tumors reflects a bias in the NIH/3T3 assay. Alternatively, activation of a member of the *ras* family may be a common step to many malignancies. Strong evidence concerning the role of *ras* oncogenes in cancer is based on chemical carcinogenesis studies.[28-33] Transforming *ras* oncogenes have been found in more than 70% of chemically induced tumors in different animal model systems. It was also shown that the H-*ras* oncogene can be activated in a 3-methylcholanthrene transformed human cell line.[34] Many studies indicate that *ras* oncogenes are activated at all three recognized stages (initiation, promotion, and progression) of cancer development. It is noteworthy that bladder, lung, and colon cancers, and acute lymphocytic leukemia have been associated with activation of more than one member of the *ras* family. Thus, activation of any one of the cellular *ras* genes may be able to set the neoplastic process in motion for these and certain other malignancies.

POINT MUTATIONS THAT ACTIVATE *ras* ONCOGENES IN HUMAN TUMORS

Nucleic acid sequencing analyses of bladder and other cancers (lung, mammary and neuroblastoma cell lines) have revealed that single base mutations (point mutations) in the coding region of the *ras* genes are responsible for acquisition of transforming activity. Moreover, it has been suggested that these mutations, leading to single amino acid substitutions in the primary sequence of the respective p21 *ras* proteins, are also critical in the development of malignancy *in vivo*. Available evidence indicates that codons 12 and 61 are the major "hot spots"[35] for genetic alterations that activate any of the 3 *ras* proto-oncogenes under natural conditions. In all cases, the activated cellular oncogenes are mutant forms of normal proto-oncogenes that differ in the structure and functions of their gene products. The finding of such oncogenes activated by somatic mutations or by DNA rearrangements in a significant fraction of human tumors clearly implicates cellular oncogene activation in the pathogenesis of human cancers.

Table 7. *In Vitro* Multistep Models for Human Cell Carcinogenesis

Cells	Immortalization	Transformation	Ref.
Epithelial cells			
Keratinocytes	Ad12-SV40	Ki-MSV	60
	Ad12-SV40	MNNG or 4NQO	61
	Ad12-SV40	x-ray	62
	Ad12-SV40	retroviruses	64
	Ad12-SV40	c-H-ras	65
	pSV$_3$ neo	Ki-MSV	66
	spontaneous	c-H-ras	67
	HPV-16	Ki-MSV	68
Bronchial	Ad12-SV40	Ki-MSV	69
	Ad12-SV40	v-H-*ras*	70
Mammary	BP	retroviruses	71
Amniotic	SV40	Ki-MSV	72
Cervical	HPV-16	v-H-*ras*	73
Urinary tract	SV40	3MC	74
	SV40	c-H-*ras*	75
Liver	SV40	-	76,77
Kidney	nickel	v-H-*ras*	78
Thyroid	adeno E1A	-	79
	SV40 Ori-	-	80
Colon	SV40 Ori-MNNG	-	81
	& sod. butyrate	-	82
Tracheal gland	Ad12-SV40	-	83
Letinal pigment	SV40	-	84
Esophagus	SV40	-	85
Melanocyte	SV40	-	86
Prostate	SV40	-	87
Fibroblasts	SV40	Ki-MSV	88
	gamma ray	H-MSV or c-H-*ras*	89
	v-*myc*	H-*ras*	90

ONCOGENE ACTIVATION BY CHROMOSOMAL TRANSLOCATION

Since genes related to retroviral *onc* genes are present in human DNA, it is possible to assess the role of proto-oncogenes in the etiology of human cancer. Cytogenetic studies showing correlations between the chromosomal locations of *onc* genes and sites of chromosomal breakage in translocation and deletion in specific malignancies constitute a second line of evidence suggesting a role for cellular oncogenes in the malignant process.[36] In human-rodent somatic cell hybrids possessing varying numbers of human chromosomes, it is possible to distinguish rodent and human proto-oncogene sequences by digestion of cellular DNAs with appropriate restriction enzymes. Where possible, cloned DNAs from human proto-oncogene loci have been used as molecular probes to enhance the signal intensity of the human proto-oncogene sequence. The results of chromosome mapping studies are summarized in Table 1.[37-55] It is apparent that proto-oncogenes are widely distributed throughout the human genome. Translocation of an oncogene to a new site or integration of a new DNA sequence either upstream or downstream of an oncogene may subject the oncogene to control by a strong transcriptional promoter or activator, resulting in enhanced expression and possibly malignancy. Alternatively, an oncogene may undergo either a small mutation or a major rearrangement during the translocation process, possibly resulting in altered expression and neoplasia. Chromosome translocations occur frequently in human cancers and clearly represent an important mechanism for oncogene activation in some tumors (Table 4). The activation of *myc* and *abl* by translocation represents the most reproducible involvement of oncogenes in human cancer, occurring in virtually all Burkitt's lymphomas and chronic myelogenous leukemias, respectively. Moreover, molecular analysis of translocations has been a powerful approach to identifying novel oncogenes (*e.g.*, *bcl*-2) with important roles in the development of human cancers.

AMPLIFICATION OF ONCOGENES IN HUMAN TUMORS

As we discussed a number of cases in which cellular oncogenes are activated by DNA rearrangements, proviral insertions and chromosome translocations which resulted increased or deregulated gene expression. Further mechanism for increasing gene expression is DNA amplification, resulting in an increased number of gene copies per cell. In this case, elevated gene expression would be expected as a direct consequence of an increase in the number of templates available for transcription, rather than from changes in transcription regulatory sequences. DNA amplification appears to be a common mode of cellular oncogene activation in a wide variety of human cancers (Table 4).[3]

TUMOR SUPPRESSOR GENES

The present decade has seen unprecedented advances in knowledge of the neoplastic process, not least in research on oncogenes and so-called anti-oncogene or tumor suppressor genes. Tumor suppressor genes are emerging as major partners in the development and progression of a variety of human neoplasms (Table 5). Loss of normal functions of tumor suppressor genes as negative regulators of cell growth is believed to lead to tumor development. Loss of function of these genes may occur in the germ line, their absence predisposing the individual to cancer. More frequently, progressive loss of function in tumor suppressor genes occurs through an accumulation of somatic mutations. Inactivation of tumor suppressor genes similar to oncogene activation is caused also by (1) chromosome loss, (2) homologous recombination, and (3) gene rearrangements.[2,3,56,57]

ONCOGENES IN MULTISTEP CARCINOGENESIS

One of the most important points emerging from investigations of oncogene activation in malignancy is that multiple steps involving alteration are involved in tumorigenesis. It has long been known from clinical observation that cancer is a

Table 8. Human osteosarcoma clonal (HOS) cells and their transformants[10]

Cell designation	Cell description	Remarks
HOS (TE85 C1-F-5)	Human osteogenic sarcoma clonal cells	P53 (low)
KHOS/NP	Nonproducer cells from KiMSV transformed(HOS) cells	Rescuable sarcoma genome (+)
MNNG/HOS	MNNG transformed HOS cells	met (+), P53 (high)
DMBA/HOS	DMBA transformed HOS cells	
Revertant 240S	Revertant from KHOS cells	
S⁺L⁻HOS	Sarcoma-positive, leukemia-negative HOS cells induced by Mo-MSV (FLV)	
RSV/HOS	RSV-SR transformed HOS cells	
Revertant 312H	3MC transformed 312H cells	cH-*ras*, 61st (+), P53 (high)
DMBA/312H	DMBA transformed 312H cells	P53 (high)

multistep process involving cooperative action of two or more separate genetic events.

The exact number of genetic alterations required for neoplastic transformation remains to be elucidated. In cell culture systems, transfection of primary rat embryo fibroblasts with 2 oncogenes, *ras* and *myc* or *ras* and EIA, an early gene of the adenovirus genome, has been shown to be sufficient to induce the malignant phenotype, whereas either transforming gene alone is unable to do so.[58,59] In a human cell culture system, it has been possible to induce malignant transformation of primary human epidermal keratinocytes by the sequential addition of viruses containing different oncogenes, a virus and a chemical carcinogen, or a virus and x-ray irradiation (Table 6). Such transformants were capable of inducing progressively growing squamous cell carcinomas in athymic nude mice, whereas untreated keratinocytes or keratinocytes treated with only one carcinogenic agent (virus, chemical carcinogen or x-ray irradiation) were not.[60-62] Such findings support a multistep model for the human neoplastic process as has been demonstrated *in vivo* in laboratory rodents and suggested in clinically observed human cancers.

Experimental models to define the role of carcinogenic agents in the development of human cancer must be established using human cells. Thus, the study of human cell transformation in culture by carcinogenic agents is of particular importance for understanding the genetic and molecular mechanisms underlying human carcinogenesis. Unlike rodent cells, normal human cells in culture do not or rarely undergo spontaneous transformation and have generally proven resistant to neoplastic transformation by carcinogens.[9,10] Since our initial report,[60] the list of successful reports on the neoplastic transformation of normal human cells including fibroblasts have been growing (Table 7).[10] Primary human cells immortalized by a variety of means (viruses, oncogenes, chemicals, irradiation or rarely spontaneous) could be transformed neoplastically by a carcinogenic agent. In all these cases, the initial event seemed to be immortalization of

Table 9. Model for colorectal tumor formation

A series of genetic alterations results in progression of tumorigenesis Vogelstein et al.,[92]

Gene	Alteration	Chromosome	
			Normal epithelium
FAP	Mutation or loss	5q ⇒	⇓
			⇓
			Hyper-proliferative epithelium
			⇓
			Early adenoma ⇐ DNA hypo-methylation
			⇓
K-ras	Mutation	12q ⇒	⇓
			Intermediate adenoma
DCC	Loss	18q ⇒	⇓
			Late adenoma
p53	Loss	17p ⇒	⇓
			Carcinoma
			⇓ ⇐ Other alterations
			⇓
			Metastasis

the cells followed by neoplastic conversion. As postulated for rodent fibroblasts,[63] the immortalization step is a critical initial step and rate limiting for in *in vitro* neoplastic transformation of human epithelial cells.

For initial human cell transformation studies, a flat nontumorigenic clonal line (TE85 clone E5), originally derived from human osteosarcoma (HOS) cells (Table 8) was used. This cell system was found to be very useful for elucidating the molecular mechanisms of carcinogenesis, since non-producer Ki-MSV transformed human cells and chemically transformed human cells have been derived using this cell system. Activation of trans-forming genes *met* and c-H-*ras* has been demonstrated in these chemically transformed HOS cell lines. Recent evidence has shown that parent HOS cells contained a missense mutation in codon 156. Each of the carcinogen-treated cell lines had acquired a second point mutation in the p53 gene . These results suggest that the codon 156 mutation in HOS cell line was not sufficient for full expression of malignant phenotype. These properties could be attributed to the second point mutation that the cell lines acquired in response to exposure to the carcinogens. Whether this is the case could be determined by transfecting HOS cells with p53 gene carrying the second mutation.[91]

SUMMARY

Since the beginning of the study of viral oncogenes two decades ago remarkable advances have been made toward understanding the biological significance of cellular oncogenes. A large number of oncogenes and tumor suppressor genes have been identified and characterized during the past decade. It is clear that both activation of oncogenes and loss of tumor suppressor genes play major roles in the pathogenesis of many human cancers. The identification of cellular oncogenes has thus opened the door to understanding the molecular basis of human cancer. The biochemical properties of many oncogene proteins have been determined and related to the mechanisms that regulate normal cell behavior. A number of proto-oncogenes have been found to encode growth factors, growth factor receptors, protein kinases, guanine nucleotide binding proteins, and transcriptional regulatory proteins that function in signal transduction pathways controlling normal cell proliferation. Thus, oncogene study has not only identified abnormalities that lead to cancer, but has also has helped to elucidate the regulatory pathways that control the growth of normal cells. Finally a number of proto-oncogenes appear to play important roles in development and differentiation. Carcinogenesis is a multistep process involving sequential genetic alterations. Neither activation of a mitogenic pathway nor inactivation of a growth-inhibitory pathway usually suffices to transform a normal cell into tumor cells. This has been well demonstrated by Vogelstein et al. for the genetic changes occurring during colon cancer formation (Table 9). In many tumors, activation of a *ras* oncogene early in tumor progression is followed later in descendant cells by inactivation of the p53 and DCC suppressor genes.[92] Cancer is a genetic disease. The molecular genetics of cancer has undergone an explosive development in the past 20 years and now it represents the cutting edge of cancer research. One end product of this research is a vast body of information on oncogenes (activated and suppressor) some of which has immediate clinical relevance.

Recent accomplishment on the immortalization and transformation of cultured human cells has far reaching implications for both cell and cancer biology. Human cell transformation studies will increase our understanding of the mechanisms underlying carcinogenesis and differentiation. The neoplastic process can now be studied in a model human cell culture system. The accompanying biochemical and genetic changes, once identified, will help define the relationship between malignancy and differentiation. Normal human cells in culture are known to be resistant to experimentally induced tumorigenicity. However, normal human cells in culture can now be transformed into tumorigenic cells.

REFERENCES

1. J. M. Bishop, Cellular oncogenes and retroviruses, Ann. Rev. Biochem. 52:350 (1983).
2. R. A. Weinberg, Oncogenes and the molecular origins of cancer, Cold Spring Harbor Laboratory Press, Cold Spring Harbor (1989).
3. G. M. Cooper, Oncogenes, Jones and Bartlett, Boston, (1990).
4. R. Sager, Genetic suppression of tumor formation, Adv. Cancer Res. 44, 44:43 (1985).

5. H. Harris, The genetic analysis of malignancy, J. Cell. Sci. Suppl. 4:431 (1986).
6. G. Klein, The approaching era of tumor suppressor genes, Nature 238:1539 (1987)
7. E. Farber and R. Cameron, The sequential analysis of cancer development, Adv. Cancer Res. 31:125 (1980).
8. G. Klein and E. Klein, Evolution of tumors and the impact of molecular oncology, Nature 315:190 (1985).
9. J. S. Rhim, Neoplastic transformation of human epithelial cells *in vitro*, Anticancer Res. 9:1345 (1989).
10. J. S. Rhim and A. Dritschilo, Neoplastic transformation in human cell culture, Mechanisms of carcinogenesis, Humana Press, Totowa (1991).
11. G. M. Cooper, S. Okenquist, and L. Silverman, Transforming activity of chemically transformed and normal cells, Nature 284:418 (1980).
12. C. Shih, B. Z. Shilo, M. P. Goldfarb, A. Dannenberg, and R. A. Weinberg, Passage of phenotypes of chemically transformed cells via transfection of DNA and chromatin, Proc. Natl. Acad. Sci. USA 76:5714 (1979).
13. M. A. Rassonlzadegan, A. Cowie, A. Carr, N. Glachenhous, and R. Kamen, The roles of individual polyoma virus early proteins in oncogenic transformation, Nature 300:713 (1982).
14. P. V. D. Elsen, A. Houweling, and A. Van der Eb, Expression of region EIb of human adenoviruses in the absence of region EIb is not sufficient for complete transformation, Virology 128:377 (1983).
15. J. S. Butel, SV40 large T-antigen:dual oncogene, Cancer Surv. 5:343 (1986).
16. D. P. Lane and L. V. Crawford, T antigen is bound to a host protein in SV40-transformed cells, Nature 298:261 (1979).
17. S. A. Courtneidge and A. E. Smith, Polyoma virus transforming protein associated with the product of the c-*src* cellular gene, Nature 303:435 (1983).
18. M. R. Green, When the products of oncogenes and anti-oncogenes meet, Cell 561:1 (1989).
19. R. A. Weinberg, Fewer and fewer oncogenes, Cell 30:3 (1982).
20. T. G. Krontiris and G. M. Cooper, Transforming activity of human tumor DNAs, Proc. Natl. Acad. Sci. USA 78:1181 (1981).
21. M. Perucho, M. Goldfarb, K. Simizu, C. Lama, J. Fogh, and M. Wigler, Human tumor-derived cell lines contain common and different transforming genes, Cell 27:467 (1981).
22. S. Pulciani, E. Santos, A. V. Lauver, L. K. Long, K. C. Robbins, and M. Barbacid, Oncogenes in human tumor cell lines:Molecular cloning of a transforming gene from human bladder carcinoma cells, Proc. Natl. Acad. Sci. USA 79:2845 (1982).
23. G. M. Cooper, Cellular transforming genes, Science 218:801 (1982).
24. Y. Yuasa, S. K. Srivastava, C. Y. Dunn, J. S. Rhim, E. P. Reddy, and S. A. Aaronson, Acquisition of transforming properties by alternative point mutations with c-*bas/has* human proto-oncogene, Nature 303:775 (1983).
25. Y. Yuasa, A. Eva, M. H. Kraus, S. K. Srivastava, S. W. Needleman, J. H. Pierce, J. S. Rhim, R. Gol, E. P. Reddy, S. R. Tronick, and S. A. Aaronson, *Ras*-related oncogenes of human tumors, in: "Cancer Cells," G. F. Vande Woude, A. J. Levine, W. C. Topp, J. D. Watson, eds., Cold Spring Harbor Laboratories (1984).
26. J. Fujita, O. Yoshida, Y. Yuasa, J. S. Rhim, M. Hatanaka, and S. A. Aaronson, Ha-*ras* oncogenes are activated by somatic alterations in human urinary tract tumors, Nature 309:464 (1984).

27. L. A. Feig, R. C. Bast, R. C. Knapp, and G. M. Cooper, somatic activation of *ras*-K gene in a human ovarian carcinoma, Science 223:698 (1984).
28. A. Balmain, M. Ramsden, G. T. Bowden, and J. Smith, Activation of the mouse cellular Harvey-*ras* gene in chemically induced benign skin papillomas, Nature 307:658 (1984).
29. E. Santos, S. Sukumar, D. Martin-Zanca, H. Zarbl, and M. Barbacid, Transforming *ras* gene, in: "Viruses and Cancer," P. W. J. Rigby and N. M. Wilkie, eds., Cambridge Univ. Press, Cambridge (1985).
30. A. Balmain and I. B. Pragnell, Mouse skin carcinomas induced *in vitro* by chemical carcinogens have a transforming Harvey-*ras* oncogene, Nature 303:72 (1983).
31. S. Sukumar, V. Notario, D. Martin-Zanca, and M. Barbacid, Induction of mammary carcinomas in rats by nitroso-methyl-urea involves the malignant activation of the H-*ras* 1 locus by single point mutation, Nature 306:658 (1983).
32. I. Guerrero, P. Corzada, A. Mayer, and A. Pellicer, Molecular approach to leukemogenesis: mouse lymphomas contain an activated c-*ras* oncogene, Proc. Natl. Acad. Sci. USA 81:202 (1984).
33. H. Zarbl, S. Sukumar, A. V. Arthur, D. Martin-Zanca, and M. Barbacid, Direct mutagenesis of Ha-*ras*-1 oncogenes by N-nitroso-N-methylurea during initiation of mammary carcinogenesis in rats, Nature 312:382 (1985).
34. J. S. Rhim, J. Fujita, and J. B. Park, Activation of H-*ras* oncogene in 3-methylcholanthrene-transformed human cell line, Carcinogenesis 8:1165 (1987).
35. S. A. Aaronson and S. R. Tronick, The role of oncogenes in human neoplasia, *in*: "Important Advances in Oncology," V. Devita, S. Helluman, and S. Rosenberg, eds., J. B. Lippincott Co., Philadelphia (1985).
36. J. J. Yunis, The chromosomal basis of human neoplasia, Science 221:227 (1984).
37. D. C. Swan, O. W. McBride, K. C. Robbins, D. A. Keithley, and S. A. Aaronson, Chromosomal mapping of the simian sarcoma virus *onc* gene analogue in human cells, Proc. Natl. Acad. Sci. USA 79:4691 (1982).
38. K. Prakash, O. W. McBride, D. C. Swan, S. G. Devare, S. R. Tronick, and S. A. Aaronson, Abelson murine leukemia virus: structural requirements for transforming gene function, Proc. Natl. Acad. Sci. USA 79:5210 (1982).
39. R. Taub, I. Kirsch, C. Morton, G. Lenoir, D. Swan, S. A. Tronick, S. A. Aaronson, and P. Leder, Translocation of the c-*myc* gene into the immunoglobulin heavy chain locus in human Burkitt's lymphoma and murine plasmacytoma cells, Proc. Natl. Acad. Sci. USA 79:7839 (1982).
40. O. W. McBride, D. C. Swan, S. R. Tronick, R. Gol, E. Klimanis, D. E. Moore, and S. A. Aaronson, Regional chromosomal location of N-*ras*-1, K-*ras*-1, K-*ras*-2 and *myb* oncogenes in human cells, Nucleic Acids Res. 11:8221 (1983).
41. O. W. McBride, D. C. Swan, E. Santos, M. Barbacid, S. R. Tronick, and S. A. Aaronson, Localization of the normal allele of T-24 human bladder carcinoma oncogene to chromosome 11, Nature 300:773 (1982).
42. R. Dalla-Favera, R. C. Gallo, A. Giallongo, and C. M. Croce, Chromosomal localization of the human homolog (c-*sis*) of the simian sarcoma virus *onc* gene, Science 218:686 (1982).
43. R. Dalla-Favera, M. Bregni, J., Erikson, D. Patterson, R. C. Gallo, and C. M. Croce, Human c-*myc onc* gene is located on the region of chromosome 8 that is translocated in Burkitt's lymphoma cells, Proc. Natl. Acad. Sci. USA 79:7824 (1982).

44. B. G. Weel, S. C. Jhanwar, R. S. Chaganti, and W. S. Hayward, Two human c-*onc* genes are located on the long arm of chromosome 8, Proc. Natl. Acad. Sci. USA 79:7842 (1982).
45. B. deMartinville, J. Giacalone, C. Shih, R. A. Weinberg, and U. Francke, Oncogene from human EJ bladder carcinoma is located on the short arm of chromosome 11, Science 219:498 (1983).
46. A. Y. Sakaguchi, S. L. Naylor, T. B. Shows, J. J. Toole, M. McCoy, and R. A. Weinberg, Human c-Ki-*ras*2 proto-oncogene on chromosome 12, Science 219:1081 (1983).
47. A. Hall, C. J. Marshall, N. I. Spurr, and R. A. Weiss, Identification of transforming gene in two human sarcoma cell lines as a new member of the *ras* gene family located on chromosome 1, Nature 303:396 (1983).
48. M. E. Harper, G. Franchini, J. Love, M. I. Simon, R. C. Gallo, and F. Wong-Staal, Chromosomal sublocalization of human c-*myb* and c-*fes* cellular *onc* genes, Nature 304:169 (1983).
49. B. deMartinville, J. M. Cunningham, J. J. Murray, and U. Francke, The N-*ras* oncogene assigned to the short arm of human chromosome 1, Nucleic Acids Res. 11:5267 (1983).
50. J. Ryan, P. E. Barker, K. Simizu, M. Wigler, and F. H. Ruddle, Chromosomal assignment of a family of human oncogenes, Proc. Natl. Acad. Sci. USA 80:4460 (1983).
51. S. J. O'Brien, W. G. Nash, J. L. Goodwin, D. R. Lowy, and E. H. Chang, Dispersion of the *ras* family of transforming genes to four different chromosomes in man, Nature 302:839 (1983).
52. T. I. Bonner, S. J. O'Brien, W. G. Nash, and U. R. Rapp, The human homologs of the *ras* (*mil*) oncogene are located on human chromosomes 3 and 4, Science 223:71 (1984).
53. J. Groffen, N. Heisterkamp, N. Spurr, S. Dana, J. J. Wasmuth, and J. R. Stephenson, Chromosomal localization of human c-*fms* oncogene, Nucleic Acids Res. 11:6331 (1983).
54. M. F. Roussel, C. I. Sherr, P. E. Barker, and F. H. Ruddle, Molecular cloning of the c-*fms* locus and its assignment to human chromosome 5, J. Virol. 48:770 (1983).
55. D. M. Glover and B. D. Hames, Oncogenes, IRL Press, Oxford.
56. M. F. Hansen and W. K. Cavenee, Genetics of cancer predisposition, Cancer Res. 47:5578 (1987).
57. A. G. Knudson, Jr., Genetics of human cancer, Ann. Rev. Gen. 20:231 (1986).
58. H. Land, L. F. Parada, and R. A. Weinberg, Tumorigenic conversion of primary embryo fibroblasts requires at least two cooperating oncogenes, Nature 304:596 (1983).
59. H. E. Ruley, Adenovirus early region IA enables viral and cellular transforming genes to transform primary cells in culture, Nature 304:602 (1983).
60. J. S. Rhim, G. Jay, P. Arnstein, F. M. Price, K. K. Sanford, and S. A. Aaronson, Neoplastic transformation of human epidermal keratinocytes by Ad12-SV40 and Kirsten sarcoma viruses, Science 227:1250 (1985).
61. J. S. Rhim, J. Fujita, P. Arnstein, and S. A. Aaronson, Neoplastic conversion of human keratinocytes by adenovirus 12-SV40 virus and chemical carcinogens, Science 232:385 (1986).

62. P. Thraves, Z. Salehi, A. Dritschilo, and J. S. Rhim, Neoplastic transformation of immortalized human epidermal keratinocytes by ionizing radiation, Proc. Natl. Acad. Sci. USA 87:1174 (1990).
63. R. F. Newbold and R. W. Overell, Fibroblast immortality is a prerequisite for transformation by EJ c-Ha-*ras* oncogene, Nature 304:648 (1983).
64. J. S. Rhim, T. Kawakami, J. Pierce, K. Sanford, and P. Arnstein, Cooperation of v-oncogenes in human epithelial cell transformation, leukemia 2:1515 (1988).
65. J. S. Rhim, J. b. Park, and G. Jay, Neoplastic transformation of human keratinocytes by polyprene-induced DNA mediated transfer of an activated oncogene, Oncogene 4:1403 (1989).
66. R. Gantt, K. K. Sanford, R. Parshad, F. M. Price, W. D. Peterson, and J. S. Rhim, Enhanced G_2 chromatid radiosensitivity, an early stage in the neoplastic transformation of human epidermal keratinocytes in culture, Cancer Res. 47:1390 (1987).
67. P. Boukamp, E. J. Stanbridge, D. Y. Foo, P. A. Ceratti, and N. E. Fusenig, c-Ha-*ras* oncogene expression immortalized human keratinocytes (HaCat) alters growth potential *in vivo* but lacks correlation with malignancy, Cancer Res. 50:2840 (1990).
68. M. Durst, D. Gallahan, G. Jay, and J. S. Rhim, Glucocorticoid-enhanced neoplastic transformation of human papilloma virus type 16 and activated *ras* oncogene, Virology 73:767 (1989).
69. R. R. Reddel, Y. Ke, E. Kaighn, L. Malan-Shibley, J. F. Lechner, J. S. Rhim, and C. C. Harris, Human bronchial epithelial cells neoplastically transformed by v-Ki-*ras* altered response to inducers of terminal squamous differentiation, Oncogene Res. 3:401 (1988).
70. P. Amstad, R. R. Reddel, A. Pfiefer, L. Malan-Shibley, G. E. Mark III, and C. C. Harris, Neoplastic transformation of a human bronchial epithelial cell line by a recombinant retrovirus encoding viral Harvey *ras*, Mol. Carcinogenesis 1:151 (1988).
71. R. Clark, M. R. Stampfer, R. Milley, E. O'Rourke, K. H. Walen, M. Kriegler, J. Kopplin, and F. McCormick, Transformation of human mammary epithelial cells by oncogenic retroviruses, Cancer Res. 48:4689 (1988).
72. K. H. Walen and P. Arnstein, Induction of tumorigenesis and chromosomal abnormalities in human amniocytes infected with simian virus 40 and Kirsten sarcoma virus, In Vitro Cell Dev. Biol. 2:57 (1986).
73. J. A. DiPaolo, C. D. Woodworth, N. C. Popeseu, V. Notario, and J. Doniger, Induction of human cervical squamous cell carcinoma by sequential transfection with human papilloma virus type 16 DNA and viral Harvey *ras* oncogene, Oncogene 4:395 (1989).
74. C. A. Reznikoff, L. J. Loretz, B. J. Christian, W. Wu, and L. F. Meisner, Neoplastic transformation of SV40 immortalized human urinary tract epithelial cells by *in vitro* exposure to 3-methylcholanthrene, Carcinogenesis 9:1427 (1988).
75. B. J. Christian, C. Kao, W. Wu, L. F. Meisner, and C. A. Rezinkoff, Transformation of SV40-immortalized human uroepithelial cells by transfection with *ras* oncogene, Cancer Res. 50:4779 (1990).
76. J. F. Lechner, D. T. Smooth, A. M. A. Pfeifer, K. H. Cole, A. Weston, J. D. Groopman, P. G. Shields, T. Tokiwa, and C. C. Harris, A non-tumorigenic human liver epithelial cell culture model for chemical and biological carcinogenesis

investigations, in: "Neoplastic Transformation in Human Cell Culture, Mechanisms of Carcinogenesis," J. S. Rhim and A. Dritschilo, eds., Humana Press, Totowa (1991).

77. M. Namba, Y., Kano, L. Y. Bai, K. Mikara, and M. Miyazaki, Establishment and characterization of SV40 T-antigen immortalized human liver cells, in: "Neoplastic Transformation in Human Cell Culture, Mechanisms of Carcinogenesis, J. S. Rhim and A. Dritschilo, eds., Humana Press, Totowa (1991).

78. A. Haugen, D. Ryberg, I. L. Hansteen, and P. Amstad, Neoplastic transformation of human kidney epithelial cell line transfected with v-Ha-*ras* oncogene, Int. J. Cancer 45:572 (1990).

79. R. D. Cone, M. Platzer, L. A. Piccinini, M. Jaramillo, and T. F. Davies, HLA-DR gene expression in a proliferating human thyroid cell clone (12S), Endocrinology 1123:2067 (1988).

80. N. R. Lemoine, E. S. Mayall, T. Jones, D. Shear, S. McDermid, P. Kendall-Taylor, and D. Wynford-Thomas, Characterization of human thyroid epithelial cells immortalized *in vitro* by simian virus 40 DNA transfection, Br. J. Cancer 60:897 (1989).

81. R. D. Berry, S. C. Powell, and C. Paraskeva, *In vitro* culture of human fetal colonic epithelial cells and their transformation with origin virus SV40 DNA, Br. J. Cancer 57:287 (1988).

82. A. C. Williams, S. J. Harper, and C. Paraskeva, Neoplastic transformation of a human colonic epithelial cell line *in vitro* evidence for the adenoma to carcinoma sequence, Cancer Res. 50:4724 (1990).

83. D. P. Chopra, A. P. Joiakin, b. Retherford, P. A. Mathieu, and J. S. Rhim, Transformation of human tracheal gland epithelial cells *in vitro*, in: "Neoplastic Transformation in Human Cell Culture, Mechanisms of Carcinogenesis," J. S. Rhim and A. Dritschilo, eds., Humana Press, Totowa (1991).

84. K. Dutt, M. Scott, M. Del Monte, N. Agarwal, P. Sternberg, S. K. Srivastava, and A Srinivastan, Establishment of human retinal pigment epithelial cell lines by oncogenes, Oncogene 5:195 (1990).

85. G. D. Stoner, M. E. Kaighn, R. R. Reddel, J. H. Resan, D. Bowman, Z. Naio, M. Matsukura, M. You, A. J. Calati, and C. C. Harris, Establishment and characterization of SV40 T-antigen immortalized human esophageal epithelial cells, Cancer Res. 51:365 (1991).

86. K. Melber, G. Zhu, and L. Diamond, SV40-transfected human melanocyte sensitivity to growth inhibition by the phorbol ester 12-0-tetradecanoyl-phorbol-13--acetate, Cancer Res. 49:3650 (1989).

87. M. E. Kaighn, R. R. Reddel, J. F. Lechner, D. M. Peehl, R. F. Camalier, E. D. Brash, U. Saffiotti, and C. C. Harris, Transformation of human neonatal prostate epithelial cells by strontium phosphate transfection with a plasmid containing SV40 early region genes, Cancer Res. 49:3050 (1989).

88. W. O'Brien, J. Stenman, and R. Sager, Suppression of tumor growth by senescence in virally transformed human fibroblasts, Proc. Natl. Acad. Sci USA 83:8659 (1986).

89. M. Namba, K. Nishitani, F. Fukushima, T. Kimoto, and K. Nose, Multistep process of neoplastic transformation of normal human fibroblasts by ^{60}Co gamma rays and Harvey sarcoma viruses, Int. J. Cancer 37:419 (1986).

90. P. H. Hurtin, V. M. Maker, and J. J. McCormick, Malignant transformation of human fibroblasts caused by expression of a transfected T24 H-*ras* oncogene, Proc. Natl. Acad. Sci. USA 86:187 (1989).

91. M. Nagarajan, B. M. Bowman, L. Rigby, J. S. Rhim, and S. Sukamar, p53, a direct target of mutational activation by chemical carcinogens, in: "Neoplastic Transformation in Human Cell Culture, Mechanisms of Carcinogenesis," J. S. Rhim and A. Dritschilo, eds, Humanan Press, Totowa (1991).
92. Vogelstein, E. R. Fearon, S. R. Hamilton, S. E. Kern, A. C. Preisinger, M. Leppert, Y. Nakamura, R. White, A. M. Smits, and J. L. Bos, Genetic alterations during colorectal-tumor development, N. Engl. J. Med. 319:525 (1988).

'IN VIVO' MODEL SYSTEMS TO STUDY ras ONCOGENE INVOLVEMENT IN CARCINOGENESIS

Ramon Mangues and Angel Pellicer

Department of Pathology and Kaplan Cancer Center, New York University Medical Center, New York, N.Y. 10016

INTRODUCTION

Correlations have been reported between exposure to chemicals, radiation, viruses or certain diets and human cancer. Some chemicals have been pointed out as causative factors in specific tumor types (IARC, 1982; U.S. Department of Health and Human Services, 1983).

Once this link became apparent, a major effort was directed to investigate the mechanisms involved in chemical carcinogenesis. The development of 'in vivo' and 'in vitro' model systems, and technical advances facilitating a molecular approach to problems in cell biology have provided the tools for progress in this area, and it has significantly expanded our understanding of the cellular and molecular changes involved in cancer development (Yuspa and Poirier, 1988).

In this article, we will review two animal models used in our laboratory to study the role of *ras* oncogenes in tumorigenesis: NMU-induced mouse thymic lymphomas and transgenic mice carrying the N-*ras* oncogene. We will also discuss the significance of using these 'in vivo' model systems in analyzing the multistage process of tumorigenesis.

It is our belief that progress in risk assessment, prevention and cancer therapy will mainly come as a consequence of deciphering the basic molecular mechanisms by which specific cancers are produced, now within our reach because of the recent developments in molecular and cellular biology.

ras ONCOGENES

The *ras* gene superfamily contains a group of three closely related genes (H-*ras*, K-*ras* and N-*ras*), ubiquitous in eukaryotes. *ras* activation appears to play a role in the etiology of human cancer, since mutated forms have been found in a significant percentage of most types of tumors (Bos, 1989).

Its importance comes from the fact that approximately 20% of human tumors of diverse origin and histological type have been shown to contain *ras* oncogenes that can transform murine cells in culture (Barbacid, 1987). They are also the most commonly detected oncogenes in carcinogen-induced animal tumors (Guerrero and Pellicer, 1987).

H-*ras* and K-*ras* were originally identified as the transforming principles of the Harvey and Kirsten strains of rat sarcoma virus, two acute transforming retroviruses generated by transduction of the rat H-*ras* and K-*ras* cellular genes, respectively (DeFeo et al. 1981; Ellis et al. 1981). N-*ras* was first found in a human neuroblastoma cell line (Shimizu et al. 1983). All of them have a similar exon-intron structure and encode strikingly homologous 21 kilodalton proteins (p21) (Bos, 1989).

The function of the *ras* proteins remains unknown despite of the intensive effort directed to decipher it. Their biochemical properties - intrinsic GTPase activity and ability to bind GDP and GTP - and their localization in the inner side of the plasma membrane have supported the hypothesis of their function as transducers in the transmembrane signaling pathways in a manner analogous to the G proteins with whose alpha subunit they share homology (Barbacid, 1987).

MODEL SYSTEMS OF CARCINOGENESIS

In Vitro Models to Study Carcinogenesis

The conjunction of two powerful techniques - DNA-mediated gene transfer (DMGT) and tissue culture - made it possible to speed up the study of the neoplastic process in the early 80s.

DMGT permitted the study and isolation of several genes which provided cells with phenotypic features that could be used to select them from the rest of the culture lacking this phenotype (Pellicer et al., 1980; Pellicer, 1986).

By using this technique with cultured mouse fibroblasts (NIH 3T3 cells), it was observed that high molecular weight DNA from several transformed cell lines and primary tumors had the ability to 'transform' those cells after exposure to the DNA in the form of calcium phosphate precipitate (Shih et al., 1979). The transformation was observed as a low frequency of foci formation, that is, clonal outgrowth of morphologically altered cells with enhanced growth potential over the background of untransformed contact inhibited NIH 3T3 cells. Transformation of recipient NIH 3T3 cells is due to uptake and incorporation of donor DNA that imparts, in dominant fashion, altered growth and morphological properties (Guerrero and Pellicer, 1987).

The detection and isolation of the gene responsible for the transformed phenotype is accomplished by tagging specific sequences to a restriction-digested DNA prior to gene transfer, which will eventually allow its retrieval by using nucleic acid hybridization techniques. The procedure requires Southern blotting of extracted DNA from the NIH 3T3 transformants, obtained from different tumors using the tagging specific sequences as a probe (Pellicer et al., 1980; Pellicer, 1986).

When the particular oncogene involved in transformation is known, the isolation is done by using the same methodology with the specific probe for this gene. The presence of extra bands, that segregate with the transformed phenotype in successive transfers, when compared to non-transfected cells, indicates that such a gene is responsible for the morphological changes in 3T3 cells and permits its cloning (Guerrero et al., 1984b; Guerrero and Pellicer, 1987).

In Vivo Models

The development of *in vivo* models of carcinogenesis, that use well-defined etiological agents and involve induction of specific tumor types in a reproducible way, offers the possibility to study the nature of the cellular factors that govern the phenotypic expression of activated oncogenes, and the role that their activation plays in the malignant origin and progression of tumors.

Chemical carcinogenesis research during the decade of the 1970s focused on the development or refinement of animal models for chemically induced tumors at specific organ sites in which the pathogenesis of benign and malignant neoplasms closely resembled tumor formation in the homologous human tissue. The use of inbred laboratory strains has provided a reproducible genetic background for these studies.

Virtually every major form of human cancer can now be reproduced in an animal species by exposure to specific chemical carcinogens - skin, liver, lung, breast, colon, pancreas, bladder, thymus, etc. - and, in many cases cancer development can be modified by additional exposure to tumor promoters (Yuspa and Poirier, 1988).

An additional *in vivo* model useful in the study of carcinogenesis has been the nude mouse assay. In this procedure, the specific DNA sequences, whose tumor induction capability is being tested, are cotransfected with a dominant selectable vector into NIH 3T3 cells. Subsequently, a pool of dominantly selected colonies are injected in mice deprived of cellular immunity, and tumor appearance observed (Blair et al., 1982; Fasano et al., 1984).

Multistage Process of Carcinogenesis

The epidemiological studies of human cancers (Knudson, 1977) and the evaluation of experimentally induced tumors in laboratory animals (Hecker et al. 1982) support the hypothesis that development of malignant tumors is a multistep process composed of a series of separable changes or events.

Geneticists consider such events to be either dominant (gain of function) or recessive (lost of function). It is likely that both types of alterations play a critical role in tumorigenesis (Weinberg, 1989).

Chemical carcinogenesis studies, moreover, have differentiated the process into three phases: initiation, promotion and progression. The first stage, initiation, is defined as the consequences of the initial interaction of a tissue with carcinogens. While pathological changes may not be apparent at this stage, the tissue is irreversibly altered and contains cells (initiated cells) which are the precursor to the future tumor.

Promotion, the second stage, is confined to those processes which facilitate the expression of the initiated phenotype at the tissue level.

Tumor progression includes further phenotypic alterations in the initiated cell population, including conversion from a premalignant to a malignant cell type. After malignant conversion progression continues with changes that give rise to tumor heterogeneity and to the metastatic phenotype (Yuspa and Poirier, 1988).

GENOTOXICITY, INITIATION AND PROMOTION

With the availability of DNA sequencing techniques, one effect common to many of the physical and chemical carcinogens has been shown to be genetic damage, by induction of point mutations, deletions or rearrangements in DNA, suggesting a crucial role for gene alteration in carcinogenesis.

For many oncogenes these mutations have been demonstrated to be somatic in nature and confined to the tumor tissue (Santos et al. 1984) supporting further the hypothesis that they arose due to the direct mutagenic action of a carcinogen or during DNA repair in a cell rendered vulnerable by various physiological factors (Sukumar, 1989).

Additionally, it has been appreciated for many years that chemical carcinogens interact with DNA to form specific adducts, some of which can lead to the introduction of mutations in the genetic material. One of the major dilemmas in studies of this kind has been the identification of the critical adduct and indeed the critical target gene that, when altered, leads to the induction of the transformed state. Although many adducts resulting from treatment of DNA or cells with chemical carcinogens have been identified, the biological significance of specific adduct formation is still not completely clear.

In terms of the defined phases in the multistage process of tumorigenesis, a major difference between initiation and promotion is that while initiators are capable of directly and irreversibly altering DNA, even if such a mechanism is not a prerequisite for initiation, 'pure' tumor promoters do not bind covalently to cellular DNA and they are not mutagenic (Dong and Jeffrey, 1990).

The linear dose-response curve for initiation of tumorigenesis, the irreversibility of the initiated state, and the persistence of initiated cells have been confirmed in a variety of target sites and indicate a genotoxic basis for initiation.

In contrast, tumor promotion is reversible, and requires repeated exposures of promoting agents, suggesting that promotion is epigenetic. Promoters are agents identified on the basis of their ability to increase the number and/or reduce the latency period of both preneoplastic and neoplastic lesions after limited exposure to initiators. Most promoting agents are tissue specific, do not require metabolic activation, and are non mutagenic as mentioned above. Experimental data suggest that the tissue-specific action of tumor promoters may play an important role in determining the target tissue for tumor formation (Yuspa and Poirier, 1988).

No single genotoxic insult appears to be sufficient to trigger cancer development. Current thought holds that multiple genetic changes are required to induce the malignant phenotype, not only at the initiation stage but for tumor progression. This argument is sustained by the fact that non-genotoxic promoters do not appear to influence neoplastic progression while repeated exposure to genotoxic initiating agents and certain tumor promoters with demonstrable genotoxicity enhances the yield of malignant tumors (Yuspa and Poirier, 1988).

ras ONCOGENE ACTIVATION IN MODEL SYSTEMS

Many authors view as an established fact *ras* oncogene involvement in human tumorigenesis, and the questions to be answered now are how these oncogenes mediate transformation and at which stage of the disease - initiation, promotion, progression, development of metastatic potential, etc. - are they activated (Sukumar, 1989). The molecular mechanisms that link oncogene activation and tumorigenesis remain to be unveiled.

ras genes acquire transforming properties by single point mutations mainly in two hot spots within their coding sequences (codons 12, 13, and 61), behaving in a dominant fashion when activated. Mutations in *ras* genes result in reduction of their products affinity for GDP and GTP or abolish their intrinsic GTPase activity losing their self-regulation by hydrolysis of GTP, and resulting in *ras* proteins constitutively activated (Barbacid, 1987).

THE MOUSE THYMIC LYMPHOMA MODEL

Among the reasons why we chose to study thymic lymphomas in mice were the long latency period that intervenes between exposure to the carcinogen at a young age and outgrowth of the malignancy many months later in middle age, that makes it an excellent model system for tumor development in humans, and the ability to produce the same type of tumor with very different agents. A defined set of protocols involving either radiation or various chemical carcinogens existed to induce, reproducibly, thymic lymphomas in treated mice with a very high frequency (Kaplan, 1967; Guerrero and Pellicer, 1987).

Our goal has been to study the possible association between *ras* gene activation and tumor development in this model system and to investigate the underlying mechanisms at a molecular level.

Tumor Induction

In our laboratory, thymic lymphomas are induced by intraperitoneal injection of

AKRxRF, RF/J, 129/J or C57BL/6 mice with N-nitrosomethylurea (NMU) (Guerrero et al., 1984b), or exposure to whole-body gamma-rays (Guerrero and Pellicer, 1987) or neutrons (Sloan et al., 1990). Five to seven weeks old mice are treated weekly either with an intraperitoneal injection of NMU for 5 weeks, whole-body irradiation of 160 rad/week for 4 weeks or 100 rads of 0.4 MeV neutrons. The NMU is given at 30 mg/Kg of body weight and the irradiation consists of gamma-rays from a ^{137}Cs source.

In both cases the animals develop tumors with a mean latency period close to 3 months for the NMU-induced tumors and 6 months for the radiation-induced tumors. Both treatments lead to thymic lymphomas, which spread to the spleen and lymph nodes, and more rarely to the liver. By histological criteria, tumors induced by the two treatments are indistinguishable (Guerrero and Pellicer, 1987).

Tumor DNA Analysis: N-*ras* Oncogene Cloning

Once the model system was established attention was focused in molecularly cloning DNA sequences with 'transforming' potential from a variety of tumors and tumor cell lines, because of the widely-held hypothesis that these genes may be responsible for the neoplastic phenotypes of the tumor cells from which they are derived (Guerrero et al., 1984b).

The DNAs extracted from the tumors were used in DNA-mediated gene transfer (DMGT) experiments to induce transformed foci in NIH 3T3 cells. These primary foci were picked, grown up in mass culture, and their DNA was extracted and used to obtain secondary foci in rat cells. Southern blot hybridizations were performed using probes for the three *ras* genes (H, K, and N-*ras*). Extra copies of the *ras* genes were present, in some cases in a different mobility than the endogenous gene.

The oncogenes that segregated with the transformed phenotype in successive transfers belonged to the *ras* gene family and were identified as N-*ras* and K-*ras* (Guerrero et al., 1984a, Guerrero et al. 1984b, Guerrero et al. 1985).

The N-*ras* activated oncogene was isolated from a rat secondary transformant obtained with DNA originally derived from an NMU-induced thymic lymphoma (the 3T3 primary transformant and this rat secondary transformant were both tumorigenic in nude mice). Using the rat secondary instead of the NIH 3T3 primary transformant permitted us to distinguish the active oncogene from endogenous homologs, due to species differences.

In order to isolate the gene in an active form, partial digestion kinetics were performed to identify a band of relative high molecular weight (approx. 14 Kb) that only appeared in the transformant and not at the same time as in the normal recipient rat cells. This band was eluted from a preparative gel and cloned in λ47.1 arms. After hybridizing the recombinant plaques with a human N-*ras* specific probe, positive clones were isolated and their restriction maps investigated.

The clone isolated from the transformant was shown to be active in the 3T3 focus forming assay and its efficiency of transformation was five orders of magnitude higher than DNA from the original tumor indicating that the gene was essentially pure (Guerrero et al., 1984b).

N-ras ACTIVATION IN THYMIC LYMPHOMAS

Once the four coding exons of the cloned gene were sequenced, only a single base change, a C->A transversion in the first base of the codon 61 of the N-*ras* gene, was found. This point mutation would change glutamine to lysine in the protein. No other alterations were found when the DNA from the transformant was compared to the DNA from a non tumoral tissue (brain) of the mouse that developed the tumor (Guerrero et al., 1985).

Other point mutations that activate the N-*ras* gene have also been found in our laboratory, in NMU induced thymomas, and are listed in Table 1 for all tumors analyzed to date, including G->A and A->T transversions in codons 12, 13 and 61 (Diamond et al., 1988; Newcomb et al., 1988).

NMU is an alkylating agent and a direct-acting carcinogen with a short half-life (approx. 20 min); that is, it forms adducts with DNA, and it does not need to be metabolized before it can exert its action on DNA (Gullino et al., 1975).

In our system NMU is behaving as a complete carcinogen, since it is able to produce malignant transformation by itself, without the cooperation of other chemicals that could act as promoters.

TRANSGENIC MODELS IN ONCOGENESIS

The ability to transfer genetic information stably into the germ line of the laboratory mouse has provided a powerful tool to investigate the consequences of gene expression on specific cells in the context of the whole organism. This technology has permitted probing and perturbation of the complex processes involved in cell proliferation or/and differentiation. In particular, it is now clear that the introduction of oncogenes in transgenic mice can be used to study oncogene actions and their role(s) in tumorigenesis.

In spite of the fact that chromosomal position and construction design contribute to define the phenotypic features associated to a specific transgene, once a transgenic line has been established, there is a stability of phenotype within it. This stability and the Mendelian inheritance of the transgene in a unique chromosomal location make the system reproducible, allowing different stages in the neoplastic process to be studied.

The heritable development of specific tumors in a highly reproducible manner in individuals with a similar genetic background permits increasingly detailed investigations into the direct and indirect consequences of oncogene expression.

Transgenic animal experiments carried out so far have shown an apparent specificity in oncoprotein action, a finding consistent with the association of many oncoproteins with specific types of cancer (Hanahan, 1988).

The consequence of the expression of oncogenes in transgenic mice is an induction of proliferation in expressing cells, which certainly supports the causality of oncogenes in cancer, and is manifested as a hyperplasia that progresses to an overt tumor in many cases. Moreover, the induction of a proliferative condition does not appear to be an immediate consequence of oncoprotein expression; thus, there is a delayed response to the oncogene signal. Rather, it often seems that the oncoprotein requires some time and the cooperation of other factors to deregulate the cell.

Table 1. Activating ras gene mutations detected in mice thymic lymphomas induced by Nitroso-methylurea.

Gene / Codon	Point mutation	Tumors with activated gene[*] Number	Percentage
K-ras			
12	GGT -> GAT Gly Asp	31	32.9%
13	GGT -> GAT Gly Asp	1	1.1%
N-ras			
12	GGT -> GAT Gly Asp	1	1.1%
13	GGT -> GAT Gly Asp	1	1.1%
61	CAA -> AAA Gln Lys	2	2.1%
	CAA -> CTA Gln Leu	1	1.1%

[*] Number of tumors examined = 94.

Furthermore, the infrequent progression of proliferating cells in hyperplastic tissue into tumor cells populating a histologically distinct tumor, suggests the necessity of rate-limiting secondary events to convert a hyperplasia into a neoplasia, implicating either genetic or epigenetic changes in the cell with tumor progression, and supporting the concept of tumor progression factors which act independently from dominant oncogenes during the process. This result is consistent with what is known about natural tumorigenesis in animals, namely that it involves a set of separable, relatively independent events (Hanahan, 1988).

One clear difference between the tumors seen in humans and those in transgenic mice is that human tumors are often invasive and highly metastatic, whereas metastasis is rare in tumors thus far seen in transgenic mice. It is conceivable that transgenic mice will permit us to identify rate-limiting events involved in metastasis, and also manipulate them genetically (Hanahan, 1988).

N-ras Transgenic Models

Once the N-*ras* oncogene was isolated in our laboratory, we wanted to see whether or not it was able to reproduce the original thymic lymphoma by introducing it in the mouse germline.

We first repeatedly injected the N-*ras* oncogene with its own promoter in the male pronucleus of fertilized mouse eggs. None of the newborn mice had integrated the transgene. We attributed this finding to the lethal effect of the high expression of the activated N-*ras* gene during embryogenesis when driven by its natural promoter, since the expected outcome should have been 20-30 % of the newborn mice being transgenics.

Consequently, we turned to a hybrid construction to make the transgenics, using regulatory elements derived from another gene and aligned to transcribe the protein coding information from the activated N-*ras* gene. This approach presumes that tissue specificity resides in the 5' flanking region of a gene where the transcriptional promoter elements, as in most cases the enhancers, are located. We decided to recombine the mouse mammary tumor virus promoter (MMTV-LTR) with the N-*ras* oncogene in order to direct the synthesis of the oncoprotein in the cell types that MMTV-LTR specify.

The tissue specificity of MMTV has been repeatedly addressed in transgenic mice experiments where MMTV-LTR was combined with several oncogenes and it was concluded that in spite of the apparent tissue specificity of the virus the MMTV-LTR has a reasonable broad transcriptional specificity, directing consistently the expression to mammary, lacrimal and submaxillary glands among other organs (Stewart et al, 1984, Leder et al., 1986, Choi et al. 1987, Choi et al. 1988, Muller et al., 1988, Bouchard et al. 1989, Tremblay et al. 1989).

MMTV/N-ras Transgenics

Transgenic mice were produced as previously described (Gordon et al., 1980; Gordon and Ruddle, 1983). Briefly, immature B6D2F1 female mice were superovulated and mated to CD-1 males. The following morning the females were examined for the presence of vaginal plugs, and that afternoon the plugged females were sacrificed. Fertilized ova were recovered and microinjected with a solution containing the MMTV/N-*ras* oncogene construct.

Screening of founder animals and offspring was performed by extraction of DNA either from spleen or tail, and Southern hybridization using the MMTV/N-*ras* as a probe.

Animals were sacrificed, tissues were taken, fixed and embedded in paraffin.

Subsequently the blocks were sectioned, stained with hematoxylin and eosin and examined for pathological findings.

Six founder animals carrying between 5 and 50 copies/cell of the transgene were identified. All animals, founders and progeny, were analyzed in detail for MMTV/N-*ras* integration, expression and associated pathology, after DNA and RNA extraction from different organs.

Transgene Expression

It was of interest to determine the expression pattern of the oncogene in different organs in order to correlate it with subsequent pathological changes.

The expression of the fusion gene was shown to be due to the induction of the MMTV promoter by Northern analysis and RNase protection studies. The RNase protection assay showed transgene transcripts initiated at the MMTV promoter in all pathological tissues, while transcripts for the endogenous gene were present in both normal and abnormal tissues (Mangues et al. 1990a).

Northern blotting showed higher levels of expression, in most of the animals, in those organs with more significant abnormalities: salivary gland, mammary gland and Harderian gland (Table 2). Intermediate levels of expression were seen in lung, prostate, testis and ovary. Some of these tissues showed also abnormalities.

The pathology we registered could be attributed to the action of the construct as none of the non-transgenic littermates showed pathological features.

Pathological Findings

The abnormalities seen in these animals were of a proliferative nature in Harderian gland, mammary gland and salivary gland.

A severe compromise of male reproductive function was also recorded. All pedigrees carrying the MMTV/N-*ras* construct manifested reduced male fertility which varied in its time of onset. (Mangues et al. 1990b).

Harderian glands showed variable degrees of pathology from hyperplasias to benign growths, evidenced by exophthalmos and histopathological analysis. Both the incidence and the degree of the pathological findings in this organ were higher in males than females. Multifocal mammary adenocarcinomas were registered in female animals only. No benign hyperplastic growth was seen in mammary tissue.

In salivary gland, we registered from focal proliferation of the ducts to overt adenocarcinoma. In lung, metastatic carcinomas were recorded (Table 2).

The latency period in Harderian gland tumors was shorter (1-4 months) than in the other two malignancies (4-9 months).

A description of the histological findings along with the level of transgene expression in pathological and normal tissues is shown in Table 2, for the two transgenic lines (4 and 7) in which enough progeny could be obtained because of a less severe compromise of the male fertility.

All tumors analyzed expressed high amounts of the transgene transcript. However, not all tissues with significant expression developed tumors. There were instances of high expression of the transgene in salivary, Harderian and mammary glands, without pathological alteration in the organ, indicating that N-*ras* oncogene

Table 2. Histopathology (HP) and transgene expression (TE) in normal and pathological tissues in MMTV/N-ras transgenic mice.

Transgenic Line & Gender		Salivary Gland	Harderian Gland	Mammary Gland	Lung
Control F		-	-	-	-
Control M		-	-	-	-
4 F	HP	Hyperplasia	normal	normal	normal
	TE	-	+	+	++
4 F[a]	HP	normal	benign growth	carcinoma	metastasis
	TE	+++	++++	++++	+
4 F	HP	normal	normal	normal	normal
	TE	+++	+	+	+
4 M	HP	normal	normal	normal	normal
	TE	++++	-	-	+
4 M	HP	carcinoma	tumor	normal	normal
	TE	++++	+++	N.D.	++++
7 F	HP	normal	hyperplasia	normal	normal
	TE	++	+++	++++	++
7 F[a]	HP	normal	hyperplasia	carcinoma	normal
	TE	-	++	++	+
7 M	HP	hyperplasia	tumor	normal	metastasis
	TE	+	+++	N.D.	+
7 M	HP	normal	tumor	normal	normal
	TE	++++	++++	-	+
7 M[a]	HP	normal	normal	normal	normal
	TE	+++	+++	-	+

[a]These animals had a carcinoma of unknown origin in the pelvic region (level of expression in tumor tissue +++, in all cases). Northern blot analysis was carried out with 15 µg of total RNA from specified organs. Crosses indicate the level of expression of the transgene in different organs. Values are normalized to 100% for the highest level of expression. The correspondence is as follows: -, 0-9%; +, 10-25%; ++, 26-50%; +++, 51-75%; ++++, 76-100%. N.D. = Not Done.

expression is not sufficient for tumor development; instead, it seems to provide a high predisposition to tumorigenesis.

The elucidation of specific pathological effects of this abnormally expressed, activated N-*ras* was facilitated by the fact that several other oncogenes and proto-oncogenes have been introduced into mice under the same MMTV regulation.

Most of the organs where we detected expression of the MMTV/N-*ras* transgene were identified as sites of expression for the other MMTV/oncogene constructs like c-*myc* (Stewart et al. 1984; Leder et al. 1986; Sinn et al., 1987), SV40 T and t antigens (Choi et al. 1987; Choi et al., 1988), c-*neu* (Muller et al. 1988; Bouchard et al. 1989) and v-H-*ras* (Sinn et al. 1987; Tremblay et al. 1989). Therefore, specific effects of N-*ras* expression in transgenic mice could be attributed to the gene product rather than to a unique pattern of expression.

The association of N-*ras* oncogene expression with the impairment of the male reproductive system seen in the transgenic mice sharply contrasts with the absence of any alteration in this system induced by H-*ras*, which encodes for a protein strikingly homologous to the N-*ras* protein, when driven by the same promoter and expressed in the same tissues. This fact and the differential expression of the *ras* proto-oncogenes during development (Leon et al., 1987) and spermatogenesis (Sorrentino et al., 1989) in mice strongly suggest a differential role for both proto-oncogenes in these processes (Mangues et al., 1990b).

The existence of three different *ras* genes coding for very similar polypeptides raises the question of whether or not they have distinct modes of regulation, and consequently, specific physiological roles for their products. This scenario seems to be supported by our data.

Since the carboxy terminal domain of the *ras* proteins is where they differ most, it is reasonable to hypothesize that this region of the protein could be responsible for their specific functions.

Our efforts are now directed to produce transgenic mice with the N-*ras* driven by a promoter that directs the expression to the thymus (Thy-1), since the MMTV promoter failed to reproduce the NMU-induced thymic lymphoma from which the activated N-*ras* oncogene was isolated.

We also want to study the effects of the NMU administration in the MMTV/N-*ras* transgenics, in terms of possible modulation of the patterns of tumor induction: shortening of the latency period of tumor appearance, reduction of the carcinogen dosage in tumor induction, and increase of metastatic potential.

IMPLICATIONS OF THE NEW MODEL SYSTEMS DEVELOPMENTS ON CANCER RISK ASSESSMENT

By using animal models, it has been possible to correlate the chemical mode of action of carcinogens with the type of mutation they induce, at least in *ras* genes, and to demonstrate in a biological system the ability of such mutations to induce transformation (Guerrero and Pellicer, 1987; Sukumar, 1989).

Many important questions however remain still unanswered; namely, Is oncogene activation the causal mechanism of cancer development or is it just a consequence of it? What is determining oncogene activation, the specific chemical or physical characteristics of the carcinogenic agent or rather the differentiation stage of the affected cell? At what stage in the carcinogenic process are oncogenes activated? Is oncogene activation necessary for initiation, promotion or tumor progression to occur? What are the effects of specific oncogene activation on the expression of the whole malignant phenotype?

The study of tumorigenesis involves attempts to understand the biochemical functions of different oncogenes, their normal roles in development and in cell function, and their deregulation in the state of disease. The question of whether oncogenes are both necessary and sufficient to trigger a particular cancer remains unresolved. The multistep nature of tumorigenesis might suggest that transcriptional or mutagenic activation of individual oncogenes represents some of the rate limiting events, but that other changes are necessary as well.

Animal models of tumor induction and the production of transgenic mice for use in combination with carcinogen or promoter exposure, will continue being instrumental in elucidating the precise role of individual genes or modifying factors in every stage of tissue-specific carcinogenesis.

Further investigations have to be conducted into the pharmacokinetics of the carcinogens, and into their specific mechanisms of DNA modification, DNA repair, cell proliferation and DNA damage fixation. A better insight into the mechanisms of cancer induced by chemicals will be provided by the study of tumor promotion at a molecular level, focusing on specific alterations in target genes: proto-oncogenes, tumor suppressor genes, and transcriptional regulating sequences. Novel methods to study the molecular epidemiology of human cancer will then be developed leading to a more accurate assessment of potential carcinogenic hazards for humans, and to a more effective prevention and treatment of human cancer.

It is possible that human tumors could be induced by chronic exposure to subcarcinogenic doses of a variety of carcinogens. Such exposures might cause the activating lesion in the oncogene; however, the expression of the neoplastic phenotype will depend on the existence of highly defined physiological conditions in the host, which in most cases may not occur during a lifetime.

The development of culture methods for cells from specific human tissues has provided the methodology required to extrapolate experimental animal data to human disease. Recently, developed expertise in the culture of human cells and explants has made it possible to study carcinogen metabolism in human tissues and to test the validity of using animal models to study this aspect of carcinogenesis. Metabolic profiles generated by human tissues from specific organ sites have been compared to profiles from the analogous site in experimental animals and found to be qualitatively similar. However, among individuals in the human population there is a 100-fold variation in the capacity to metabolize a specific carcinogen. These studies are important, both, for predicting the likelihood that compounds known to cause tumors in other species will be human carcinogens, and for identifying cohorts of potentially high cancer susceptibility within the human population (Yuspa and Poirier, 1988).

The development of new cancer risk-assessment models in humans should integrate all relevant developments that from different fields now converge in the study of oncogenesis. These models have to consider:
1. The exposure to carcinogens as it occurs in people (low-dose, chronic exposures).
2. The molecular, biochemical, cellular (cytotoxic or non-cytotoxic), and physiological mechanisms of the carcinogenic process, for genotoxic as well as non-genotoxic agents, and
3. A basic mechanistic understanding of chemical interactions in defined physiological conditions (genetic/sex/developmental stage differences).

Our progress in this field will heavily depend upon the development of new assays to detect genetic as well as epigenetic mechanisms of toxicity.

ACKNOWLEDGEMENTS

R.M. acknowledges support from NATO Science Fellowship Programme (7.30.-89), Spanish Fondo de Investigaciones Sanitarias de la Seguridad Social (90/1185) and Spanish Ministry of Education and Science (FPU-91 40859166). These studies were conducted with support from CA 36327, CA 50434, and ACS PDT392 to A.P.. A.P. is a Leukemia Society Scholar.

REFERENCES

Barbacid, M. (1987). *ras* genes. Ann. Rev. Biochem. 56:779-827.
Blair, D.G., C.S. Cooper, M.K. Oskarsson, L.A. Eader and G.F. Vande Woude (1982). New method for detecting cellular transforming genes. Science 218:1122-1125.
Bos, J.L. (1989). *ras* oncogenes in human cancer: a review. Cancer Res. 49:4682−4689.
Bouchard, L., L. Lamarre, P.J. Tremblay, and P. Jolicoeur. (1989). Stochastic appearance of mammary tumors in transgenic mice carrying the MMTV/c-neu oncogene. Cell. 57:931-936.
Choi, Y., D. Henrard, I. Lee, and S.R. Ross. (1987). The mouse mammary tumor virus long terminal repeat directs expression in epithelial and lymphoid cells of different tissues in transgenic mice. J. Virol. 61:3013-3019.
Choi, Y., I. Lee, and S.R. Ross. (1988). Requirement for simian virus 40 small tumor antigen in tumorigenesis in transgenic mice. Mol Cell Biol. 8(8):3382−3390.
DeFeo, D., M.A. Gonda, H.A. Young, E.H. Chang, D.R. Lowy, E.M. Scolnick, and R.W. Ellis. (1981). Analysis of two divergent rat genomic clones homologous to the transforming gene of Harvey murine sarcoma virus. Proc. Natl. Acad. Sci. USA. 78:3328-3332.
Diamond, L., I. Guerrero, and A. Pellicer. (1988). Concomitant K- and N-*ras* gene point mutations in clonal murine lymphoma. Mol. Cell. Biol. 8(5):2233-2236.

Dong, Z., and A.M. Jeffrey. (1990). Mechanisms of organ specificity in chemical carcinogenesis. Cancer Investigation 8(5):523-533.

Ellis, R.W., D. DeFeo, T.Y. Shih, M.A. Gonda, H.A. Young, N. Tsuchida, D.R. Lowy, and E. Scolnick. (1981). The p21 src genes of Harvey and Kirsten sarcoma viruses originate from divergent members of a family of normal vertebrate genes. Nature. 292:506-511.

Fasano, O., D. Birnbaum, K. Edlund, J. Fogh and M. Wigler (1984). New human transforming genes detected by a tumorigenicity assay. Mol. Cell. Biol. 4:1695-1705.

Gordon, J.W., and F.H. Ruddle. (1983). Gene transfer into mouse embryos: production of transgenic mice by pronuclear injection. Meth. Enzymol. 101:411-433.

Gordon, J.W., G.A. Scangos, D.J. Plotkin, J.A. Barbosa, and F.H. Ruddle. (1980). Genetic transformation of mouse embryos by microinjection of purified DNA. Proc. Natl. Acad. Sci. USA 77:7380-7384.

Guerrero, I., P. Calzada, A. Mayer, and A. Pellicer. (1984a). A molecular approach to leukemogenesis: mouse lymphomas contain an activated c- *ras* oncogene. Proc. Natl. Acad. Sci. USA. 81:202-205.

Guerrero, I., A. Villasante, P. D'Eustachio, and A. Pellicer. (1984b). Isolation, characterization, and chromosome assignment of mouse N-*ras* gene from carcinogen-induced thymic lymphoma. Science 225 (4666):1041-1043.

Guerrero, I., and A. Pellicer (1987). Mutational activation of oncogenes in animal model systems of carcinogenesis. Mut. Res. 185:293-308.

Guerrero, I., A. Villasante, V. Corces, and A. Pellicer. (1985). Loss of the normal N-*ras* allele in a mouse thymic lymphoma induced by a chemical carcinogen. Proc. Natl. Acad. Sci. (U.S.A.), 82:7810-7814.

Gullino, P.M., Pettigrew, H.M., and F.M. Grantham. (1975). N-Nitroso-methylurea, a mammary gland carcinogen in rats. J. Natl. Cancer Inst. 54:401-414.

Hanahan, D. (1988). Dissecting multistep tumorigenesis in transgenic mice. Annu. Rev. Genet. 22:479-519.

Hecker, E., N.E. Fusenig, W. Kunz, F. Marks, and H.W. Thielman. (1982). Cocarcinogenesis and biological effects of tumor promoters. Raven Press. New York

IARC. (1982). IARC Monographs on the evaluation of the carcinogenic risk of chemicals to humans. Supplement 4.

Kaplan, A.S. (1967). On the natural history of the murine leukemias. Cancer Res. 27:1325-1329.

Knudson, A.G. Jr. (1977). Genetic predisposition to cancer. In: Origins of Human Cancer. Hiatt, H.H., Watson, J.D. and Winsten, J.A. (eds.). Cold Spring Harbor Laboratory. New York. pp. 45-52.

Leder, A., K. Pattengale, A. Kuo, T.A. Stewart, and P. Leder. (1986). Consequence of widespread deregulation of the c-myc gene in transgenic mice: multiple neoplasm and normal development. Cell 45:485-495.

Leon, J., I. Guerrero, and A. Pellicer. (1987). Differential expression of the *ras* gene family in mice. Mol. Cell Biol. 7:1535-1540.

Mangues, R., I. Seidman, A. Pellicer, and J.W. Gordon. (1990a). Tumorigenesis and male sterility in transgenic mice expressing a MMTV/N-*ras* oncogene. Oncogene 5(9):1491-1497.

Mangues, R., I. Seidman, A. Pellicer, and J.W. Gordon. (1990b). MMTV/N-*ras* transgenic mice as a model for altered capacitation, male sterility and tumorigenesis. In: Advances in assisted reproductive technologies. Mashiach et al. (eds.). Plenum Press, New York. pp. 939-957.

Muller, W.J., E. Sinn, P.K. Pattengale, R. Wallace, and P. Leder. (1988). Single-step induction of mammary adenocarcinoma in transgenic mice bearing the activated c-neu oncogene. Cell. 54:105-115.

Newcomb, E.W., J.J. Steinberg, and A. Pellicer. (1988). ras oncogenes and phenotypic staging in N-Methylnitrosourea and gamma irradiation induced thymic lymphomas in C57BL/6J mice. Cancer Res. 48:5514-5521.

Pellicer, A. (1986). Gene purification by transfection methods. In: Gene transfer. Kucherlapati, R. (ed.). Plenum Publishing Co. pp. 263-287.

Pellicer, A., D. Robins, B. Wold, R. Sweet, J. Jackson, I. Lowy, J.M. Roberts, G.K. Sim, S. Silverstein, and R. Axel. (1980). Altering genotype and phenotype by DNA-mediated gene transfer. Science 209:1414-1422.

Santos, E., D. Martin-Zanca, E.P. Reddy, M.A. Pierotti, G. Della Porta, and M. Barbacid. (1984). Malignant activation of a K-ras oncogene in lung carcinoma but not in normal tissue of the same patient. Science 223:661- 664.

Shih, C., B-Z. Shilo, M.P. Goldfarb, A. Dannenberg, and R.A. Weinberg. (1979). Passages of phenotypes of chemically transformed cells via transfection of DNA and chromatin. Proc. Natl. Acad. Sci. (U.S.A.), 76:5714-5718.

Shimizu, K., M. Goldfarb, M. Perucho, and M. Wigler. (1983). Isolation and preliminary characterization of the transforming gene of a human neuroblastoma cell line. Proc. Natl. Acad. Sci. USA. 80:383-387.

Sinn, E., W. Muller, P. Pattengale, I. Tepler, R. Wallace, and P. Leder. 1987. Co-expression of MMTV/v-Ha-ras and MMTV/c-myc genes in transgenic mice: synergistic action of oncogenes in vivo. Cell. 49:465-475.

Sloan S.R., E.W. Newcomb, and A. Pellicer. (1990). Neutron radiation can activate K-ras via a point mutation in codon 146 and induces a different spectrum of ras mutations that does gamma radiation. Mol. Cell. Biol. 10(1):405-408.

Sorrentino, V., M.D. McKinney, M. Giorgi, R. Geremia, and E. Fleissner. (1988). Expression of cellular protooncogenes in the mouse male germ line: A distinctive 2.4 kilobase pim-1 transcript is expressed in haploid postmeiotic cells. Proc. Natl. Acad. Sci. USA. 85:2191-2195.

Stewart, T.A., P.K. Pattengale, and P. Leder. (1984). Spontaneous mammary adenocarcinomas in transgenic mice that carry and express MTV/myc fusion genes. Cell. 38:627-637.

Sukumar, S. (1989). ras oncogenes in chemical carcinogenesis. Current Topics in Microbiology and Immunology 148:93-114.

Tremblay, P.J., F. Pothier, T. Hoang, G. Tremblay, S. Brownstein, A. Liszauer, and P. Jolicoeur. (1989). Transgenic mice carrying the mouse mammary tumor virus ras fusion gene: distinct effects in various tissues. Mol. Cell Biol. 9:854-859.

U.S. Department of Health and Human Services. Public Health Service. (1983). Third Annual Report on Carcinogens. Summary. NTP 83-010. Springfield, Virginia.

Weinberg, R.A. (1989). Oncogenes, antioncogenes, and the molecular bases of multistep carcinogenesis. Cancer Res. 49:3713-3721.

Yuspa, S.H., and M.C. Poirier. (1988). Chemical carcinogenesis: from animal models to molecular models in one decade. Adv. Cancer Res. 50:25-70.

K-ras ONCOGENE ACTIVATION IN NEOPLASIAS OF PATIENTS WITH FAMILIAL ADENOMATOUS POLYPOSIS OR KIDNEY TRANSPLANT

A. Haliassos[1] and D. A. Spandidos[1,2]

[1]Institute of Biological Research and Biotechnology
National Hellenic Research Foundation
48, Vas. Constantinou Ave. 116 35 Athens, Greece
[2]Medical school, University of Crete, Heraklion, Greece

INTRODUCTION

Oncogenes are highly conserved genes in any living organism and have important functions in cellular proliferation and differentiation. Loss of regulation in this complex system resulting from structural modification of a normal gene or of its controlling regions, is called oncogene activation. However, no specific correlation has been demonstrated that was relevant to this process, during the carcinogenesis, until the detection of point mutations in total genomic DNA became feasible (1).

The Polymerase Chain Reaction (PCR) is an *in vitro* method for the primer directed enzymatic amplification of specific DNA sequences (2). This technique leads to a very sensitive method for the detection of point mutations in genes already sequenced. Until now, the amplified DNA product is analyzed for screening purposes by a dot blot hybridization, with synthetic radiolabeled oligodeoxynucleotide probes, under high stringency conditions (3). The probes are specific of normal DNA sequence or of each possible mutation of a given codon. The results of the hybridization are revealed by auto-radiography. The sensitivity of this method is 5% to 10% i.e. cells carrying a mutation can be detected, among a population of non carrying cells, if they represent at least 5% of the total cell population.

However tumors are composed of heterogeneous cell populations (neoplastic cells, infiltrated inflammatory cells, normal tissue cells). This heterogeneity increases the background "noise" elevating the number of amplified non mutated sequences and thus decreases the sensitivity of the detection. Therefore, it is often important to detect point mutations at this level of sensitivity in many samples of the same patient, especially after a treatment, in order to evaluate the efficiency of the therapeutic action and the level (if any) of the residual disease. The detection of a point mutation in the adjacent apparently "normal" tissue provides a prognostic marker, relevant to the identification or the prevention of further tumor progression. These factors are of great importance in the understanding of the commitment of cells to neoplasia and malignancy.

ras oncogene activation is known to occur during the multiple step process of carcinogenesis in many forms of human cancers. The activation of K-*ras* oncogene by point mutation at its 12th codon has been reported as a molecular marker of tumor progression in various neoplasias (4,5). Although the majority of these neoplasias are sporadic, there is a special category of cancers occurring in immunosuppressed patients after kidney transplantation. In these cases the frequency of K-*ras* oncogene activation, which is established for the sporadic forms, is not fully studied.

The same activation of the K-*ras* oncogene has been reported (6) as a molecular marker of tumor progression in colonic neoplasias (early event, correlated with the increasing size of adenomas, but not with dysplasia). Although the majority of colorectal cancers are sporadic, there are hereditary forms such as Familial Adenomatous Polyposis (FAP) were the same transition is believed to be true.

The overall frequency of K-*ras* point mutations has probably been underestimated by the usual former techniques. We described a new sensitive method for an easier detection of this mutation, based upon the introduction of an artificial restriction site in a modified primer for selective *in vitro* amplification (7). This sensitive method, combined with the feasibility of polymerase chain reaction (PCR) amplification of DNA in formalin-fixed and paraffin embedded tissues, prompted us to evaluate the frequency of K-*ras* point mutations at codon 12 in both sporadic and familial colonic adenomas and in neoplasias occurring in patients after kidney transplantation.

MATERIALS AND METHODS

DNA samples

DNA from histological slides of tissues from various rare forms of human neoplasias, fixed by formol and embedded in paraffin, was extracted by boiling in a lysis mixture after the dissolution of paraffin in chloroform.

Slices of paraffin with no embedded tissues, empty vials and normal lymphocytes were processed together with the DNA specimens and used as negative tests for PCR amplification and K-*ras* mutations.

As a positive control, we used DNA from the SW 80 cell line, which is known (3) to carry 2 mutated alleles at codon 12 of the K-*ras* oncogene.

DNA extracted by the simplified guanidium method (8) from peripheral lymphocytes (50A), which are homozygous wild type at this same position as proved by allele specific oligodeoxynucleotide (ASO) probe hybridization, was used as a negative control.

Oligonucleotides primers and probes

The oligodeoxynucleotides were synthesized by the solid phase triester method in an Applied Biosystems DNA synthesizer. The primers were designed to introduce base substitution in the amplified fragments.

```
                   K-ras sequence with
                   wild codon 12 (GGT)
5'..TAAACTTGTGGTAGTTGGAGCTGGTGGC......GACGAATATGATCCAACAATAGA..3'
   5'TAAACTTGTGGTAGTTGGAGCC 3'  primers  3'CTTATACTAGGTTGTTATCT 5'
      K12Nm                                              KB12
                              PCR
                               ▼
5'TAAACTTGTGGTAGTTGGAGCCGGTGGC..........GACGAATATGATCCAACAATAGA 3'
              Msp I site       (99 bp)

                         Msp I digestion
                               ▼
TAAACTTGTGGTAGTTGGAGCCG and GTGGC..........GACGAATATGATCCAACAATAGA
         (21 pb)                                     (78 pb)
```

Fig. 1. P.C.R. with modified primer creating artificial RFLP.

Polymerase chain reaction

In vitro enzymatic DNA amplification (PCR) was performed on an automated apparatus (DNA thermal Cycler from Perkin Elmer Cetus).

We performed 35 cycles of amplification. Each cycle included 3 steps:
- denaturation of DNA at 94°C for 20 sec.
- annealing of the primers at 53°C for 30 sec.
- enzymatic extension at 72°C for 1 min.

Modified primers were designed to introduce a base substitution adjacent to the codon of interest in order to create an artificial restriction site with only one allelic form (wild type or mutated). We performed PCR with a modified primer creating a Msp I recognition site only if codon 12 was of the wild type (Figure 1). This approach allowed us to screen for point mutations at codon 12 of the K-*ras* oncogene (7).

An aliquot of the PCR product was controlled in a 2-3% Low Melting Point (LMP) agarose gel for the presence of the amplified fragment (99 bp). The PCR product was digested by Msp I enzyme which gave two fragments of 21 and 78 bp in the case of a wild type sequence or left the product undigested in the case of a mutated sequence. The fragments were analyzed by electrophoresis on LMP agarose gels.

The visualization of the results in the gel under U.V. light, without the use of radioactive probes, gives a sensitivity analogous to the previous method with the allele specific radiolabeled probes (i.e. ~ 5%).

RESULTS AND DISCUSSION

We used the above technique to detect the mutations of codon 12 of the K-*ras* oncogene in cell populations from formalin-fixed and paraffin-embedded tissues from various rare forms of human neoplasias

We studied 17 DNA samples from 14 patients with sporadic adenomas (3 with synchronous colon cancers), and we found 14 mutations of the K-*ras* oncogene at codon 12 and 11 DNA samples from 11 patients with FAP (2 with synchronous colon cancers) and we found only one mutation in an adenoma from a patient with an atypical form of FAP (Lynch family). This patient had a colectomy some years ago after the diagnosis of a colon adenocarcinoma (9).

We also studied 8 DNA samples extracted from formalin fixed and paraffin-embedded tissues from various neoplasias of 8 patients after kidney transplantation. We found 4 mutations in 2 Kaposi sarcomas, one infiltrating breast cancer and one cancer of common bile duct and pancreas from 4 patients with kidney transplantation. In contrast, in 4 other similar cases with a Kaposi sarcoma, an *in situ* cervical cancer, a stomach cancer and a prostatic gland cancer no lesion sample was positive for the studied mutation (10).

Attempts were made to establish a correlation between the presence of an activated K-*ras* allele and Dukes stages in the above cases but our results do not permit it because our sample was statistically insufficient.

CONCLUSIONS

We used an approach which proved sensitive and specific for the detection of point mutations despite the presence of many copies of wild type alleles. It is about 1000 times more sensitive than all the previous methods, and many times faster.

According to our preliminary results, the occurrence of K-*ras* point mutations at its codon 12 appeared to be far more common in sporadic adenomas than previously reported. They also revealed a possible difference in the occurrence of this molecular event during tumor progression of familial polyposis as the activation of *ras* oncogene is a rare event among the FAP patients, even in the very aggressive forms of the disease, contrary to sporadic adenomas.

This activation of K-*ras* oncogene is frequent among kidney transplant patients presenting a neoplasia, even in the least aggressive forms of the disease.

Definitive correlations between K-*ras* activation and these rare forms of human neoplasias could be established only after study of a large number of patients using a sensitive method of detection in DNA samples extracted by a standard protocol.

Finally, this approach will permit correlating clinical stages with data from histology and molecular biology in these rare and interesting cases of carcinogenesis.

REFERENCES

1. Spandidos, D.A., and Anderson, M.L.M. (1989). Oncogenes and oncosuppressor genes: their involvement in cancer. J. Pathol., 157, 1-10.
2. Saiki, R.K., Scharf, S., Faloona, F., Mullis, K.B., Horn, G.T., Erlich, H.A., Arnheim,N. (1985). Enzymatic Amplification of β-Globin Genomic Sequences and Restriction Site Analysis for Diagnosis of Sickle Cell Anemia. Science, 230, 1350-1354.
3. Verlan-de Vries , M ., Bogaard, M . E ., Elst , H ., Boom, J . H., Eb, A. J., & Bos, J. L. (1986) . A dot-blot screening procedure for mutated ras oncogenes using synthetic oligodeoxynucleotides. Gene, 50, 313-320.
4. Field, J.K., Spandidos. D.A. (1990). The role of ras and myc oncogenes in human solid tumors and their relevance in diagnosis and prognosis. Anticancer Res., 10, 1-22.
5. Vogelstein, B., Fearon, E.R., Hamilton, S.R., Kern, S.E., Preisinger, A.C., Leppert, M., Nakamura, Y., Whyte, R., Smits, A.M.M. and Bos, J.L. (1988) . Genetic alterations during colorectal-tumor development. N. Engl. J. Med., 319, 525-532.
6. Burmer, G.C. and Loeb, L.A. (1989) . Mutations in the K-ras2 oncogene during progressive stages of human colon carcinoma. Proc. Natl. Acad. Sci. USA, 86, 2403-2407.
7. Haliassos, A., Chomel, J.C., Tesson, L., Baudis, M., Kruh, J., Kaplan, J.C. and Kitzis, A. (1989) . Artificial modification of enzymatically amplified DNA for the detection of point mutations. Nucl. Acids Res., 17, 3606.
8. Jeanpierre, M. (1987) . A rapid method for purification of DNA from blood. Nucl. Acids Res., 15, 9611.
9. Chomel, J.C., Grandjouan, S., Spandidos, D.A., Tuilliez, M., Bognel, C., Kitzis, A., Kaplan, J.C. and Haliassos A. Search for correlations between K-ras oncogene activation and pathology in sporadic and familial colonic adenomas. In the Proceedings of NATO ARW on the super-family of related genes (1991). Plenum Publishing Corporation, New York, USA, 153-161.
10. Skalkeas, Gr. D., Spandidos, D. A., Kostakis, A., Balafouta-Tseleni, S., Choremi, E., Eliopoulos, D., Haliassos, A.(1991) K-ras oncogene mutations in Neoplasias of kidney transplanted patients: Preliminary results with a new technique. Anticancer Res., 11, 2091-2093.

APPENDIX

Cell lysis solution:
0.1M NaOH, 2M NaCl, 0.1% sodium dodecylsulfate
PCR mixture:
400ng of each primer, 1mM of each dNTP, 67mM Tris HCl pH 8.8, 6.7μm EDTA 6.7mM $MgCl_2$, 10mM β-mercaptoethanol, 16.6mM ammonium sulfate, 10% DMSO 1μg-100ng DNA to amplify, 1 unit of Taq polymerase, and H_2O to 100 μl

AN OVERVIEW OF TRANSGENIC MICE AS FUTURE MODELS OF HUMAN DISEASES, IN DRUG DEVELOPMENT, AND FOR GENOTOXICITY AND CARCINOGENESIS STUDIES

J.C. Cordaro

Office of Research Resources
Center for Drug Evaluation and Research
Food and Drug Administration
Washington, DC 20204, USA

INTRODUCTION AND BACKGROUND

Many of the recent advances and remarkable benefits afforded by modern medicine derive from the numerous contributions of an emerging biotechnology - a discipline drawing upon the strategies and techniques which form the cornerstones of both basic and applied research from animal husbandry to zoology, the knowledge base in organic chemistry and biochemistry, the vast literature on the genetics, molecular, and cell biology of both prokaryotes and eukaryotes, and the current and rapidly accumulating databases on HIV, oncogenes, and the human genome.

The last two decades have witnessed seminal discoveries in and effective integration among such areas as DNA restriction enzyme specificities employed in gene cloning strategies and recombinant DNA protocols; techniques in nucleotide sequence analyses used to determine gene structure-function relationships and the molecular mechanisms of DNA-protein interactions; retroviruses both as natural agents of disease and as vectors specially constructed for gene transfer experiments; oncogenes and tumor suppressor genes as the genetic determinants which contribute to a variety of neoplasias; a litany of ingenious variations of the polymerase chain reaction (PCR) whose utility both as a research and diagnostic tool for *in vitro* gene amplification and DNA sequence analysis are already legion; and transgenics, the technology which allows the introduction of foreign genes into animal embryos, all of which continue to expand and intensify the exciting and explosive strides in biotechnology. Sustained progress in the above individual areas is noteworthy and unique to be sure, but it is the points of emphasis where these achievements, strategies, and technologies merge and integrate that will greatly amplify both resolution and utility involving critical questions in biotechnology and biomedicine. Such focal points will include specific questions and strategies uniquely defined and answered in transgenic mice. More than just summarizing some advances in key areas using such animals, this narrative will propose future lines of experimentation using transgenic mice which were inconceivable merely a decade ago as potential models for human disease, drug development, genotoxicity and carcinogenesis.

Many recent and excellent reviews summarize various aspects of creating and using transgenic animals including rabbits, sheep, pigs, and mice which contain deliberately introduced exogenous or foreign genes that are both stably integrated into their genomes and reliably transmitted to their progeny. Although construction of the larger transgenic animals lends itself well to specific commercial applications such as pigs with greater muscle to fat ratios, calves that grow faster and come to market earlier, cows that produce more milk per pound of feed consumed, and goats that secrete in their milk human hormones, growth factors, and activator proteins encoded by transgene constructs, transgenic mice (and possibly hamsters and guinea pigs) are more ideal for research purposes because they are small and easier to handle and maintain. The choices of genetic backgrounds of mutant mouse egg donors characterized and maintained over decades will prove critical to the construction of potential transgenic derivatives of both whole animals and unique cell lines sure to be derived from such mice for future *in vivo* and *in vitro* determinations of the molecular mechanisms of action of drugs, biologicals, carcinogens and other xenobiotics.

The knowledge base created by the classic literature over the last six decades involving prokaryotic systems which utilize bacteria and phage describes gene transfer techniques as conceptual and historic precursors to the creation of transgenic animals. These techniques have included transformation using naked DNA, conjugation depending upon cell-to-cell contact, and transduction relying upon generalized and specialized bacteriophage viral vectors to encapsidate regions of the donor cell chromosome for transfer to and integration into the recipient cell genome under the appropriate selective conditions. Such advances in genetics and molecular biology, which engender understanding of genetic regulatory circuitry and gene structure and function, coupled with using retroviruses and oncogenes, DNA restriction enzyme specificities, gene cloning and DNA sequencing strategies, and microinjection techniques will now allow virtually limitless questions to be asked and answered in both basic and applied research areas involving genetics, molecular biology, biochemistry, pharmacology, physiology, cell biology, immunology, embryology/developmental biology, endocrinology, virology, toxicology, and carcinogenesis/oncogenesis using transgenic mice as the paradigm.

Two major techniques are currently utilized in the creation of transgenic mice. The first involves infection of 4-8 cell stage embryos by either direct exposure to retroviral virions or co-cultivation with fibroblast cell monolayers constitutively producing retroviruses. This technology results in a transgenic mouse containing either unaltered proviral DNA derived from intact retroviruses or the integration of specially constructed retroviruses that have been genetically engineered to contain exogenous or foreign genes.

The second technique employs microinjection of desired DNA fragments generated by either correct restriction enzyme digestion, gene cloning, or PCR gene amplification, into the male pronuclei of fertilized eggs after one-day gestation using a smooth fire-polished holding pipette under gentle suction and a narrow injection pipette extruded to 1 μm at the tip. Successful gene transfer usually occurs following a noticeable but transient swelling of the male pronucleus. Re-implantation of these putative transgenic mouse embryos in the uteri of pseudopregnant foster mothers is followed by birth of the pups in 20 days. Presence of the introduced transgene can be verified by restriction enzyme digestion of total mouse genomic DNA obtained from 1-2 cm tail clippings in Southern blot hybridization using either the original DNA segments or shorter oligodeoxynucleotide probes synthesized in the laboratory which represent DNA sequences unique to the transgene. This strategy both insures survival of the founder mice for further study and allows their participation in matings to produce unique and valuable progeny which can transmit the introduced transgene to future offspring as a stable Mendelian trait. Tissue-specific fidelity of transcription and translation of the transgene is detected by Northern blots and Western blots (and 2-D gels) for the transgene-specific production of mRNA and protein, respectively, in the correct cells and tissues of such animals.

Gene transfer using either retroviral vectors or DNA microinjection to construct transgenic mouse embryos share important similarities and exhibit at least one critical difference. The similarities between the two techniques include: (i) random integration of the transgene within the host genome; (ii) a 50-75% viability for treated eggs prior to their introduction into pseudopregnant foster mothers; (iii) a 10-30% survival rate for implanted eggs which develop into live mice; and (iv) an overall rate of 1-50% for integration and stabilization of the transgene affected by such variables as the molecular nature of the vector-transgene construct and the mouse strain from which the eggs are derived.

Once the transgene is stably integrated into the genome of the founder mice subsequent matings then provide a mouse colony containing the desired construct. The difference between the two methods is both important and interesting: only one copy of the transgene occurs at its integration site when a genetically engineered retrovirus is used as the vector contrasted to the presence of multiple copies in head-to-tail tandem array of transgene cassettes resulting from micro-injection of naked DNA fragments. The molecular mechanism(s) responsible for this difference in arrangement and copy number of transgenes depending upon which gene transfer method is used is not understood and therefore remains a challenging and fascinating research question waiting to be answered.

The advantages and disadvantages of using either retroviral transfer or DNA microinjection should be weighed carefully in designing the construction of each transgenic mouse system. The use of specially designed retroviral vectors allows ease of both infection and resulting introduction of the foreign gene(s) selected and flanked by the retroviral LTRs usually insuring high-level expression of the desired exogenous transgene inserts they contain. DNA microinjection must be preceded by *in vitro* construction of the proper tissue-specific promoter-enhancer combination in each case. On balance, however, DNA microinjection strategies are usually preferable over using retroviral vectors because the latter method has disadvantages which include additional steps in genetic design required to produce the optimal vectors carrying the transgenes of interest; a size limit of the foreign DNA insert; gene transcript processing constraints due to the need to have the proper splicing and termination signals present; subsequent mosaicism prevalent in the founder animals because successful retroviral infection depends upon cell division; and by potential interference with desired expression of the transgene due to the presence of retroviral vector DNA sequences. In each case, the final choice between the two methods should be determined by the experimental goals and protocols, the regulatory circuitry required for expression of the transgene construct, and the genetic characteristics of the mouse strains to serve as egg donors.

MODELS OF HUMAN DISEASE

The catalogue containing thousands of human diseases and the chromosomal locations of their genetic determinants is currently maintained and continuously extended by the laudable efforts of OMIM (On-line Mendelian Inheritance in Man) and GDB (Human Genome Database). Only a few human genetic diseases will be mentioned here to illustrate some of the strategic uses of transgenic mice as unique tools in the analysis of the genetic and molecular mechanisms which underlie their etiologies. Cystic fibrosis (CF), one of the most common genetic diseases in man, affects 1 in 2,000 Caucasians with a carrier frequency of 1 in 22 defining individuals in the population that are heterozygous at this locus. The signs and symptoms which define the clinical pathophysiology of CF are readily identifiable and include elevated chloride ion in sweat, reduced sodium ion influx, and a thick, chronic pulmonary mucous secretion which fosters subsequent and frequent pulmonary infections. CF mutant proteins are defective in their interaction(s) with the chloride ion channel in cells lining the airways of the lungs. The CF gene is located on human chromosome 7q31-32 (the long arm of chromosome 7 between bands 31 and 32) and its alleles exhibit autosomal recessive inheritance. Pulsed-field gel electrophoresis (PFGE) and Southern blot analysis of DNA fragments from this region isolated by cosmid sub-cloning locate the CF gene within 1 cM (Centimorgan = approx. 1 megabase) of hu-c-*met* (the human *met* protooncogene). RFLPs (restriction fragment length polymorphisms) generated by EcoRI, PstI, and BamHI digestion which define individual molecular differences between normal and mutant alleles of the CF gene can detect both common consensus sequences and individual base alterations which are unique to each CF DNA sequence. These facts warrant construction of specific strains of transgenic mice to implement the following experimental strategies. First, a transgenic mouse strain containing a normal human CF gene could be used to calibrate the detection in Northern blots of CF mRNA and protein expression in such animals because CF mRNA is usually the same in abundance and size among normal *vs*. affected individuals; second, the creation of transgenic mice containing the CF mutant allele which results from a deletion of the CTT codon causing the loss of a single amino acid, phe-508, thereby affecting the ATP-binding domain of CFTR (cystic fibrosis transmembrane regulator protein), would provide an excellent animal model to examine the effect(s) of this mutation on the interaction of this altered CFTR protein with other critical components of the chloride channel *in vivo*; and third, since exon 10 of the CF gene exhibits 70% of all genetic defects observed in the human population, transgenic mice containing selected polymorphs in this region would be invaluable to examine how the molecular characteristics of any CF mutant allele correlates with cystic fibrosis of various degrees of severity. Using the proper DNA probes in Southern and Northern blot analyses of these transgenic mice should reveal which mutant forms of the CF gene result from deletions, point mutations, rearrangements, or insertions, and how each contributes to the symptoms of the disease by critical alterations in CFTR protein function and in mRNA splicing patterns, respectively. Finally, the effect(s) of such genetic alterations can be exactly defined by both DNA sequence analysis and the affected amino acid(s) localized to the hydrophobic or ATP binding sites of the mutant human CF protein(s).

Alzheimer's disease (AD), occurring in at least 25% of people over 85 years of age, is a particularly loathsome and protracted clinical condition with profoundly debilitating effects resulting from neuronal degeneration and subsequent nerve cell destruction. The cardinal features of cognition, including learning, memory, and creative abilities, are drastically compromised along with increased personality disintegration as the disease progresses. As collateral damage, AD both causes extreme emotional stress among mates and family members of affected individuals and places overwhelming demands on health care delivery resources. The brains of AD patients at autopsy reveal neuronal amyloidosis in the hippocampus caused by intracellular accumulations of tubular aggregates of beta-amyloid transmembrane glycoprotein normally possessing both intra- and extracellular domains. The location of the AD gene is on human chromosome 21q11.2- 22.2 and the abnormal 4 kD proteins which finally accumulate in the dendrites are truncated versions of the amyloid precursor protein (APP) which result from various splicing events of the primary transcript and the generation of one of five AD-associated isoforms in each case. Transgenic mice containing genetically polymorphic alleles of the AD gene region should be readily constructed using only 4-5 Kb of coding human AD:beta-amyloid DNA. Different disease-producing AD alleles will generate at least two sets of informative experimental findings: some disease polymorphs might contain mutations which code for an APP that is cleaved more frequently by protease to enhance accumulation of the truncated protein whereas other alleles may inhibit turnover and clearance of the abnormal beta-amyloid versions thereby increasing their persistence; other polymorphs may affect the phosphorylation of the protein and consequently change the affinity for their intracellular anchors. Studies which correlate the particular polymorphic disease alleles introduced into transgenic mice with genetically determined alterations in the splicing patterns of the transcripts or with point mutations, deletions, or insertions, altering the activities of the proteins and/or their relative affinities for the protease(s) which either create or destroy them, can result in understanding the exact genetic and molecular mechanisms underlying the factors which contribute to the progression and severity of the disease. The X-chromosome offers extremely fertile experimental material for the construction of key strains of transgenic mice because there exist virtually hundreds of genes which specify disease states that are located on that component of the human genome. Pulse-field gel electrophoretic analyses demonstrates that among the plethora of X-linked genes, many loci occurring within 8 Mb (1 megabase $=10^6$ bases) effect mental retardation. One such locus determines the fragile X syndrome (Fra X) which presents with grotesque facial features and severely diminished mental capacity. Affecting mainly males at a frequency of 1 in 1,500, Fra X exhibits perplexing characteristics in its phenotypic expression: greater than 20% of males with the Fra X genetic determinant do not express the phenotype whereas 50% of carrier females exhibit symptoms of the disease. Variations in incomplete penetrance of fragile site expression is evident among families examined where the mutant allele shows low penetrance in brothers of normal transmitting males and high penetrance in brothers of affected males. The chromosomal region of Fra X is Xq27-28, along with the markers of other genetic diseases located in this region, with most Fra X alleles occurring at Xq27.3. Critical CpG islands occur within Fra X which result in definitive methylation differences contributing to the collection of many other informative polymorphisms including deletions and translocations of this locus. In addition a VNTR site is located distally to Fra X which is useful as an outside marker for independent identification of DNA fragments isolated from this region of the X-chromosome. Construction of the proper strains of transgenic mice containing informative Fra X alleles should allow examination of the genetic and molecular factors contributing to the unusual phenotypic expression, incomplete penetrance, and complex inheritance of the disease. An experimental approach to understanding these characteristics could involve isolating each Fra X allele from parents and offspring from a given family carefully tracking whether the allele existed in the females or males showing either carrier status or degrees of phenotypic expression. At least two sets of experiments can be envisioned: First, to examine a given Fra X allele for its exact molecular identity when the parental source is compared with the allele present in sons, daughters, and grandchildren; penetrance of the allele may increase through mechanisms dependent upon either maternal or paternal inheritance should mutational changes occur during gametogenesis. Additions to or losses from the original Fra X allele could be detected by DNA hybridization experiments detecting new or different fragments created by base changes, deletions, duplications, insertions, inversions, or a combination of genetic events of each allele cloned in transgenic mouse strains.

Second, these transgenic mice containing stabilized human Fra X polymorphs can now be selectively mated with appropriate males and females so that the contribution of the genetic background of the parent strain to alterations in each Fra X site can be monitored by Southern blot hybridization and DNA sequencing. The pattern of changes in DNA sequences of the original alleles from male or female grandparent, or parent, to offspring can be compared to the molecular changes in the original alleles in each case as a function of subsequent matings of the transgenic mice in which they reside. Such analyses should provide important insight into the mechanisms which determine variations in penetrance and the genetic basis of the changing pattern of fragile X inheritance.

The genetic locus for Duchenne's muscular dystrophy (DMD) is on the X-chromosome at Xp21.1-21.3 (p=short arm of the chromosome). The DMD locus provides a wealth of information because the region is very large being composed of 60 exons over 2 Mb of DNA which codes for a final protein of 427 kD. Many different kinds of mutations affecting the DMD region in humans can be identified, isolated, and cloned into transgenic mice to allow detailed molecular characterization of each mutation and its correlation with the mechanism by which it causes the disease in varying degrees of severity both in the transgenic progeny and in the original patients from which it was isolated. Most DMD mutations occur mainly by rearrangements, deletions, and duplications as contrasted to point mutations and one-third of DMD cases result from new mutations among which deletions and duplications comprise 65%. Endoduplications within the DMD locus occur with substantial frequency and the mechanism of their formation can be examined as the precursor alleles pass through the germline of transgenic mice. Each new RFLP appearing among the progeny should be readily detectable in Southern blot hybridizations which monitor additions (or losses) of kilobases of DNA representing the presence of each new allele. Rapid detection and sequencing of these new alleles should be facilitated by current DNA multiplexing technologies. Detailed molecular characterization and corresponding pathological behavior of both pre-existing and newly-arising DMD alleles using transgenic mice as whole animal models should ultimately yield critical information regarding the genetic mechanisms responsible for the onset, progression, and severity of Duchenne's muscular dystrophy.

One of the most disturbing, nettlesome, and perplexing diseases arising in the human population during the past two decades is acquired immune deficiency syndrome (AIDS) which is caused by an infectious retrovirus called human immunodeficiency virus (HIV). In the early 1990s, AIDS is pandemic with large populations in Africa, Southeast Asia, Europe, and the United States suffering and dying from the disease. Estimates include 10-20 million HIV infections worldwide with 1-2 million people currently symptomatic and a conservative death toll of one million. In the US 1-2 million infections include at least 300,000 people with various stages of the disease and a death toll of greater than 100,000. The signs and symptoms of AIDS include weakness and malaise; loss of appetite and weight; swollen lymph nodes; skin abnormalities including Kaposi's sarcoma; a pronounced predisposition to other cancers; diminished cognitive functions affecting powers of concentration, memory, and reasoning; and pronounced immune suppression resulting in increased susceptibility to chronic and severe secondary opportunistic infections including cytomegalovirus (CMV)-induced retinitis, hepatitis, and pneumocystis carinii pneumonia. HIV replicates in and thereby reduces the number of $CD4^+$ lymphocytes which form critical components of the immune system and usually accounts for the sequelae of diseases leading to death. Extensive knowledge accumulated thus far about both the genetic composition and the mechanism of HIV replication is crucial to achieving a full understanding of the disease and devising promising strategies for its treatment, prevention, and possible cure. The relatively small size of the HIV genome (10 Kb) efficiently contains many genes which specify viral infection, replication, and latency. The U3, R, and U5 regions of the long terminal (direct) repeats (LTRs) which bracket the proviral genome code for powerful and precise enhancer, promoter, initiator, and terminator signals controlling the transcription of HIV genes; *gag* contains the open reading frames for the p15, p17, and p24 core proteins of intact HIV virions which maintain the internal structure and define the serological groupings of the virus; *pol* encodes the viral reverse transciptase which converts virion-encapsidated HIV single-stranded RNA into double-stranded proviral DNA; *env* specifies gp160 which is cleaved by HIV protease into two the gp41 and gp120. The gp41 stalk protein attaches gp120 to the virion coat, allowing exposure of the gp120 glycoprotein to recognize the $CD4^+$ receptor and facilitate entry of HIV virions into T4 lymphocytes and macrophages; *tat* codes for the key 12 kD protein, resulting from translation of a double-spliced mRNA, whose richness in basic amino acids facilitates DNA binding and therefore controls the transcription rates of HIV genes that increase virus production; *rev* codes for a protein acting as a switch between early and late replication steps; *vpr* specifies a protein that plays a significant role in cell killing yet when over-expressed not enough progeny virus can be made from a productive infection before cell death occurs so that consequently the *vpr* gene is often mutated or deleted among progeny virus; *vpu* encodes a protein which allows rapid exit of newly assembled virus from productive infections; and *nef* mutants emerge as *nef*$^+$ revertants among the viral progeny.

Construction of key strains of transgenic mice containing HIV as provirus would provide ideal animal model systems to study key genetic and molecular events contributing to AIDS. The founder mice although containing the entire HIV genome as provirus would not be viremic and therefore should be safe to handle with little or no risk of infectious spread to other animals and research personnel. Crosses between founder mice to produce an HIV-containing F1 generation and/or subsequent matings with F1 individuals to produce F2 progeny should allow *in vivo* production of intact HIV virions from the proviral transgene under controlled conditions for many key experiments. An additional strategy would encompass breeding either the founder mice and/or individuals of the F1 generation with other mouse strains lacking the HIV provirus to identify the endogenous genetic determinants specific to each parent which potentiate or attenuate virus production. A list of experimental areas enhanced by the development of appropriate HIV-containing transgenic mouse systems would include (a) infectivity, how well and under what genetic conditions and environmental circumstances HIV virions penetrate lymphocytes and macrophages, the comparative susceptibility to different routes of infection, and the contribution of syncytial cell formation to various steps in virus production; (b) replication, the genetic determinants that define both HIV mutants and strains of mice which augment or diminish each critical stage of virus synthesis and maturation; (c) cytotoxicity, the contribution of various different viral mutants to cell killing which would contribute to understanding the mechanisms of viral latency; and (d) the genetic instability of HIV, the emergence of new viral mutants following productive infections *vs* proviral inductions which alter virulence, pathogenesis, latency, and the mechanisms of HIV contribution to Kaposi's sarcoma.

Many experimental strategies using transgenic mice carrying HIV as a provirus can be designed to elucidate the steps leading from viral infectivity to provirus insertion through virus assembly finally ending in the release of progeny virions. Only a few among many possible approaches will be mentioned.

One such strategy could determine whether HIV has preferred sites of integration or whether proviral insertion occurs randomly throughout the resident genome. The creation of HIV-containing mice with different genetic backgrounds (e.g., C3H vs. Bl/6 vs. BALB/c, etc.) would give these experiments greater resolving power than if only one mouse strain were used in their design. Digestion of genomic DNA isolated from each HIV-transgenic derivative with the appropriate restriction enzymes followed by Southern blot hybridization of the fragments with HIV probes would detect differences in HIV integration sites. Analysis of these HIV-containing DNA fragments by inverse PCR followed by DNA sequencing would identify (a) how many different HIV integration sites exist both within a given strain and among different strains of transgenic mice containing HIV as the foreign insert and (b) whether such HIV insertion sites are defined by identical or degenerate host DNA consensus sequences. Correlations between sites of HIV integration and the ability to produce HIV virions when crosses are made between either the founder mice or the F1 generations derived from them would indicate the contribution of HIV integration sites to viral infectivity, latency, and titer and where those sites are distributed in the genomes of each host mouse mutant strain.

A second strategy would involve studying genetic recombination between mutants of a given strain of HIV. Beginning with an HIV-transgenic mouse containing one mutant of the virus, when either the adults or their embryos are exposed to another mutant of the same virus, do recombinants emerge and if so how do their latency periods, pathogenicities and other phenotypic characteristics differ from the parent viruses from which they were derived? The putative recombinants can be isolated and their molecular characterization should identify which HIV genes (e.g., *tat*, *vpu*, *vpr*, *nef*, *sor*, *lor*, *gag*, *pol*, *env*, etc.) are predominantly affected and whether their frequencies in the population vary with the particular strains of transgenic mice in which they were selected.

Analogous experiments which monitor genetic recombination between different strains of HIV (e.g., HIV-1 and HIV-2) can be performed to determine the conditions for creating new HIV strains which begin from parental viral isolates with different origins.

In aggregate, data from the molecular analysis of recombinant HIV virions from such an assay system would increase the understanding of the genetic mechanisms involved both in the progression and latency of AIDS among high-risk individuals in the human population that are behaviorally predisposed to multiple HIV infections and the contribution of viral and host factors affecting the role of genetic recombination in the molecular evolution of HIV.

A third strategy could be employed to analyze whether different strains of HIV either activate cellular proto-oncogenes to oncogenes and/or inactivate tumor suppressor (TSGs) genes by insertional mutagenesis. This question is important because the mechanisms by which HIV-infected individuals exhibit greater frequencies of malignancies including leukemias and lymphomas may involve more than a generally compromised immune surveillance. Analysis of HIV-containing transgenic mice should include examination of the endogenous c-*abl* and c-*myc* genes by HIV insertion events which place the proviral LTRs close enough to activate these (and possibly other) oncogenes by abnormally up-regulating their transcription. Positive results using Southern blot analyses of HIV/transgenic mouse genomic DNA with cDNA probes unique to HIV, *abl*, and *myc* would provide reasons for Northern blot analyses to detect over-production of either *abl* or *myc* mRNAs.

By analogy, the p53 and other TSGs could be examined for inactivation caused by HIV-promoted insertional mutagenesis. In these cases, cDNA probes of HIV would reveal the presence of provirus within such genes interrupting their reading frames. More detailed molecular analyses of the host DNA sequences as putative preferential insertion sites could follow by generating PCR products from the HIV-TSG fusion sites and their subsequent DNA sequencing. A fourth strategy would attempt to determine the mutation rate of *pol* (and other HIV genes of choice) beginning with a pure viral clone isolated by PCR-mediated amplification and subsequent recovery of the virus from HIV-containing macrophages of transgenic mice. Such HIV clones would be re-introduced into embryos of selected mouse mutant strains creating HIV/transgenic animals harboring a pure viral isolate instead of deriving such animals from samples possibly containing a heterogeneous mixture of HIV virus originating in clinical sources of pooled sera from infected individuals. Subsequent matings which produce viremic mice would generate progeny viruses for analysis of mutations occurring in the *pol* gene. Once this system is standardized, a variety of putative anti-HIV compounds can be tested and screened for the conditions under which *pol*-encoded reverse transcriptase (RT)-resistant clones appear. Understanding the mechanisms of anti-HIV drug resistance would be invaluable in the rational design of compounds intended for use either singly or in combination with other antiviral agents depending upon their sites of action from HIV infection to release.

RATIONALES IN DRUG DEVELOPMENT

The construction of key strains of transgenic mice followed by their genetic and molecular characterization should provide an invaluable resource for rational drug design to create compounds including both drugs and biologicals for the prevention, treatment, and possible cure of human diseases. Such mice would provide unique animal models containing the exact human disease-producing targets for research and development of many classes of efficacious therapeutic agents.

Among the myriad of transgenic systems envisioned for development, only four specific models will be discussed here to illustrate sample strategies. One transgenic mouse system ideally suited for drug development would have individual animals that contain both normal and disease-producing alleles of Alzheimer's disease (AD). The phenotypic expression of human AD polymorphs should be readily identifiable in affected transgenic mice by aberrant behavior such as disorientation and compromised muscular movement. AD alleles isolated from human 21q chromosomes at position 11.2-22.2 would produce the disease in transgenic constructs for developing candidate drugs which ameliorate the signs and symptoms of the disease. For instance, tacrine and other model compounds could be analyzed in such animals for inhibition of beta-amyloid protein production of each AD-associated polymorph and isoform.

Studies of positive effects by model compounds could be further extended to using radio-immunoassays (RIAs) to measure reductions in levels of AD allele-specific beta-amyloid proteins from transgenic animals responding to such drugs. Additional congeners could be synthesized with increased specificities for particular disease isoforms associated with the more rapid progression of Alzheimer's disease. Transgenic mice harboring normal AD alleles, or any non-disease producing polymorphs from unaffected individuals in the human population, would serve as controls for possible adverse effects from therapies against the disease. Those compounds which show promise can be further analyzed for their molecular mechanisms of action in these animals by studies that detect drug-induced alterations in the splicing pattern of the primary AD transcripts vs. compounds that either modify or retard the enzymatic activities of cellular proteases which recognize mutant amyloid precursor proteins (APP) during conversion to their beta-amyloid derivatives. Other compounds may be synthesized to restore phosphorylation of AD mutant proteins to create drug-induced phenocopies of normal AD alleles; still others could enhance proteolytic clearance of mutant AD isoforms and polymorphs. Such approaches would finally distinguish among and extend the development by rational design of promising compounds for use against Alzheimer's disease based upon both their molecular pharmacology and mechanisms of attack on specific AD-associated target sites.

A second set of strategies in rational drug development would utilize key strains of transgenic mice in which crosses between F1 animals containing HIV as provirus triggers viral replication generating viremic F2 animals. A variety of compounds with potential antiviral activities could be tested in this system as inhibitors of virtually any step in the pathway leading from integrated provirus to release of infectious virions. Small antisense (AS) DNA and RNA oligomers synthesized with deliberate complementarity to unique viral coding or regulatory sequences would act against HIV proviral target sites ranging from the LTRs through to transcription termination signals. Antisense technology would include the synthesis of such sequence specific oligomers for obvious HIV targets with the following results. Oligomers directed against (a) *gag* sequences would inhibit the synthesis of viral products required for assembly of critical internal components of infectious HIV virions; (b) transcription or translation of the *pol* gene thereby preventing synthesis of either HIV protease, reverse transcriptase, and ribonuclease H which are needed for proteolytic cleavage of gp160 required for assembly of the viral envelope, for transforming single stranded HIV RNA into double-stranded DNA copies of the viral genome, and nucleolytic digestion of the RNA strand of viral RNA-DNA early hybrids of nascent HIV genomes, respectively; and (c) *env* sequences thereby preventing the synthesis of gp160, the source of the gp41 stalk and gp120 protein, the latter being the critical component on the surface of mature infectious virions required to recognize the $CD4^+$ receptor for HIV entry into lymphocytes and macrophages. Various other drugs that specifically inhibit HIV proviral DNA synthesis can be developed using HIV-transgenic mice. Chain terminators that target HIV reverse transcriptase activity, analogous to but possibly more potent than *azidothymidine* (AZT) and other dideoxyribonucleosides such as ddC, ddI, ddA and their derivatives, act through conversion to the corresponding triphosphates catalyzed by host-encoded intracellular kinases followed by their subsequent incorporation into nascent HIV genomes signalling end of chain. Newly developed chain terminators can be tested in HIV viremic mice for their *in vivo* capabilities to inhibit viral DNA synthesis. Derivatives of key compounds (e.g., ddC) such as hydroxymethyl-ddC which inhibit replication of DNA viruses including CMV and HSV may work at different steps in the reverse transcription pathway. Drugs which inhibit more efficiently the conversion of single-stranded viral RNA to the RNA-DNA hybrid intermediate may be tested in combination with compounds that prevent formation of the final DNA-DNA duplex which becomes incorporated into the host cell genome as provirus. Candidate compounds with anti-HIV protease activity can be monitored by measuring decreases in serum levels of HIV p24 protein or reductions in the formation of critical myristylation protein products which are required for auto- proteolytic activation to generate an active HIV protease from its precursor which, in turn, catalyzes the proteolytic conversion of virally encoded proteins required for self-assembly of infectious HIV virions.

The third strategy would begin by utilizing the existence of transgenic mice in which skin lesions mimic the development of Kaposi's sarcoma (KS) in humans as part of the AIDS disease sequelae. The appearance of KS-like lesions in transgenic mice affords ideal whole animal models in which to test compounds that either inhibit development, cause remission, or prevent the occurrence of this form of cancer. Transgenic mice already constructed as prototypes contain an HIV-LTR-*tat* construct and exhibit skin lesions with histopathologies of KS in human AIDS patients. Experimental protocols using these strains would allow development and testing of compounds that inhibit (a) angiogenesis which results in blood supply required for growth of KS lesions; (b) cytokine production which facilitates cell division and an increase in lesion size; (c) HIV-LTR-driven gene expression within the epidermis required for transcriptional activation of cellular genes encoding factors for angiogenesis and cytokine synthesis measured by Northern blots and SDS gels which monitor reductions in *tat* mRNA and TAT protein production, respectively; (d) production of KS-like lesions from HIV-LTR-*tat*-containing affected epidermal cells of the original transgenic host caused by their transplantation into *nu/nu* mice; and (e) occurrence of both epidermal hyperplasty and dermal cell proliferation so that KS tumors never develop. Detailed genetic and molecular analysis of certain HIV-LTR-*tat*-containing strains may reveal that the transgene cassette (a) is located next to a protooncogene which becomes transcriptionally activated to oncogene status; (b) has inserted within the reading frame of a tumor suppressor gene; (c) has caused truncation of a protooncogene (e.g., *raf*) converting it to active status; or (d) has rearranged by translocation other cellular protooncogenes which deregulates their normal transcriptional control mechanisms.

The collateral benefits of such strains would increase understanding of the mechanism(s) required for production of KS lesions and provide additional murine transgenic systems to develop antineoplastic agents using rational drug design. The rational design of antineoplastic agents for use against a variety of cancers represents the fourth strategic area in which transgenic mice can provide unique experimental models. The current spectra of compounds in use against leukemias, lymphomas, and solid tumors exhibit wide variations in effectiveness primarily because such chemotherapies are usually based upon differences in cytotoxicities between malignant cells and their normal progenitors, not upon a detailed understanding of their molecular mechanisms of action against tumor-specific target sites. The rational design of antitumor agents must involve human oncogenes, oncoproteins, tumor suppressor genes and their products in transgenic mice properly constructed to test such drugs and biologicals directed against these specific targets. Cancer is a collection of diseases caused by a multistep process which results in normal to malignant cell transformation. A cascade of at least five kinds of events is required for neoplastic cell transformation which includes (a) initiation, the occurrence of critical changes which begins the process; (b) promotion, the increase in cell division due to loss of growth control; (c) invasiveness, the establishment of a distinct tumor cell identity required for tissue penetration; (d) metastasis, the expansion to and colonization of distant organs from the site of the primary tumor; and (e) immortality, the escape of primary and secondary neoplasias from immune surveillance. The oncogene theory of cancer describes the existence of two major classes of genes as components of normal cells, the protooncogenes and the tumor suppressor genes (antioncogenes), which participate in the induction and maintenance of the neoplastic state. Cancer results from critical combinations of protooncogenes activated to oncogenes and tumor suppressor genes converted to their inactive counterparts as the individual signature of each tumor type. Genetic events which can convert protooncogenes to oncogenes and inactivate tumor suppressor genes include (a) point mutations caused by DNA base pair alterations resulting in missense or nonsense events; (b) deletions which remove only a few nucleotides or include kilobase lengths of DNA; (c) translocations resulting in chromosome rearrangements mediated by genetic recombination mechanisms; and (d) amplifications which create multiple copies of activated oncogenes. Currently there exist more than sixty protooncogenes and tumor suppressor genes whose chromosomal positions and cellular locations of their protein products are determined. Oncogenes which specify products located in the nucleus include *fos*, *myb*, and *myc* whose normal locations in the human genome are chromosomes 14q21-31, 6q22-24, and 8q24, respectively; those oncogenes producing cytoplasmic proteins include *fms*, *mos*, and *raf* located on chromosomes 5q34, 8q22, and 3p25, respectively; and those oncogenes which produce proteins localized in the plasma membrane include *src*, *abl*, and *fes*, whose normal locations are chromosome 20, 9q34, and 15q24-25, respectively, and three members of the *ras* super-family whose locations are H-*ras* on chromosome 11p14-15, K-*ras* on chromosome 12p12, and N-*ras* on chromosome 1. Possible strategies for constructing critical strains of transgenic mice containing human oncogenes as specific targets of rational drug design might be developed as follows. First, insert the disease-producing polymorphs of each oncogene, either singly or in combination, isolated from human tumors of a specific type; for instance, certain breast tumors, leukemias, and lymphomas exhibit activated *myc*, *ras*, *fos*, and *fms* oncogenes. Second, testing drugs which inhibit transcription of *myc* in combination with compounds that restore *ras*-mediated GTP hydrolysis and signal transduction might result in shrinkage and disappearance of these activated human oncogene-driven tumors in transgenic mice.

Compounds that inhibit amplification of human N-*myc* or c-*myc* in select strains of such animals might prevent further progression of human neuroblastomas and oat cell lung carcinomas, respectively. Specific inhibitors of human c-*myc* transcription resulting from its translocation to chromosomes containing high-efficiency promoter-enhancer combinations could cause remission of human c-*myc*-driven lymphomas in such animals.

These and similar strategies depend upon the utilization of specific interactions between the drug and the particular human target in transgenic mice thus providing a deliberate and rational basis for designing the next generation of antineoplastic agents.

STRATEGIES FOR GENOTOXICITY AND CARCINOGENESIS

A major goal of carcinogenesis risk assessment is the ability to calculate the probabilities and therefore predict with reasonable certainty which individuals within a population are more likely to contract cancer based upon understanding the genetic and environmental causation of the disease. The vast literature from the current repertoire of tests in genetic toxicology contributes to this understanding but usually fails to describe exactly how foreign compounds such as by-products of cooking certain foods, endogenous compounds derived from intermediary metabolism, drugs, pesticides, pollutants, and other organic compounds are both activated to carcinogens and how they finally act on and convert discrete human genetic targets required for neoplastic transformation. The current literature in genetic toxicology contains a myriad of examples using prokaryotic and eukaryotic systems that detect important classes of heritable changes induced by a variety of compounds. Such events include point mutations, both missense and nonsense events caused by transitions and transversions, and frameshifts resulting from adding or losing 1 or 2 base pairs of DNA. Still other systems can detect amplifications, deletions, duplications, insertions, and inversions of genetic material which alter the phenotypic properties determined by the affected genes.

The Salmonella/Ames test has unequivocally demonstrated the causal and in part the mechanistic connection among compounds that are both carcinogens and mutagens but neither consistently nor reliably detects deletionogens, recombinogens, and agents that cause such genetic events as gene amplifications, duplications, insertions, or inversions.

Certain yeast systems can assay by measuring both forward and reverse mutations the induction or correction of ochre mutations to detect transversions, agents that cause mitotic recombination between alleles on the same chromosome, and compounds which cause chromosome loss resulting in aneuploidy.

Mammalian cell assay systems using Chinese hamster lung (V79) and ovary (CHO) cells measure the induction of forward mutations in the HGPRT genes located on the X-chromosome causing 6-thioguanine resistance clonogenicity and a mouse lymphoma cell line (L5178Y tk^+/tk^-) detects forward mutations which inactivate the autosomal thymidine kinase gene allowing colonies to form that are resistant to trifluorothymidine because they lack the ability to convert the analogue to cytotoxic metabolites. The induction of chromosome aberrations by test compounds can be crudely measured using CHO cells and human blood lymphocytes.

Gross measures of chromosomal aberrations in CHO cells and human peripheral blood lymphocytes induced by genotoxicants record effects on mitotic index, chromosome number, chromosome and chromatid deletions and exchanges, and endoreduplication when chromosomes double in number but remain inseparable. In response to compounds with genotoxic activity, unscheduled DNA synthesis (UDS) supposedly monitors induction of DNA repair by counting grains appearing autoradiographically following exposure to ^{3}H-thymidine as precursor, while in the micronucleus test (MNT), expulsion of normal nuclei and retention of DNA fragments or lagging chromosomes are observed in daughter nuclei of immature mouse erythrocytes. In the fruit fly, Drosophila, the sex-linked recessive lethal test (SLRL) can detect the induction of X-linked mutations in back-crosses using virgin females derived from and then mated with males treated with a suspected genotoxicant by measuring the reduction in or complete absence of wild-type males among the progeny. An analogous assay which measures embryonic lethality in the mouse or rat induced by suspected genotoxicants is called the dominant lethal test.

One popular whole animal test for detecting compounds with carcinogenic activity uses B6C3F1 male mice which are characteristically prone toward developing spontaneous and chemically induced hepatomas. This assay can detect hepatocarcinogens but criticisms regarding the 30% spontaneous background rate (murine H-*ras* is activated in both spontaneous and induced liver tumors in B6C3F1 male mice), minimal efforts to identify the specific mechanisms by which protooncogenes are activated to oncogenes and/or which tumor suppressor genes are mutationally inactivated, along with important differences between mouse and man in both the metabolism of putative (pro)carcinogens and the DNA sequences of crucial genes upon which they act, severely weakens conclusions derived from B6C3F1 carcinogenesis data regarding the mechanisms of cancer causation and their relevance to and predictive value in man.

Recent successes have created strains of transgenic mice containing a 1.1 Kb *lac*I gene as target linked to a functional *lac*Z gene as reporter in which mutagens such as N-ethyl-N-nitrosourea, cyclophosphamide, and benzo(a)pyrene that inactivate *lac*I (coding for the repressor protein of the *lac* operon) allow expression of beta-galactosidase which cleaves X-gal to release a blue chromophore in plaques generated from an *in vitro* system which packages target (and reporter) DNA treated with mutagens in whole animals using lambda phage and a sensitive bacterial indicator. By analogy, forward mutations induced by mutagens which inactivate the *lac*Z gene directly in other strains of transgenic mice therefore do not release the blue chromophore from X-gal and are detected as clear lambda phage plaques on solid media. Both assays can monitor a subset of mutagenic events caused by potential human carcinogens but are severely limited in their detection of larger deletions and insertions due to constraints caused by lower and upper bounds of the final fragment sizes of *lac* DNA to be encapsidated by lambda phage heads in the *in vitro* packaging system. A further criticism of both transgenic mouse systems is that the *lac* targets acted upon by putative carcinogens are derived from a bacterium and are not human sequences which makes the resolution and relevancy of such data only slightly more informative than if the mutational spectrum were done on compounds using the battery of bacterial strains in the Salmonella/Ames microsome test. In general the most valuable transgenic mice for studying the mechanisms of action of compounds from therapeutic drugs and biologicals to suspected carcinogens will be those strains containing human genes and their products as the final targets. Strategies for the construction of such key strains will depend upon the exact nature of either the research goals intended to develop these therapeutic agents or the definition of the carcinogenicity tests to be devised. The number of transgenic mouse strains that can be created to test the carcinogenic potential of any class of compounds is virtually limitless. In each case, individual strain construction should focus on the suspected genetic activity of the compounds and the carcinogenic end point of each assay system. The further derivation of mouse strains which should contain appropriate combinations of human transgenes can be achieved either by crosses between founder mice containing the original transgene(s) of interest or by deliberate insertion of additional transgene cassettes into embryos derived from the original transgenic progenitor strains.

Genetic polymorphisms should be the first among many variables considered in devising strategies for constructing key strains of transgenic mice to test the mechanisms of action of putative human carcinogens. Various enzyme systems rich in polymorphic content exist in the human population which are involved in metabolic activation and inactivation of endogenous and foreign compounds as potential carcinogens. A brief list of such human enzyme systems would include (a) the cytochrome P450s located in microsomal fractions of human liver, lung, kidney, intestines, skin, and brain. The P450s, composed of 50-60 kD proteins containing noncovalently bound protoporphyrin IX at the carboxy terminal end which acts as receptor for electrons donated from the NADPH-P450 oxidoreductase, participate in the generation of highly reactive metabolic intermediates created by hydroxylation and epoxidation of N, C, and O groups and from dealkyation, deamination, reductive dehalogenation and conjugation reactions with sulfate, glutathione, and glucuronic acid. Eight separate families of P450s are located on seven different chromosomes and contain almost 50 distinct forms each one capable of metabolizing a large number of substrates such as drugs (e.g. benzphetamine, hexobarbitol, tolbutamide, retinoic acids and alcohols, etc.), tranquilizers (midazolam), immunosuppressants (cyclosporine), antibiotics (erythromycin), and calcium channel blockers (nifedipine) and important classes of human carcinogens including arylamines, heterocyclic arylamines, and polycyclic aromatic hydrocarbons (e.g. benzo(a)pyrene, benz(a)anthracene, dioxins, etc.); and (b) other enzyme systems such as flavin monooxygenases, carboxylesterases, dihydrodiol dehydrogenases, peroxidases, acyl-, aryl-, acetyl-, and methyltransferases, DT diaphorases, epoxide hydrolases, glutathione S-transferases, carboxymethyl L-cysteine sulfoxidases, and UDP-glucuronyl transferases. The genetic design of transgenic mice containing selective combinations of human polymorphs of these enzymes should be determined from an intimate knowledge of the pathways of metabolic activation required for each class of putative carcinogen. These judiciously constructed progenitor strains can then be further genetically modified to create the cadre of final transgenic mice for testing parent compounds and their metabolic derivatives which act against specific genetic targets defined by the presence of chosen human protooncogenes and tumor suppressor genes. The following specific examples suggest only a few of many rationales for creating strains of transgenic mice intended to generate unequivocal data describing the molecular mechanisms of action of genotoxicants and carcinogens, their specific genetic targets, and the human oncogene-driven tumors produced in such animals. Some desirable candidates among the many human protooncogenes and tumor suppressor genes selected for placement as definitive targets for carcinogens in transgenic mice would include the following choices: (a) to detect the induction of point mutations caused by transitions and transversions resulting in missense events, the protooncogenes H-*ras*, K-*ras*, and N-*ras* are chosen for their involvement in bladder cancers, liver and lung tumors, and neuroblastomas, respectively, and *erb*B2 for its role in mammary carcinomas. The p53 tumor suppressor gene occupies a central role in tumorigenesis because its mutationally inactivated forms which can occur as missense mutations in over 30 different codons, as well as small or extended deletions, are involved in at least 50% of all human cancers including lymphoid leukemias, hepatocellular carcinomas, colon cancers, Li-Fraumeni syndrome, esophageal cancers, myeloid leukemias, small cell lung cancers, osteosarcomas, and chronic myelogenous leukemias in either the presence or absence of *abl* translocations. The characteristics of *p53* include synergism with *ras* and *myc*, an involvement in transcriptional control of selected genes affecting cell proliferation, and the presence of important restriction fragment length polymorphisms (RFLPs) which can be detected diagnostically in normal and mutated forms of the gene by BamHI, BglII, SccII and MspI *p53* DNA digestions; (b) to monitor deletion formation, protooncogenes including *raf*, involved in cell proliferation, *met* detected in osteosarcomas, and *trk*, activated in colon carcinomas, requires the loss of certain sequences resulting in truncations caused by suspected carcinogens such as benzo(a)pyrene and its metabolites. The DNA sequences upon which a given carcinogen prefers to act can cause a deletion to form and to thereby activate a protooncogene by at least two mechanisms: (1) at a direct tandem repeat or when a unique sequence is bracketed by non-tandem direct repeats causing a deletion mediated through by-pass replication, or (2) in chromosomal regions lacking any repeats or palindromes but whose DNA sequences are more susceptible to cleavage by endonucleases and topoisomerases causing a deletion by excision followed by blunt-end ligation; (c) to assay genetic recombination events resulting in translocations which fuse high-efficiency promoter-enhancer sequences to activate quiescent protooncogenes, the selection of human *abl* and *myc* which cause chronic myelogenous leukemia and Burkitt's lymphoma, respectively, would supply definitive targets for recombinogens; and (d) to detect compounds that cause oncogene amplifications, *erb*B2/*neu* involved in progressive mammary carcinomas, N-*myc* in aggressive neuroblastomas, and c-*myc*, L-*myc*, and N-*myc* in metastatic small cell lung carcinomas would be among the human protooncogenes of choice. For example, transgenic mouse strains derived from C57Bl/6xCBA F1 eggs containing the human protooncogenes *ras*, *abl*, *myc*, and the p53 tumor suppressor gene might be ideal to analyze carcinogens that (a) induce missense mutations in *ras* analyzed by synthetic oligodeoxyribonucleotide probes that detect codon 12, 13, and 61 alterations (e.g., 4-aminobiphenyl forms guanosine adducts which cause K-*ras* codon 13 GGC to CGC transversions and GGC to GAC transitions, and H-*ras* codon 61 CAA to AAA and CAA to CTA transversions) and changes in codons 59, 63, 116, and 119 in transgenic mice that develop *ras*-driven bladder, liver, and lung tumors; (b) cause translocations of either *abl* or *myc* resulting in leukemias and lymphomas, respectively, the *abl* recombinants identified in Northern blots which detect chimeric *abl*-encoded mRNA transcript(s) coding for an abnormal tyrosine kinase(s) activity causing lethality in such mice within two months, and *myc* fusions to new chromosomal locations detected in Southern blots using the appropriate *myc* probes; (c) alter the p53 tumor suppressor gene by induction of either point mutations or deletions which cooperate with *ras* and/or *myc* alleles to form leukemias, lymphomas, and/or solid tumors of the liver and lung. Transgenic mice produced from BALB/cxC3H F1 hybrids might express p53-driven Li-Fraumeni-like syndrome with either a mutated H- or K-*ras* allele in the presence of a c-*myc* translocation whereas H-*ras*- and -*myc*-activated C57Bl/6 transgenic mice would express liver tumors and/or B-cell lymphomas compared to a K-*ras*- and -*myc*-activated C3H/HEJ transgenic strain which could develop pancreatic tumors and/or T-cell lymphomas. Molecular analyses of *ras*-activating point mutations would reveal the existence of DNA sequences that constitute preferred "hot spots" within the *ras* genes which may differ among classes of mutagens and carcinogens. Lymphomas driven by c-*myc* translocations could occur by a variety of mechanisms including deletions of normal *myc* transcriptional control regions substituted by up-promoter sequences due either to its recombination and fusion to a new chromosomal location(s) or to alterations in stability and/or turnover of the resulting *myc* mRNA. Inverse PCR protocols followed by DNA sequencing would reveal which regions are responsible for each characteristic *myc* translocation and might even reveal which carcinogen-induced *myc* translocations result from legitimate vs. illegitimate genetic recombination: if legitimate (homologous) genetic recombination is involved, then the region(s) in which *myc* originally resided and those to which it translocates should exhibit significant DNA sequence homologies; if the carcinogen-induced *myc* translocation events occur by illegitimate (non-homologous) genetic recombination, then little or no DNA sequence identity should be detected between the original *myc* sites of residence and its newly created translocation site(s). These findings would discriminate between possible pathways preferred by classes of carcinogens that are recombinogens using *myc* (or *myb* or *abl*) translocations: either such compounds prefer to act through homologous (legitimate) genetic recombination which uses sequences that possess long-track DNA homologies [or (degenerate) consensus sequences] or non-homologous (illegitimate) genetic recombination which does not depend on DNA sites with detectable sequence identity; and (d) cause insertional mutagenesis which can be detected using two separate but related strategies. The p53 tumor suppressor gene may become inactivated by insertion of a foreign piece of DNA caused by treatment with carcinogens that prefer to fragment certain regions of the chromosome. Myriad derivatives of benzene derived and isolated from microsomal metabolism are notorious for their clastogenic abilities and are implicated in the etiologies of both acute and myelogenous leukemias. DNA fragments generated by such compounds could insert into either the reading frame and/or the promoter region of a normally active p53 tumor suppressor gene thereby preventing its function. Correct PCR analyses and Northern blots would detect the occurrence of insertion mutations whose presence would be observed by DNA sequences fused to p53 and in longer p53 mRNA which also contain transcripts of the insertions, respectively. In addition the *erb*B2 protooncogene may become activated by carcinogens due to insertion of high-efficiency promoter sequences which alter and/or increase *erb*B2-encoded receptor protein production resulting in highly metastatic cell populations contributing to either aggressive leukemias, glioblastomas, squamous cell carcinomas, and estrogen-dependent mammary and ovarian tumors.

In the future when the polymorphic alleles causing cancer-prone hereditary conditions are cloned and isolated from humans, analogous combinations of human protooncogenes and tumor suppressor genes can be introduced into strains of transgenic mice containing the mutant genes from such diseases. These hereditary defects in processing damaged DNA include (a) at least two forms of xeroderma pigmentosum (XP) both of which exhibit increased sensitivity to carcinogens, the classic form being deficient in an early step of DNA excision repair, and an XP variant unable to use efficiently a damaged template for DNA replication; (b) ataxia telangiectasia (AT) which exhibits pronounced chromosomal aberrations and a reduced ability to repair complex DNA strand breaks; (c) Bloom's syndrome (BS) which shows a reduction in the rate of DNA synthesis, an increase in both sister chromatid exchange (SCE) and chromosome breaks, and a DNA ligase defect; and (d) Fanconi's anemia (FA) whose unknown genetic defect results in an increased sensitivity to DNA cross-linking agents and a high frequency of chromosomal aberrations.

Comparisons between strains of transgenic mice containing normal and mutant forms of these determinants for predispositions to cancer may reveal additional mechanisms of action of carcinogens on human target protooncogenes and tumor suppressor genes.

CONCLUSIONS AND SUMMARY OF FUTURE DIRECTIONS

The rationales and strategies for creating and using critical strains of transgenic mice are limited only by imagination, creativity, and available technology. As illustrated by the specific examples developed in the previous sections, transgenic mice can serve as unique and informative models of human disease, drug development, and carcinogenesis. Future transgenic systems can be devised to examine the recombination mechanisms resulting in new human light and heavy chain combinations which determine B-cell antibody response to specific immunogens, the genetic determinants which specify antigen processing in macrophages and lymphocytes, the cellular factors that influence formation, secretion, and specificity of IgE immunoglobulins that define the spectrum of allergic reactions, and the factors that determine the cytotoxic activities of natural killer cells.

Transgenic mice are invaluable tools to study genetic determinants in tissue-specific expression, the regulatory circuitry which governs the interactions between genes, and the characteristics of gametogenesis which affect genome stability including replication, insertion, movement, and excision of human transposable elements. That future strains of transgenic mice have enormous potential to change the study of genotoxicity and carcinogenesis is a safe prediction. The isolation and chemical characterization of DNA adducts formed by reactions between carcinogens and critical human oncogene DNA sequences in transgenic mice will facilitate research on the enzymatic pathways which fix the final mutational lesions be they point mutations, deletions, insertions, inversions, duplications, or translocations. In future transgenic systems designed to measure translocational activation of human protooncogenes, specific carcinogen-sensitive cleavage sites can be spliced *in vitro* adjacent to chosen human protooncogenes followed by their introduction into transgenic mice. Since some carcinogens exhibit regional preferences for causing chromosome scissions, precisely defined GC- vs. AT-rich DNA sequences deliberately placed upstream of either human *myc* or *abl* would provide ideal sites of action used by such compounds for cleavage, translocation, and subsequent transcriptional activation of these oncogenes.

Suppose a neuropharmacologic agent is being developed which shows promise in the control and management of schizophrenia. A preliminary genotoxic screen reveals that the parent compound exhibits moderate missense mutagenic activity but some of its congeners which result from intermediary metabolism and those synthesized in the laboratory possess both elevated missense and frameshift mutagenesis. Since the agent is essentially intended for long-term therapy over a lifetime, definitive transgenic mouse systems, which detect mutations that activate *ras*, *erb*B2, *fos*, and inactivate p53, coupled with translocation assays involving *abl* and *myc*, would be critical in selecting which congeners possess the highest therapeutic to carcinogenic index to be included in expanded clinical trials. These systems can also be adapted to assay the mechanisms of action by which anticarcinogens, such as those isolated from cruciferous vegetables, either inhibit the production of carcinogens through the cytochromes or prevent carcinogen recognition of specific DNA target sequences.

Recall the fascinating difference between arrangement and copy number of transgenes introduced by retroviral vector (RV) vs. DNA microinjection: that in the mouse genome only one copy of the transgene occurs when an RV is used whereas multiple copies of the transgene in head-to-tail tandem array occurs with DNA microinjection. An experimental strategy to examine the molecular basis for this phenomenon could be designed around the notion that insertion of a given transgene into the host genome usually occurs by illegitimate genetic recombination. Suppose the 5' and 3' (direct repeat) murine-derived RV LTRs possess (degenerate consensus) sequences with enough common homology contained in endogenous DNA tracts of the host genome. One expectation might be that an exogenous transgene(s) carried by the RV would preferentially recombine into such host regions without prior alteration of the vector-transgene construct; in this case, only one copy would be inserted into the host genome at a given albeit somewhat slightly preferred integration site.

Conversely, naked foreign DNA fragments might be subjected to partial degradation by cellular exonucleases causing transient single-strandedness; acted upon by the DNA replication/repair machinery of the cell, these structures could then be converted into concatenates. Since significant homologies usually do not exist between the mouse genome and foreign DNA sequences, the resolution of such catenates would then result, allowing the final structure to recombine with and insert into the host genome at random locations as tandem transgene cassettes.

Strategies for the construction of transgenic mice have pitfalls. Since the integration sites of foreign genes introduced into mouse embryos cannot be controlled at present, certain experiments may produce inviable founder mice depending upon both the nature and structure of the exogenous genes and the genetic characteristics of the intended host mouse mutant strain. Furthermore progeny resulting from crosses between the founder mice either with each other or with other strains of mice may result in embryonic lethality or death while approaching adulthood. The study of such mechanisms may reveal the occurrence of insertional mutagenesis into genes required for normal stages of development, activation of endogenous viral sequences, or genomic destabilization by hybrid dysgenesis. Appearing initially as obstacles, these observations can be developed as model systems of value to study the possible mechanisms by which lethal birth defects occur in both animals and humans. The creation of key strains of transgenic mice affords many benefits for both research and development purposes which include the use and sacrifice of smaller numbers of animals; the potential to shorten the latency periods in both acute and chronic studies requiring whole animals; critical increases in the sensitivities and specificities of particular assays of the test articles; eliminating the necessity to make high-dose to low-dose extrapolations; reduction in species to species uncertainty; increased accuracy of threshhold dose determination; comparisons which correlate with mechanisms of action of drugs, biologicals, and carcinogens; improved understanding of the mechansims of action of carcinogens using specific human genetic targets; databases from crucial experiments derived from unambiguous fold increases and decreases instead of percentage differences; new options for drug and carcinogen administration and first-pass metabolism studies which affect important pharmacodynamic and pharmacokinetic parameters affected by key human transgenic polymorphisms; and the derivation of cell lines from specific tissues containing human targets for developing rapid *in vitro* screens. Transgenic mice will therefore provide invaluable laboratory tools for the phrasing of and solutions to experimental problems well into the 21st century.

SUGGESTED REFERENCES FOR FURTHER READING

1. Adams JM, Harris AW, Pinckert CA, Corcoran LM, Alexander WS, Cory S, Palmiter RD, Brinster RL (1985) The c-myc oncogene driven by immunoglobulin enhancers induces lymphoid malignancy in transgenic mice. Nature 318: 533-538
2. Baker SJ, Markowitz S, Fearon ER, Willson JKV, Vogelstein B (1990) Suppression of human colorectal carcinoma cell growth by wild-type p53. Science 249: 912-915
3. Bishop JM (1989) Oncogenes and clinical cancer. In Oncogenes and the Molecular Origins of Cancer, Cold Spring Harbor Laboratory, Chap 13, pp 327-358
4. Brinster RL, Chen HY, Trumbauer ME, Yagle MK, Palmiter RD (1985) Factors affecting the efficiency of introducing foreign DNA into mice by microinjecting eggs. Proc Natl Acad Sci 82: 4438-4442
5. Cooper GM (1990) Oncogenes. Jones and Barlett, Boston, pp 1-323
6. Dionne CA, Kaplan R, Seuanez H, O'Brien SJ, Jaye M (1990) Chromosome assignment by polymerase chain reaction techniques: assignment of the oncogene FGF-5 to human chromosome 4. BioTechniques 8: 190-194
7. Druker BJ, Mamon HJ, Roberts TM (1989) Oncogenes, growth factors, and signal transduction. N Engl J Med 321: 1383-1391
8. Dyson PJ, Rabbitts TH (1985) Chromatin structure around the c-myc gene in Burkitt's lymphomas with upstream and downstream translocation points. Proc Natl Acad Sci 82: 1984-1988
9. Forrester K, Almoguera C, Han K, Grizzle WE, Perucho M (1987) Detection of high incidence of K-ras oncogenes during human colon carcinogenesis. Nature 327: 298-303
10. Fromont-Racine M, Bucchini D, Madsen O, Desbois P, Linde S, Nielsen JH, Saulnier C, Ripoche M-A, Jami J, Pictet R (1990) Effect of 5'-flanking sequence deletions on expression of the human insulin gene in transgenic mice. Molec Endocrinol 4: 669-677
11. Guengerich FP (1988) Roles of cytochrome P-450 enzymes in chemical carcinogenesis and cancer chemotherapy. Cancer Res 48: 2946-2954
12. Hanahan D (1988) Dissecting multistep tumorigenesis in transgenic mice. Ann Rev Genet 22: 479-519
13. Heisterkamp N, Jenster G, ten Hoeve J, Zovich D, Pattengale PK, Groffen J (1990) Acute leukemia in bcl/abl transgenic mice. Nature 344: 251-253
14. Jones PA (1986) DNA methylation and cancer. Cancer Res 46: 461-466
15. McCormick F (1989) ras oncogenes. In Oncogenes and the Molecular Origins of Cancer, Cold Spring Harbor Laboratory, Chap 7, pp 125-145
16. Pattengale PK, Stewart TA, Leder A, Sinn E, Muller W, Tepler I, Schmidt E, Leder P (1989) Animal models of human disease: pathology and molecular biology of spontaneous neoplasms occurring in transgenic mice carrying and expressing activated cellular oncogenes. Amer J Pathol 135: 39-61
17. Roldan-Arjona T, Luque-Romero LL, Ruiz-Rubio M, Pueyo C (1990) Quantitative

relationship between mutagenic potency in the Ara test of Salmonella typhimurium and carcinogenic potency in rodents. A study of 11 direct-acting monofunctional alkylating agents. Carcinogenesis 11: 975-980
18. St George-Hyslop PH, Tanzi RE, Polinsky RJ, Haines JL, Nee L, Watkins PL, Myers RH, Feldman RG, Pollen D, Drachman D, Growdon J, Bruni A, Foncin J-F, Salmon D, Frommelt P, Amaducci L, Sorbi S, Piacentini S, Stewart GD, Hobbs WJ, Connelly M, Gusella JF (1987) The genetic defect causing familial Alzheimer's disease maps on chromosome 21. Science 235: 885-890
19. Stanbridge EJ (1990) Human tumor suppressor genes. Ann Rev Genet 24: 615657
20. Sukumar S (1989) ras oncogenes in chemical carcinogenesis. Curr Topics Microbiol Immunol 148: 93-114
21. TNO Protocols in Genetic Toxicology (1990) Version 1.2
22. Tommerup N (1989) Cytogenetics of the fragile site at Xq27. In The Fragile X Syndrome, Ed. Davies KE, Oxford University Press, NY, pp 102-135
23. Vogel J, Hinrichs SH, Reynolds RK, Luciw PA, Jay G (1988) The HIV tat gene induces dermal lesions resembling Kaposi's sarcoma in transgenic mice. Nature 335: 606-611
24. Windle JJ, Albert DM, O'Brien JM, Marcus DM, Disteche CM, Bernards R, Mellon PL (1990) Retinoblastoma in transgenic mice. Nature 343: 665-669

REGULATING GENE EXPRESSION IN MAMMALIAN CELL CULTURE AND TRANSGENIC MICE WITH YEAST GAL4/UAS CONTROL ELEMENTS

David M. Ornitz, Radek Skoda, Randall W. Moreadith and
Philip Leder

Howard Hughes Medical Institute and Department of Genetics
Harvard Medical School 25 Shattuck Street
Boston, Massachusetts 02115

SUMMARY

To overcome the difficulties in maintaining strains of mice with decreased reproductive potential due to the expression of toxic genes, we have developed a binary system that will allow one to activate an otherwise silent transgene. This system requires two types of transgenic mice, target strains and transactivator strains. The target strain is engineered to contain a transgene controlled by yeast regulatory sequences that respond only to the yeast transcription factor, GAL4, or a derivative of this protein. The transactivator strains are engineered to express a GAL4 gene that can be driven by any enhancer/promoter combination. In this manuscript we review studies designed to test this binary system in tissue culture cells and in mice. We also discuss several target genes currently being transactivated in mice, and review some of the published data on *int*-2 expression in GAL4 mice.

INTRODUCTION

Deleterious effects of expressed transgenes have often complicated the study of transgenic mice. Transgenic strains expressing an oncogene may develop tumors before they reach reproductive age (Quaife et al. 1987), while mice expressing potent lymphokines may succumb to disease resulting from immunodeficiency (Tepper et al. 1990). In order to overcome these difficulties we wished to create an optimized binary transgenic system in which the deleterious gene could be carried in a silent, unexpressed form and then be activated by breeding with a second transgenic mouse containing an activating gene.

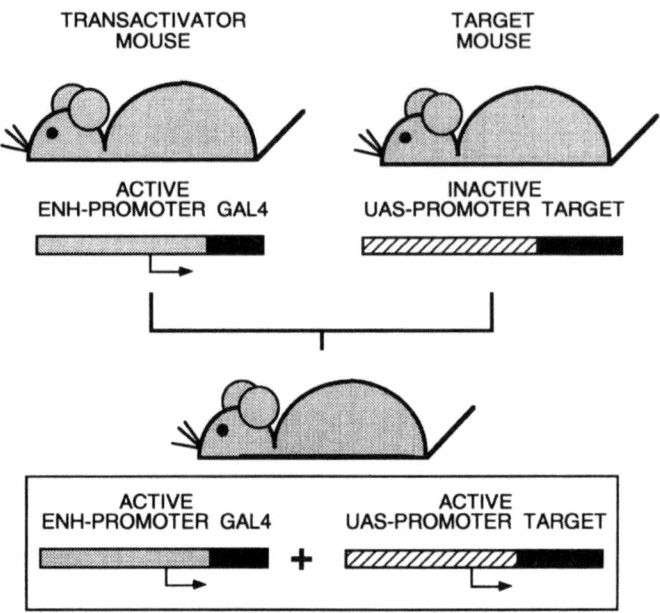

Fig. 1. The binary transgenic system in which transcriptionally active transactivator mice are mated with transcriptionally silent target mice to yield bigenic offspring in which the target gene is transactivated

Recently we have developed a binary system that allows the activation of a transgene in an animal by mating two lines of mice (Ornitz et al. 1991). The general scheme of this binary system is diagrammed in Figure 1. The gene of interest is engineered to contain regulatory elements that respond only to the yeast transcriptional activator GAL4 (Ginger et al. 1985). These regulatory elements, in combination with a specific mammalian or viral promoter element, are silent in mammalian cells and in transgenic animals. When such an engineered gene is introduced into the germline of an animal it should be active only in the presence of its transcriptional activator GAL4. We refer to this as the target gene and target animal.

The second component of this system requires the directed expression of the yeast transcriptional activator GAL4. The GAL4 gene is engineered to contain viral or mammalian transcriptional regulatory elements and polyadenylation signals. These elements can be chosen to regulate expression in specific cell types, in a limited number of tissues, in a broad range of tissues, and at various times throughout development. Such a gene, referred to as the transactivator gene, can be introduced into the germline of an animal. This animal, referred to as the transactivator animal, when mated to a target animal will produce offspring in which the target gene is transactivated. The pattern of transactivation is determined by the pattern of GAL4 expression in the transactivator animal.

Table 1.   Genes being tested in the transgenic binary system.

Transactivator gene	Target gene	Phenotype
MMTV-GAL4/236	UAS INT2	Hyperplasia
	UAS NEU	Neoplasia
	UAS LACZ	Stains for lac Z
	UAS DT	ND
	UAS MYC	ND
	UAS RAS	ND
	UAS LIF	ND
PF4-GAL4/VP16	UAS LACZ	Stains for lac Z
	UAS INT2	ND
	UAS NEU	ND
	UAS DT	ND
	UAS LIF	ND
EuPuGAL4/236	UAS INT2	No phenotype
CMV GAL4/236	UAS INT2	No phenotype

ND, not determined; DT, diphtheria toxin; LIF, leukemia inhibitory factor

GAL4 is a 881 amino acid transcription factor in the yeast, Saccharomyces cerevisae (Laughon and Gesteland, 1984). GAL4 binds to four similar 17 bp sequences located in the upstream activating region (UAS) of several yeast genes. A consensus 17 bp oligonucleotide will confer GAL4 regulation to heterologous genes in yeast (Giniger et. al., 1985), plants (Ma et al., 1988), and flies (Fisher et al. 1988). Webster et al. (1988) and Kakidani and Ptashne (1988) using transient transfection methods, have demonstrated that either a synthetic 17 bp oligonucleotide or an intact UAS will confer GAL4 regulation on a reporter gene in mammalian cells, respectively. Because the background expression in these experiments was low and the induction by GAL4 was comparable to other mammalian enhancers it seemed that GAL4 might function well in a transgenic system. However, it remained to be demonstrated whether GAL4 could transactivate a stably integrated UAS in a mammalian cell.

Deletion mapping and functional analysis of the GAL4 protein has defined several important regions: amino acids 1-147 includes the DNA binding domain of the GAL4 protein (Keegan et al. 1986); amino acids 148-196 (Region I) and 768-881 (Region II) can function as transcriptional activators when linked to a DNA binding domain (Ma and Ptashne, 1987). Several GAL4 mutants containing various combinations of these domains were tested by transient transfection into mammalian cells. One mutant, pAG236, containing the DNA binding domain and region II functioned better than the wild type GAL4 protein in this assay (Kakidani and Ptashne, 1988). We chose to test this GAL4 mutant (GAL4/236) in stable transfection experiments and in transgenic mice.

Table 2. Transactivator, target and bigenic mice.*

Line	Transgene	Expression[a]						Phenotype[b] Tissue Hyperplasia					
		A	B	C	D	E	F	A	B	C	G	H	I
RV	MMTV GAL4	+++		+++		+++		-	-	-	no	no	no
RW	MMTV GAL4	-	-	-	-	-	-	no	no	bo	no	yes	fertile
RX	MMTV GAL4	-		-	-	-	-	no	no	no	no	yes	fertile
DG	UAS INT-2							no	no	no	no	yes	fertile
DH	UAS INT-2							no	no	no	no	yes	fertile
DX	UAS INT-2							no	no	no	no	yes	fertile
DY	UAS INT-2	-						no	no	no	no	yes	fertile
DZ	UAS INT-2	-						no	no	no	no	yes	fertile
OA	UAS INT-2	-		-	-	-	-	no	no	no	no	yes	fertile
DGxRV	INT-2, GAL4	+	++					moderate			yes	yes	yes
DHxRV	INT-2, GAL4	++						variable		no	no	yes	fertile
DXxRV	INT-2, GAL4	±		-				mild		no	no	yes	fertile
DYxRV	INT-2, GAL4	+++			+++				severe	yes	severe	yes	yes
DZxRV	INT-2, GAL4	+++		+++		+++		-	-	-	severe	yes	no	sterile
OAxRV	INT-2, GAL4	+++		+++		+++		-	-	-	severe	yes	no	sterile

A=Breast; B=Salivary Gland; C=Epididymis; D=Spleen; E=Kidney; F=Liver; G=Prostate; H=Ability; I=Male Fertility. *Reprinted from Ornitz et al. 1991. [a] +++, band visible on northern blot after two hours. +, band visible on Northern blot after 12 hours. Blanks are data not determined. [b]Tissue hyperplasia based on gross and/or histologic analysis. Blanks are data not determined.

Transgenic mice were constructed using the mouse mammary tumor virus long terminal repeat (MMTV LTR; Huang et. al. 1981) fused to the GAL4/236 gene (Ornitz et al. 1991) as described below. The MMTV LTR was chosen because it is a reasonably strong promoter/enhancer element in mice (Stewart et al. 1984; Leder et al. 1986; Muller et al. 1988), and has the advantage of being expressed in multiple tissues and cell-types. This extends the potential generality of this binary system.

Murine *Int*-2 was chosen as the target gene (Dickson et al. 1984; Moore et al. 1986; Dixon et al. 1989) in the first test of this binary system. This gene, originally implicated in mouse mammary tumorigenesis (Morris et al. 1990), is also expressed in specific patterns during mouse development (Jakobovits et al. 1986; Wilkinson et al., 1988). *Int*-2 has a known phenotype when overexpressed in breast and prostate tissues of transgenic mice (Muller et al. 1990), however its phenotype when overexpressed in other tissues is unknown. *Int*-2 is also of interest because it is amplified in human breast tumors (Ali et. al. 1989; Lidereau et al. 1988; Zhou et al. 1988).

During the course of our experiments we demonstrated that transiently transfected GAL4/236 could transactivate a stably integrated target gene. These experiments demonstrate the feasibility of a binary transgenic system. We then demonstrate that a transgenic line expressing an MMTV GAL4/236 transgene in several tissues faithfully transactivates *Int*-2 target genes in appropriate tissues. Most recently we have extended the repertoire of target genes to include the *neu* oncogene and a LacZ reporter gene, other target genes currently being produced include the *myc* and *ras* oncogenes and the Leukemia Inhibitory Factor (LIF) gene (Table 1).

RESULTS

Transactivation in stably transfected cells

In order to model a potential GAL4 binary genetic system it was necessary to determine whether transfected GAL4/236 could activate a silent stably integrated target gene. The plasmacytoma cell line, J558L was co-electroporated with pSV2NEO and a UAS E1B/CAT reporter gene (Fisher et al. 1988, Giniger et al. 1985). Three independent stable pools of cells were selected and then transiently electroporated with either a promoterless GAL4/236 plasmid (pGEM GAL4), a cytomegalovirus enhancer/promoter fused to GAL4/236 (CMV GAL4), or an immunoglobulin enhancer/promoter fused to GAL4/236 (EuPuGAL4). Figure 2 shows no detectable CAT activity in the control transfections, a strong signal with CMV GAL4, and a weak signal with EuPuGAL4. Thus, *in vitro* a stably integrated UAS containing reporter gene is not expressed yet can be activated *in trans* by GAL4/236.

Transactivator Mice

We constructed several lines of mice containing the MMTV GAL4/236 transgene (Figure 3, Table 2). One line (RV, Table 2) expressed the transgene at relatively high levels in several tissues. To assess the pattern and level of transgene expression in the RV line, RNA from several tissues were analyzed by Northern blot and by RNase protection. Figure 5A demonstrates GAL4/236 gene expression in mammary gland, salivary gland, and epididymis of RV transgenic mice. No expression was observed in liver, kidney, spleen, or pancreas of these animals. This pattern of expression is

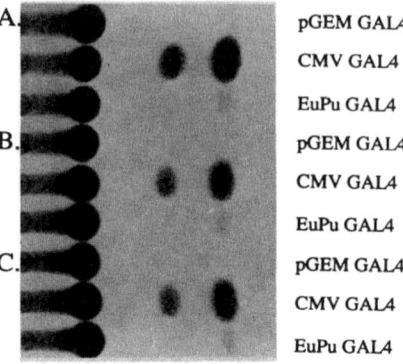

Fig. 2. Chloramphenicol acetyl transferase assays (Gorman et al., 1982) in J558L plasmacytoma cells stably transfected with a UAS/CAT reporter gene and then transiently transfected with control or GAL4 expressing transactivator genes. A, B, and C represent independent pools of stably transfected cells. CMV, cytomegalovirus LTR (Boshart et al., 1985; EuPu, immunoglobulin enhancer/ promoter.)

consistent with the pattern of MMTV LTR regulated expression in other lines of transgenic mice previously established in this lab and in other labs (Stewart et al. 1984, Leder et al. 1986).

We have also placed GAL4 under the control of several other enhancer/promoter combinations (Table 1). The cytomegalovirus LTR was used to provide a very wide range of expression in transgenic mice and in tissue culture cells. The immunoglobulin enhancer/promoter was used to direct GAL4 expression to the lymphoid system and the platelet factor 4 promoter was used to direct expression to megakaryocytes and platelets.

Target Mice

The target transgene contains four synthetic GAL4 binding sites fused at position -72 of the rat elastase I promoter (Figure 4C). This target promoter was placed 5' to the human growth hormone gene (Figure 4A) with an intervening polylinker for cloning cDNAs. The hGH gene in this target vector provides splice and polyadenylation sequence for the targeted cDNA. In our primary test of the binary system (Ornitz et al. 1991) the *int*-2 cDNA was cloned into this vector (Figure 4B). Six lines of target transgenic mice containing the UAS *Int*-2 transgene were established (Table 2). No expression of the *int*-2 target gene could be detected in any of these transgenic lines by Northern blot or by a sensitive RNase protection assay (Figure 5 and Ornitz et al. 1991). Furthermore, there is no noticeable phenotype grossly or histologically in breast tissue or salivary gland tissue of both virgin and pregnant female animals (Figure 6, 7, and data not shown). These target transgenic lines were then bred to the RV line and bigenic offspring were analyzed for *Int*-2 gene expression and *Int*-2-induced phenotypes.

Int-2 Expression in Bigenic Mice

Northern blot analysis demonstrates that the pattern of *int*-2 mRNA expression in the bigenic mouse is identical to that of GAL4 (Figure 5). GAL4/236 is expressed at

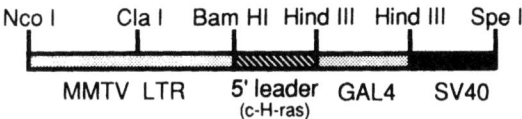

Fig. 3. MMTV transactivator construct tested in transgenic mice: The GAL4/236 transactivator gene fused downstream of the MMTV LTR (Huang et al. 1981) followed by SV40 splice and polyadenylation sequences (Seed 1987).

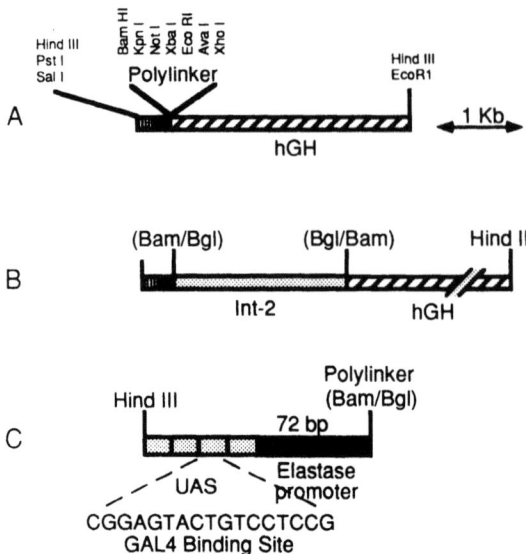

Fig. 4. Targeting genes responsive to GAL4. A. UAS/elastase promoter upstream of the human growth hormone gene (Seeburg 1982). Polylinker sequences are provided for insertion of cloned cDNAs. The hGH gene provides splice and polyadenylation sequences. B. *int*-2 cDNA inserted between the UAS/Elastase promoter and the human growth hormone gene. C. Four high affinity GAL4 binding sites (Sca I version, sequence shown) fused upstream of the otherwise silent elastase promoter (Ornitz et al. 1987) constituting the transgenic GAL4 target sequence

moderately high levels in breast, salivary glands and epididymis. Similarly *int*-2 mRNA is present at high levels in these tissues. RNase protection also demonstrates proper initiation of transcription (Ornitz et al. 1991). Thus the GAL4/236 transcription factor interacts with mammalian factors responsible for proper mRNA synthesis.

Hormone Dependent Mammary Hyperplasia in *Int*-2 Bigenic Mice

Control monogenic mice containing only the *int*-2 target gene or the GAL4/236 transactivator gene are normal (Figure 6A, 6C). Whole mount preparations from mammary tissue of *int*-2 mice are indistinguishable from that of normal age matched FVB mice (Figure 7A, and data not shown). Histologically both virgin and pregnant *int*-2 target mice are identical to non-transgenic FVB mice (Figure 7C, and data not shown). Similarly the RV transgenic mice which express GAL4 have no evidence of pathology. Mammary tissue in virgin females is not palpable, whereas pregnant mice develop massively enlarged mammary glands (Figure 6B, 6D). The mammary glands regress substantially in the first two weeks following delivery in the absence of suckling pups. However, they still remain markedly hyperplastic compared to a mono-transgenic mouse post weaning (data not shown). Whole mount preparations of mammary fat pads reveal abnormal ductal structures in mammary tissue of virgin bigenic mice (Figure 7B). Compared to a normal control mouse the duct's ability to grow into the fat pad is impaired. *Int*-2 expressing ducts form multiple budding structures and, in some bigenic mice, form dilated multi-cystic vesicles.

Histologic sections from mammary tissue of virgin bigenic females demonstrate mammary ductal structures that are dilated, and some tissue contains hyperplastic multilayered epithelia (Figure 7D). Some epithelial cells contain lipid droplets characteristic of lactating cells. The mammary tissue of pregnant mice is massively hyperplastic (Figure 7F). The lumina are small, and the epithelial cells do not produce secretory lipid droplets characteristic of lactating mammary epithelium. The ducts are surrounded by a dense connective tissue with increased numbers of fibroblasts. In most cases, the females are unable to nurse, however lactation can begin in some of the bigenic mice if the litter is maintained by a foster female mouse for several days. Of six independent target transgenic lines examined three have similar severe phenotypes when mated to the MMTV/GAL4/236 transactivator mouse RV, while two of the lines appear to have a milder form of mammary hyperplasia (Table 2). One of the lines, DH x RV has variable phenotypes ranging from mild in some mammary fat pads to severe in other fat pads within an individual animal.

Int-2 Induced Pathology in Male Bigenic Mice

Male mice from several of the target lines develop epididymal and prostate hyperplasia (Table 1, Figure 6F). Although these mice are able to mate they fail to sire any offspring even after multiple confirmed vaginal plugs. Autopsy on RV x DY bigenic males reveals an enlarged and dilated epididymal structure, normal appearing vas deferens, and a cystic enlarged prostate gland. Histologic examination of the prostate demonstrates a glandular tissue that closely resembles the disease benign prostatic hyperplasia common in human males. The epididymis of these bigenic mice is distended with sperm. The distal end of this tissue demonstrates interstitial and epithelial hyperplasia. Several of these specimens contain sperm-granulomas, consistent with obstruction and normal sperm production (data not shown).

Salivary Gland Hyperplasia in Bigenic Mice

In contrast to mice containing either a transactivator gene or a target gene the bigenic animals demonstrate parotid and sublingual gland enlargement (data not shown). Histologic analysis reveals hyperplastic salivary gland tissue in bigenic animals (data not shown).

Other Target Genes and Target Mice

Recently we have developed several more target transgenic mice (Table 1). The *Neu* oncogene, also implicated in human mammary neoplasia (van de Vijver et al. 1987) was placed in the UAS/hGH targeting vector. Bigenic offspring from one UAS/*Neu* line mated to the RV line develop multiple solitary mammary adenocarcinomas in virgin females at approximately 4 to 7 months of age. Male mice develop severe epididymal and prostate hyperplasia similar to that seen in bigenic *int*-2 males (D.M.O., R.W.M., and P.L., unpublished data)

A third target line of mice containing a β-galactosidase (lac Z) gene was developed to allow rapid screening of GAL4 transactivator lines. Preliminary data indicates that a GAL4/VP16 gene expressed in the hematopoetic system efficiently transactivates the UAS/LacZ gene in bigenic offspring (R.S., D.M.O., and P.L., unpublished data). Additionally this target transgene can be activated by the RV transactivator mouse in the male genitourinary tract. This data now extends further the generality of the binary system. The lacZ target mouse will not only allow rapid screening of future transactivator mice, but will also allow the tissue-specific pattern of gene expression and transactivation to be determined with relative ease.

DISCUSSION

Several other groups have developed binary systems in mice. These approaches involve transcriptional regulatory elements from Herpes Simplex Virus (HSV-1), from the Human Immunodeficiency Virus (HIV), and from the Human T-Cell Leukemia Virus (HTLV 1) (Byrne and Ruddle, 1989; Khillan et al. 1988; Nerenberg et al. 1987). The HIV and HTLV 1 LTRs both exhibit basal expression in mice (Khillan et al., 1988; D.M.O. and P.L., unpublished data). Khillan et al. were able to demonstrate transactivation by the HIV transactivator protein however the level of induction never exceeded 6 fold. Byrne and Ruddle, using the HSV system have demonstrated very low basal expression, and tissue-specific transactivation by the VP16 protein. However, they indicated that VP16 may be deleterious when expressed in some tissues. Although the viral binary system may have some applications in animals, it is unlikely that it will be broadly applicable. The viral transactivating proteins interact with host proteins and viral target sequences, and are more likely to have basal levels of expression in animal tissues.

We anticipate that this binary system will be used to express genes that are purposefully toxic. For example, one may want to eliminate a particular cell lineage with a toxic gene in order to determine its developmental consequences (see Palmiter et al., 1987). For such an experiment it is essential that the target gene is silent in the absence of its transactivator, GAL4. We have thus far used five different target genes, in twelve lines of transgenic mice, and have not yet observed leaky expression (Table 1). No epithelial phenotype has been observed in any of our six *int*-2 target animals nor in our *neu* target line. Two lacZ target lines and two diphtheria toxin target lines also appear inactive. Using our most sensitive RNAse protection assays, we have not been able to detect any *int*-2 mRNA in any tissue tested in our mono-transgenic target mice. Nevertheless, it is possible that the random integration of the target gene will occur in the vicinity of an active genomic locus resulting in leaky expression. Therefore several target strains may be required to find one best suited for the binary system. From our cumulative experience the UAS/Elastase promoter system is quite inactive in the absence of the GAL4 transactivator, both in tissue culture and in transgenic mice.

The level of expression of the *int*-2 and *neu* target genes in the presence of the

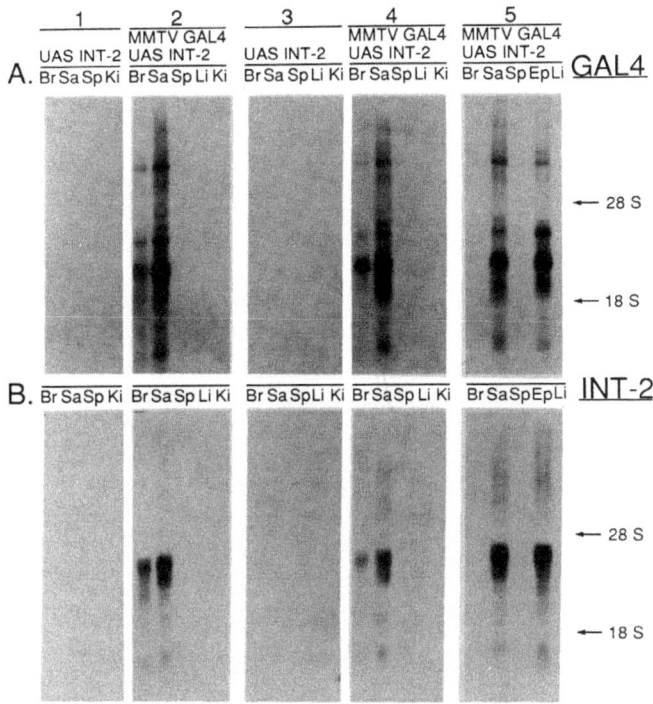

Fig. 5. Northern blot analysis of *Int*-2 target and bigenic mice. Br, Breast; Sa, salivary gland; Sp, spleen; Li, liver; Ki, kidney; Ep, epididymis. 1, 24 week virgin mouse (DZ). 2. 8 wk virgin bigenic mouse (DZ x RV). 3. 24 weeks pregnant mouse (DZ). 4. 14 week pregnant mouse (DZ x RV). 5. 14 week male mouse (DZ x RV). A. Blot hybridized with a GAL4 specific DNA probe. B. The same blot stripped and rehybridized with an *Int*-2 specific DNA probe. Ten mg total RNA was run from each tissue. Reprinted from Ornitz et al. 1991.

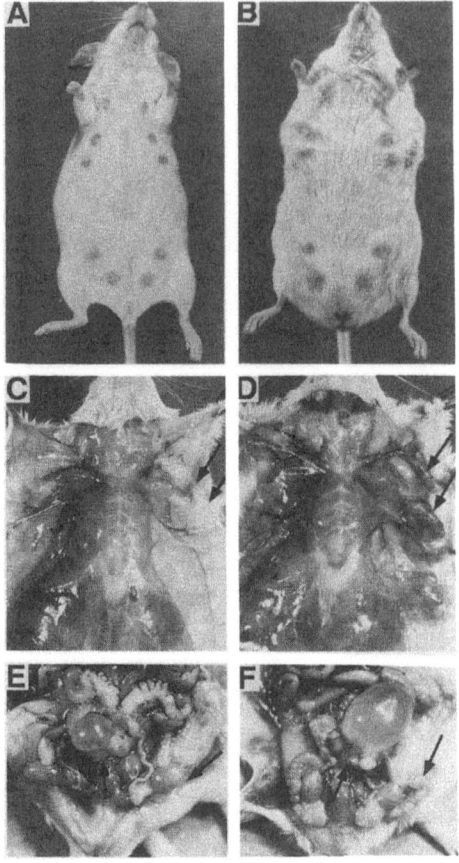

Fig. 6. Gross pathology of *Int*-2 target mice. Top. day one postpartum mice. A. 14 week mono-transgenic mouse (DZ). B. 19 week old bigenic mouse (DZ x RV). Middle. autopsy of a day one postpartum mouse. Arrow indicates mammary tissue. C. 14 week mono-transgenic mouse (DZ). D. 26 wk bigenic mouse (RV x DZ). Bottom. Autopsy of male mice. Arrows indicate prostate and epididymis. E. 22 week mono-transgenic mouse (DY). F. 22 week bigenic mouse (DY x RV). Reprinted from Ornitz et al. 1991.

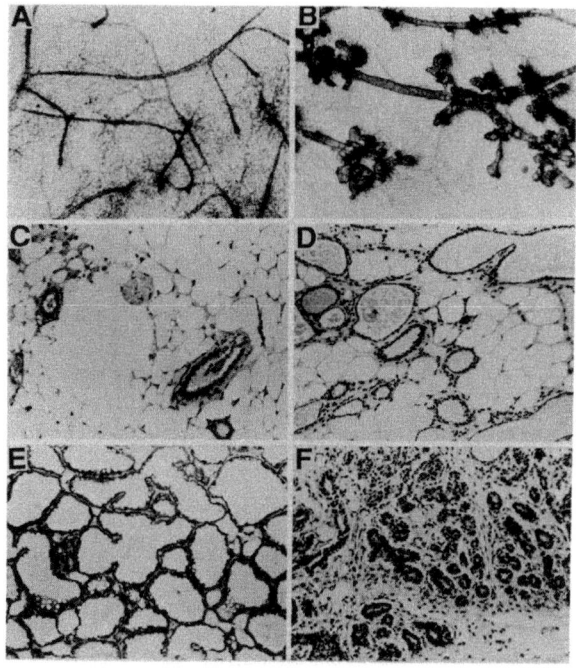

Fig. 7. Whole mount and histologic analysis of *int*-2 target mice. Top. Whole mount preparations of mammary fat pads from virgin mice (magnification 25X). A. 12 wk control mouse (RV). B. 12 wk bigenic mouse (RV x OA). Middle. Mammary histology from virgin mice (magnification 250X). C. 24 wk mono-transgenic mouse (DZ). D. 9 wk bigenic mouse (RV x DZ). Bottom. Mammary histology from pregnant mice (magnification 250X). E. 22 wk mono-transgenic mouse (OA). F. 20 wk bigenic mouse (RV x OA). Reprinted from Ornitz et al. 1991.

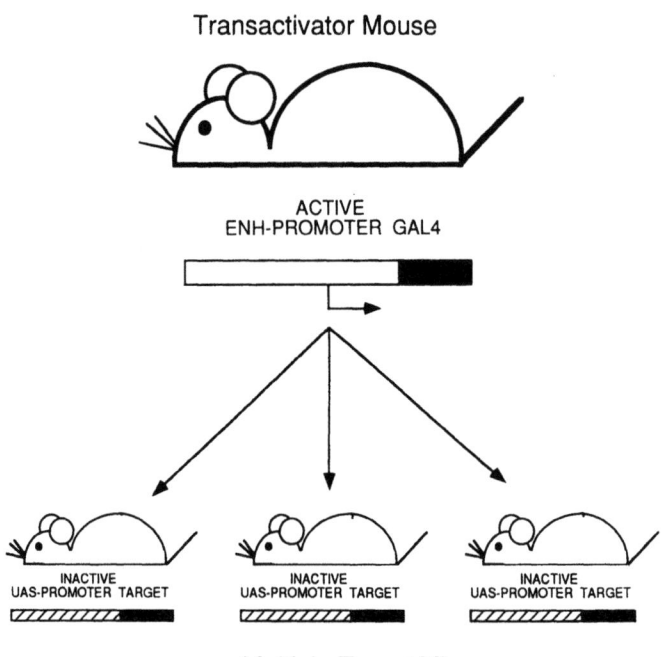

Fig. 8. Multiplex use of the binary system in which one transactivator mouse can be used to confer identical tissue-specific patterns of gene expression on multiple target mice. Alternatively, multiple transactivator mice can be mated with a single target mouse.

transcriptional activator GAL4 is high. The level of transcriptional activation must be extremely high, however, it cannot be accurately estimated because the denominator (expression in monogenic lines) in this calculation approaches zero. *Int-2* and *neu* mRNA are easily detected on total mRNA Northern blots. Furthermore, by RNase protection analysis, the mRNA is properly initiated. The pattern of target gene expression parallels that of the transactivator gene GAL4. The phenotype resulting from *int*-2 and *neu* expression is dramatic. Most tissues expressing *int*-2 develop a hyperplastic response. Mammary epithelium, a tissue that is normally under hormonal control develops an exaggerated response during pregnancy, followed by partial regression postpartum.

The utility of this system is demonstrated by two of our bigenic lines, DY and DZ x RV (see Table 1), in which both the males are sterile and the females fail to nurse their pups. The GAL4 binary genetic system allows these animals to be maintained and studied with ease. In comparison to MMTV-*int*-2 transgenic mice previously described which fail to express *int*-2 in mammary tissue from virgin female mice (Muller et al. 1990), several bigenic animals express both the GAL4 transactivator gene and consequently the *int*-2 gene in virginal mammary tissue. The mice described in this report and in Ornitz et al. (1991) develop a profound defect in mammary duct development even in mammary tissue that has not undergone the hormonal stimulation associated with pregnancy. This phenotype in virgin female mice affects ductal growth characteristics, whereas the phenotype in pregnant mice results in acinar cell hyperplasia. Overall the phenotype observed in several of these bigenic lines is similar, yet more severe than that previously observed in MMTV-*int*-2 transgenic mice. This may correlate with substantially higher levels of *int*-2 expression in some bigenic mice.

Comparing the histology from multiparous *int*-2 bigenic mice with that of transplantable mammary plaques resulting from the insertion of MMTV proximal to the *int*-2 gene in the GR strain of mice reveals a striking similarity (Morris 1990). Both the acinar and stromal hyperplastic phenotype are indistinguishable. Thus our bigenic mice are an excellent model of the *in vivo* disease caused by MMTV insertional activation of the *int*-2 locus.

In addition to the differences in mammary expression, several of these bigenic animals exhibit epididymal hyperplasia, an effect not previously associated with *int*-2 expression. By developing additional lines of transactivator mice with different patterns and levels of GAL4 expression we should be able to define a spectrum of tissues that respond to the putative growth factor *int*-2.

In summary, we have developed a bigenic system in which the GAL4/236 gene linked to a tissue-specific regulatory sequence transactivates target genes linked to GAL4 binding sequences. The result of this transactivation in the case of *int*-2 is essentially the same as that of an MMTV-*int*-2 fusion gene. The pattern of expression of the target gene under the control of the UAS/elastase promoter is the same as that of the transactivator gene, and the level of expression is qualitatively the same.

A further advantage of the bigenic system, pointed out by Byrne and Ruddle (1989), is that it provides an opportunity to combine different target genes with different promoters by use of a simple mating protocol (Figure 8). Once a target gene has been made, it can be easily maintained for use in conjunction with any GAL4 transactivator strain bearing any promoter. Thus far we have demonstrated the utility of this system with two different transactivator lines and several reporter lines (Table 1).

ACKNOWLEDGMENTS

We are grateful to Steve Sansing for outstanding help with our mouse colony, Ann Kuo, Cathy Daugherty and Peter Gentile for excellent technical assistance, Robert Cardiff for whole mount preparations and helpful discussions on mammary pathology, and members of the Leder lab for their helpful discussions. This research was supported in part by a grant from E.I. Dupont deNemours & Co., Inc.

REFERENCES

1. Ali, I.U., Merlo, G., Callahan, R., Lidereau, R., (1989) The Amplification Unit on Chromosome 11q13 in Aggressive Primary Human Breast Tumors Entails the bcl-1, *int*-2 and hst loci. Oncogene 4, 89-92.
2. Boshart, M., Weber, F., Jahn, G., Dorsch-Hasler, K., Fleckenstein, B., and Schaffner, W., (1985) A Very Strong Enhancer is Located Upstream of an Immediate Early Gene of Human Cytomegalovirus. Cell 41, 521-530.
3. Byrne, G.W., and Ruddle, F.H., (1989) Multiplex Gene Regulation: A Two-Tiered Approach to Transgene Regulation in Transgenic Mice. Proc. Natl. Acad. Sci. USA 86, 5473- 5477.
4. Dickson, C., Smith, R., Brookes, S., Peters, G., (1984) Tumorigenesis by Mouse Mammary Tumor Virus: Proviral Activation of a Cellular Gene in the Common Integration Region *Int*-2. Cell 37, 529-536.
5. Dixon, M., Deed, R., Acland, P., Moore, M., Whyte, A., Peters, G., Dickson, (1989) Detection and Characterization of the Fibroblast Growth Factor-Related Oncoprotein *Int*-2. Molec. Cell. Biol. 9, 4896-4902.
6. Fisher, J.A., Giniger, E., Maniatis, T., Ptashne, M., (1988) GAL4 Activates Transcription in Drosophila. Nature 332, 853-856.
7. Giniger, E., Varnum, S.M., Ptashne, M. (1985) Specific DNA Binding of GAL4, A Positive Regulatory Protein of Yeast. Cell 767-774.
8. Gorman, C.M., Moffat, L.F., and Howard, B.H., (1982) Recombinant Genomes which Express Chloramphenicol Acetyl Transferase in Mammalian Cells. Molec. Cell. Biol. 2, 1044-1051.
9. Huang, A. L., Ostrowski, M. C., Berard, D., Hager, G. L., (1981) Glucocorticoid Regulation of the Ha-MuSV p21 gene conferred by sequences from Mouse Mammary Tumor Virus. Cell 27, 245-255.
10. Jakobovits, A., Shackleford, G. M., Varmus, H. E., Martin, G. R., (1986) Two Proto-Oncogenes Implicated in Mammary Carcinogenesis, *Int*-1 and *Int*-2, are Independently Regulated During Mouse Development. Proc. Natl. Acad. Sci. USA 83, 7806-7810.
11. Kakidani, H., Ptashne, M. (1988) GAL4 Activates Gene Expression in Mammalian Cells. Cell 52, 161-167.
12. Keegan, L., Gill, G., Ptashne, M., (1986) Separation of DNA Binding from the Transcription-Activating Function of a Eukaryotic Regulatory Protein. Science 231, 699-704.

13. Khillan, J.S., Deen K.C., Yu, S., Sweet, R.W., Rosenberg, M., Westphal, H. (1988) Gene Transactivation Mediated by the TAT Gene of Human Immunodeficiency Virus in Transgenic Mice. NAR 16, 1423-1430.
14. Laughon, A., Gesteland, R.E. (1984) Primary structure of the Saccharomyces cerevisiae GAL4 Gene. Molec. Cell. Biol. 4, 260-267.
15. Leder, A., Pattengale, P.K., Kuo, A., Stewart, T., Leder, P. (1986) Consequence of Widespread Deregulation of the c-*Myc* Gene in Transgenic Mice: Multiple Neoplasms and Normal Development. Cell 45, 485-495.
16. Lidereau, R., Callahan, R., Dickson, C., Peters, G., Escot, C., Ali, I.U. (1988) Amplification of the *Int*-2 Gene in Primary Human Breast Tumors. Oncogene Res. 2, 285-291.
17. Ma, J., Przibilla, E., Hu, J., Bogorad, L., Ptashne, M. (1988) Yeast Activators Stimulate Plant Gene Expression. Nature 334, 631-633.
18. Ma, J., Ptashne, M., (1987) Deletion Analysis of GAL4 Defines Two Transcriptional Activating Segments. Cell 48, 847-853.
19. Medina, D. (1973) Preoeplastic Lesions in Mouse Mammary Tumorigenesis. Methods Cancer Res. 7, 3-53.
20. Moore, R., Casey, G., Brookes, S., Dixon, M., Peters, G., Dickson, C. (1986) Sequence, Topography and Protein Coding Potential of Mouse *Int*-2: A Putative Oncogene Activated by Mouse Mammary Tumor Virus. EMBO J. 5, 919-924.
21. Morris, D.W., Barry, P.A., Bradshaw, H.D. Jr., Cardiff, R.D. (1990) Insertion Mutation of the *Int*-1 and *Int*-2 Loci by Mouse Mammary Tumor Virus in Premalignant and Malignant Neoplasms from the GR Mouse Strain. J. Virol. 64, 1794-1802.
22. Muller, W. J., Lee, F. S., Dickson, C., Peters, G., Pattengale, P., Leder, P (1990) The *Int*-2 Gene Product Behaves as an Epithelial Growth Factor in Transgenic Mice. EMBO J. 9, 907-913.
23. Muller, W. J., Sinn, E., Pattengale, P., Wallace, R., Leder, P. (1988) Single-Step Induction of Mammary Adenocarcinoma in Transgenic Mice Bearing an Activated C-*Neu* Oncogene. Cell 54, 105-115.
24. Nerenberg, M., Hinrichs, S.H., Reynolds, R.K., Khoury, G., Jay, G. (1987) The TAT Gene of Human T-Lymphotropic Virus Type I Induces Mesenchymal Tumors in Transgenic Mice. Science 237, 1324-1329.
25. Ornitz, D.M., Hammer, R.E., Davison, B.L., Brinster, R.L., Palmiter, R.D. (1987) Promoter and Enhancer Elements from the Rat Elastase I Gene Function Independently of Each Other and of Heterologous Enhancers. Molec. Cell. Biol. 7, 3466-3472.
26. Ornitz, D.M., Moreadith, R.W., Leder, P. (1991) Binary System for Regulating Transgene Expression in Mice: Targeting *int*-2 Gene Expression with Yeast GAL4/UAS Control Elements. Proc. Natl. Acad. Sci. USA 88, 698-702.
27. Palmiter, R.D., Behringer, R.R., Quaife, C.J., Maxwell, F., Maxwell, I.H., Brinster, R.L. (1987) Cell Lineage Ablation in Transgenic Mice by Cell-Specific Expression of a Toxin Gene. Cell 50, 435-443.
28. Quaife, C.J., Pinkert, C.A. Ornitz, D.M., Palmiter, R.D., Brinster, R.L. (1987) Pancreatic Neoplasia Induced by *Ras* Expression in Acinar Cells of Transgenic Mice. Cell 48, 1023-1034.
29. Seeburg, P.H., (1982) The Human Growth Hormone Gene Family: Nucleotide Sequences Show Recent Divergence and Predict a New Polypeptide Hormone. DNA 1, 239-249.

30. Seed, B. (1987) An LFA-3 cDNA Encodes a Phospholipid-Linked Membrane Protein Homologous to its Receptor CD2. Nature 329, 840-842.
31. Stewart, T.A., Pattengale, P.K., Leder, P. (1984) Spontaneous Mammary Adenocarcinomas in Transgenic Mice that Carry and Express MTV/*myc* Fusion Genes. Cell, 38, 627-637.
32. Tepper, R.I., Levinson, D.A., Stanger, B.Z., Campos-Torres, J., Abbas, A.K., Leder, P. (1990) IL4 Induces Allergic-like Inflamatory Disease and Alters T Cell Development in Transgenic Mice. Cell 62, 457-467.
33. Vijver van de, M., Bersselaar van de, R., Devilee, P., Cornelisse, C., Peterse, J., Nusse, R. (1987) Amplification of the *Neu* (c-erbB-2) Oncogene in Human Mammary Tumors is Relatively Frequent and is Often Accompanied by Amplification of the Linked c-erbA Oncogene. Molec. Cell. Biol. 7, 2019-2023.
34. Webster, N., Jin, J.R., Green, S., Hollis, M., Chambon, P. (1988) The Yeast UAS_G is a Transcriptional Enhancer in Human HeLa Cells in the Presence of the GAL4 Trans-Activator. Cell 52, 169-178.
35. Wilkinson, D. G., Peters, G., Dickson, C., McMahon, A. P. (1988) Expression of the FGF-Related proto-oncogene *Int*-2 During Gastrulation and Neurulation in the Mouse. EMBO J. 7, 691-695.
36. Zhou, D.J., Casey, G., Cline, M.J. (1988) Amplification of Human *Int*-2 in Breast Cancers and Squamous Carcinomas. Oncogene 2, 279-282.

GENES AND ONCOGENES OF THE *ras* SUPERFAMILY

Armand Tavitian

INSERM U.248
10, avenue de Verdun, 75010
Paris, France

INTRODUCTION

The search for oncogenic sequences in human tumor DNA, using the DNA gene transfer technique for transformation of recipient cells, have repeatedly led to the rediscovery of one of the three closely related genes : H-*ras*, K-*ras* (first found as the cellular genomic sequences (oncogenes) transduced in the Harvey and Kirsten sarcoma viruses) and N-*ras*. The three *ras* genes code for 21kD GTP/GDP binding proteins; they are transiently anchored at the inner face of the plasma membrane upon reversible C-terminal farnesylation. Acquisition of a transforming potential is due to a point mutation that results in an amino acid substitution that most frequently occurs at positions 12, 13 or 61. Expression of such activated proteins causes dramatic changes (membrane ruffling, cell motility, etc...) and loss of contact inhibition. Ras proteins are highly conserved throughout evolution and are found in all eucaryotic organisms where they must exert essential functions. In spite of intense investigation the precise biochemical role of the ras proteins remains elusive(1).

The fortuitous findings of two proteins: YPT (from yeast)(2) and rho (from the marine snail Aplysia)(3), whose encoded proteins share between themselves and with the ras proteins about 30% homologies, did suggest the existence of a multigene family. In all the members of the family, from yeast to higher eucaryotes, a sequence of six a.a. was noticeably conserved.

This conserved sequence was taken advantage of, to isolate several additional *ras*-related genes. Using the corresponding oligonucleotides, we first isolated the cDNA of the ral protein (*ral* for "*ras* like") sharing some 50% a.a. identity with the ras proteins(4) and, later, the cDNAs of the four rab proteins sharing ~30% a.a. identity with the ras proteins and 75% to 40% identities with the yeast YPT protein (5); the most homologous member: *rab1*, being the mammalian counterpart of YPT. Other *ras*-related genes have also been found in various organisms by low stringency annealing with the H-*ras* probe: R-*ras* in mammals (6), Apl-*ras* in Aplysia(7), D-*ras*2 and D-*ras*3 in Drosophila (8); or with the YPT probe: *ryh1* and *ryh2* in the fission yeast (9). Meanwhile, it appeared that a cDNA isolated from a yeast secretion mutant: SEC4 did belong to the *rab*/YPT family (10).

THE *ras* SUBFAMILIES

Presently, the *ras* superfamily might be divided into three main branches that can be denoted: *ras*, *rho* and *rab*.

In mammals, the first branch (*ras* proper) includes the three oncoproteins ras: , K-ras and N-ras ($\sim$85% a. a. identity between one another) and, in addition, the ralA and ralB proteins (11) (also $\sim$85% a.a. identical to each other and about 50% homology with the ras proteins); the R-ras protein (50% identity with ras or ral) (6); the rap1A, rap1B and rap2 proteins ($\sim$ 55% identity with ras or ral) which are the human counterparts to the drosophila D ras3 protein (12-13-14).

In mammals, the second subfamily (*rho*) comprises the rho proteins: rhoA, rhoB, and rhoC ($\sim$85% and $\sim$30% identity when compared to ras) (15-16) . The yeast counterpart of these human proteins is RHO1; a second protein, RHO2, sharing $\sim$50% identity with RHO1, was also found in yeast and should be considered as a yeast rho-like protein (3). No real counterpart of RHO2 has been found, as yet, in mammals. It would be surprising if such RHO2-like proteins do not exist. However, additional rho-like cDNAs have been cloned and characterized in human cDNA libraries, more recently. They are the *rac1* and *rac2* (17), two genes for two closely related forms of G25K which are the mammalian homologs of the yeast CDC42 protein (18) and the TC10 cDNA which was isolated from a human teratocarcinoma library (19).

The third branch was given the acronym "*rab*", first used in 1987 when four additional members of the *ras* gene family were isolated from a rat brain cDNA library(5). It was observed that these mammalian *rab* genes constituted a distinct branch of the *ras* superfamily, more closely related to two genes discovered in saccharomyces cerevisiae: the YPT gene found serendipitously, and the SEC4 gene, involved in yeast secretion(2-10).

rab genes have been the focus of intense research for the last four years and many additional members have been characterized by several groups: the YPT cDNA of the mouse, which is capable of complementing that of yeast, confirmed that *rab1* (*rab1A*) was the homolog of the yeast YPT gene (20). G-proteins purified from bovine brain permitted the design of oligonucleotide that revealed additional members related to *rab3* (denoted smg 25A, B and C) (21). The search for human counterparts of the rat *rab* cDNAs in human cDNA libraries revealed the same four (rab1, 2, 3, 4) and additional rab3B,5 and 6 (22). BRL-*ras*, isolated from a rat liver cell line cDNA library, is referred to as *rab*7(23). An other clone, very close to *rab*1A was isolated from a rat brain cDNA library and denoted *rab*1B(24) . *sas*1 and *sas*2 were isolated from Dictyostelium discoideum (25). More recently, additional *rab*8, 9, 10 and 11 were characterized in canine cells (26). The fission yeast has three genes YPT1 , YPT2 (that may be functionally equivalent to SEC4 and *sas*1 and RHY, for which the human counterpart is *rab*6 (27-28).

At present there are some 25 members, characterized in mammals, that pertain to the *rab* branch of the family. This figure may represent only a fraction of the total number of the existing rab proteins, even though different approaches have often led to the rediscovery of previously known members. One may foretell that there are, perhaps, as many as 40 different genes in this *rab* family.

STRUCTURAL FEATURES OF THE PROTEINS OF THE *ras* SUPERFAMILY

The minimal homology observed, in terms of amino-acid identity, between any one member of one branch and an other member of an other branch is about 28-30%. The 30% identity is not scattered along the protein but rather clustered in the four regions that constitute the GTP/GDP binding site. These regions are around residues 10-17 (GXXXXGKS/T); 57-62 (DTAGQE); 115-120 (GNKXDL/M), and 143-147 (EXSAK/L). In addition some amino acids are almost always conserved at given positions such as lysine-5; phenylalanine-28; Threonine-35; tyrosine-71; phenylalanine-156, etc. Residues between lysine-5 and position 164 constitute the central core of the protein, very much alike in all the members, with very few insertions or deletions needed to align the proteins, except in those of the *rho* branch where there are insertional sequences of 12 amino acids between position 122 and 123. Upstream from position 5 and downstream from position 164, the sequences are termed N-terminal and C-terminal extensions respectively. They vary in size between individual members. However, the rho proteins have the shortest C-terminal extensions, whereas the rab proteins have the longest. There is always (at least) one cysteine at the C-terminus (or at the fourth position before the C-terminal end); the cysteine is needed for fatty acylation (farnesylation or geranylgeranylation) and subsequent anchoring to the plasma or internal membranes. It is most probable, also, that the C-terminal part, from residue 165 to the end, constitutes a region that interacts with specific membrane sites (29).

An other important region of the protein resides between amino acid 32 and 40, known as the effector region. This region was found essential in active transforming *ras* since it was the only one (besides the C-terminal cysteine) that could not be deleted nor mutated without loss of transforming activity. It was found that this effector region was the one interacting with "GAP," the GTPase activating protein that might be considered as a target or, rather, as a component of a target complex. An other type of protein, that at present has only been found and described as rho and rab protein is the GDI (for: GDP-Dissociation Inhibitor) (30).

POSSIBLE FUNCTIONS OF THE PROTEINS OF THE *ras* FAMILY

Being submerged by this abundance, this surfeit, of proteins of the *ras* superfamily, many questions arise. The oncologists, considering that the *ras* proto-oncogenes are potential transforming genes, would wonder whether some other small molecular weight G-proteins of the *ras* superfamily might be putative oncogenes or, perhaps, tumor suppressor genes. It would also be interesting to know the subcellular localization of the ras proteins. Moreover, an obvious question is: are the small molecular weight G-proteins ubiquitously expressed in all cell types or, are there some qualitative and/or quantitative differences in diverse differentiated cells? Is there a specific phenotype that correlates with the expression, or the lack of expression of a given protein of the ras superfamily? I will try to answer these questions while reviewing each branch of the superfamily and listing briefly their members.

THE *ras* BRANCH OF THE FAMILY

In addition to the three "leaders": H-, K-, and N-*ras*, several *ras*-like genes were discovered, *ral*A and B, *rap*1A and B, *rap*2A and B, R-*ras*, TC21, etc... The central core, from amino-acid-5 to amino-acid 164, is very well conserved in all the proteins with no insertion nor deletion. Only conservative amino acid substitutions are found in the important structural domains. The N-terminal extensions vary from 4 for *ras*, to 30 for R-*ras* in which this extension represents an additional domain very rich in glycine and proline. C-terminal extensions vary from 18 in rap2, to 30 in ralA, and always terminate with the CAAX motif for isoprenylation, clipping and carboxy-methylation.

Is any member of this branch, a putative oncogene or antioncogene? It seems logical to ask this question here, for members of this branch since they share more homology with the *ras* proto-oncogene proteins, than do the other members of the *rho* or the *rab* branches.

Experiments were done, for instance, where the *ral*A cDNA was mutagenized in order to gain the most potent mutation at the critical positions like valine-12 and/or arginine-61. The same kinds of experiments were performed more recently with activated *rap*1A and *rap*2A cDNAs. Tranfections with these mutated genes on recipient NIH-3T3 cells always yielded negative results (unpublished: P. Chardin, V. Pizon, J. de Gunzburg, in our laboratory).

Can they act as tumor suppressor genes? The discovery of the K-*rev*1 gene (31), capable of reverting the phenotype of cells transformed by K-*ras* was quite striking. It was the same gene that was found before (*rap*1A), as a human homolog of Drosophila D-*ras*3 gene(12). In our laboratory, homologs were searched in human cDNA libraries because D-*ras*3 was the only intriguing exception that had no glutamine, but a threonine instead, at the critical activating position 61.

The antagonistic effect of *rap*1A/K-*rev* on ras transformed cells, however, is not complete, nor well understood either, since it was clearly shown that *rap*1a is not located at the plasma membrane but in the Golgi (32). It was also shown, meanwhile, that *rap*1 GTPase was not activated by *ras*-GAP, even though *rap*1A, in the GTP-bound form, was capable of interacting with *ras*-GAP and inhibiting GAP activity on *ras*-GTP *in vitro* (33).

Interacting molecules - GAP - exchange factor - GDI

The only proteins clearly and undoubtedly interacting with *ras* are the ras-GAP proteins and the protein encoded by the neurofibromatosis type 1 locus (34). Several groups have shown that the ras-GAP proteins interact with region 32-40 of *ras*, but it was also shown that a tryptophan substitution in position 64 did not impair the interaction with GAP, but abolished the ability for GAP to activate GTP hydrolysis. This suggests that the region 60-68 interacts also with ras-GAP and might explain why rap1A interferes with ras-GAP without stimulation of its GTP hydrolysis activity. We know also that rap1 has a specific GAP (35). Besides, I. Lerosey has found a specific GAP active for rap2, different from GAP-rap1 (unpublished). Several groups have also described *exchange factor activities* for ras and rap1B, that have not been fully characterized, as yet, but it is probably difficult to isolate such factors which might acquire their full efficiency in response to upstream signals that are unknown. No GDI has been characterized yet for the *ras* branch.

RHO AND RHO-LIKE PROTEINS

The first three *rho* genes characterized in mammals were *rho*A, B and C (15-16). In addition an other subgroup has been recently characterized comprising the *rac*1 and *rac*2 genes. Finally, two other genes have been characterized. One codes for two closely related forms of G25K which are the mammalian homologs of the Yeast CDC42 proteins. The TC10 gene was obtained from teratocarcinoma cells. Proteins of the *rho* branch are distinct. They are characterized by a clear difference in the central G-domain which consists in an insertion of 12 residues between position 122 and 123 (numbering of ras). N-terminal extensions vary from 4 to 18 residues and the C-terminals are the shortest of the superfamily varying from 14 to 22 residues (in TC10). These small C-extensions compensate for the G-domain insertions, so that the molecular weights of these proteins are close to 21-22 Kd (with the exception of TC10); like ras, rho proteins always terminate with the consensus CAAX sequences.

Rho proteins are substrates for ADP ribosylation by exo enzyme C3 of Clostridium Botulinium strains C and D (36). The ADP ribosylation modifies asparagine 41 of rho, corresponding to serine 39 of ras. This ADP ribosylation seems to impair the normal function of rho but does not impair GTP hydrolysis after interaction with rho-GAP. This is surprising since position 39 is believed to be an important site for interaction of ras with ras-GAP.

As mentioned, a rho-GAP has been described that seems mainly cytosolic and has a M.W. of 30 KDa. A rho-GDI has also been purified and sequenced by Takai's group (37).

The cellular localization

The proteins are mainly localized in the Golgi or in vesicles close to the Golgi. A significant fraction is also found in the cytosol and the proportion seems to depend on the cell type. Other proteins of this family are also found associated with membrane but their precise localization have not been defined.

Possible function of the rho protein

Micro injection of activated rho induces dramatic changes in cell morphology. That was shown by Alan Hall and colleagues. Also, when introduced into cells, the C3 exoenzyme induces cell rounding and disorganization of actin microfilaments, probably by inactivating *rho*. This suggests an interaction of rho with the organization of actin (38).

Potential involvement in transformation ?

Overexpression of *rho*A in fibroblasts results in a partially transformed phenotype and reduces the cell dependance on serum. It seems also that cells overexpressing rhoA are tumorigenic in nude mice. Although it seems premature to draw any conclusions and assert that rho is a classical oncogene, the overexpression of rho is clearly able to induce a partially transformed phenotype.

Recently, it was shown that *rho*B was found to be an early gene induced by stimulation of resting cells to reenter the cell cycle by growth factors such as EGF or tyrosine kinases such as v-fps(39).

RAB AND RAB-LIKE PROTEINS

This branch has become now the largest of the family (40). It is also the most heterogenous. It comprises more than two dozen proteins named rab1A and B, rab2, rab3A, B and C (also named smg p25) rab4A and B, rab5A and B, rab6, rab7 previously named BRL-ras by Bucci et al, rab8, 9, 10 and 11, etc... and TC4 (5-19-22-23-24-40). The TC4 is placed provisionally in this subfamily, but it might represent, a fourth branch of the superfamily, since it is the only known member lacking a terminal cysteine (19).

General features

As for the ras or rho proteins, the central GTP-GDP binding domain is very well conserved; in contrast, N-terminal extensions vary from 3 residues in rab7 to 23 residues in rab3. The rab contrasts with the rho or ras branches. The C-terminal extensions are larger than in the ras branch and much larger than in the rho branch and comprise from 31 residues in rab1B up to 47 in rab2. The consensus CAAX sequences are not found in most of the rab proteins except for rab5, 8 and 10. However, one cysteine, at least (except in TC4) is always present close to, or at the C- terminus. The significance of these differences at the C-termini are not known. They might confer specificity for post translational modifications and targeting to specific membranes. Moreover, there seems to exist a selection pressure since some C-terminal GGGCC sequences, for instance, are identical in YPT, rab1A, rab1B and rab2.

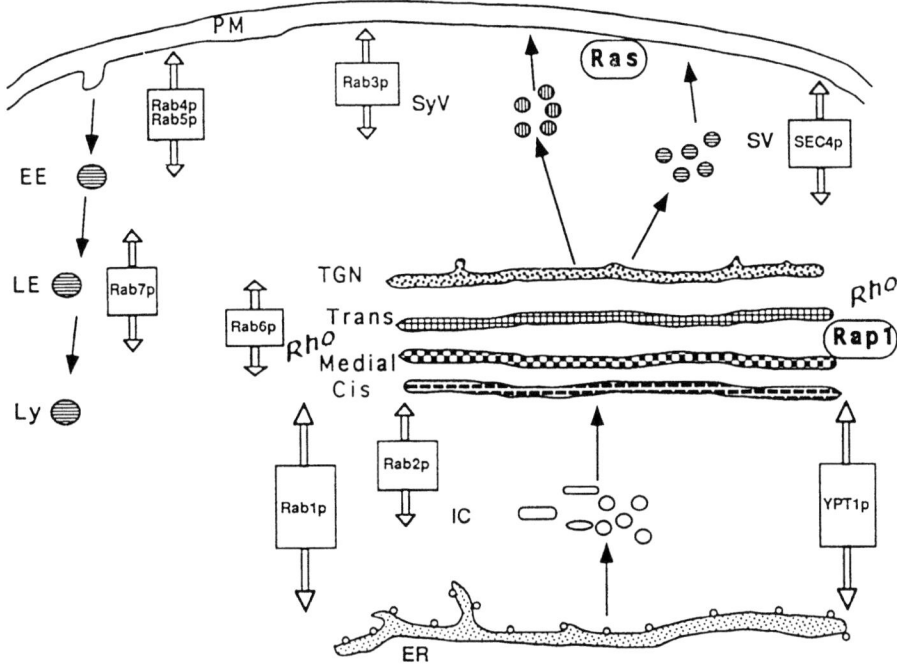

Fig. 1. The figure schematically represents the subcellular localization of the Rab/YPT1/SEC4 protein family. A scheme of different exocytic and endocytic compartments is shown. The figure displays the distribution of the Rab/YPT1/SEC4 proteins along the secretory and endocytic pathways. ER: endoplasmic reticulum; IC: Intermediate Compartment; Cis-Medial and Trans Golgi; TGN: Trans Golgi Network; SV: Secretory Vesicles; SyV: Synaptic Vesicles; PM: Plasma Membrane; EE: Early Endosome; LE: Late Endosome and Ly: Lysosome. In addition, Rho and Rap1 proteins are also tentatively localized.

Interacting proteins GAP, Exchange Factor, GDI

GAP activities are not well defined although they have been reported for SEC4, mammalian rab3 and YPT1. Presently, a cytosolic exchange factor has been reported for rab3. A GDI activity has also been reported for rab3A (also called smg p25) by the group of Takai.

Cellular localization

Rab proteins have been reported in many different compartments in the cell. Most are implicated in vesicular traffic as shown in the schematic figure .

Cellular functions

Because yeast is more amenable to genetic studies than mammalian cells, the real true and direct genetic evidence of implication in secretion of the rab proteins is given by the yeast genes YPT and SEC4. At present a similar role for the rab proteins in mammalian cells comes from indirect experiments with synthetic peptides, for instance, of the rab effector region that inhibit transport from the Endoplasmic Reticulum to the Golgi and intra Golgi compartments(41). Besides, a large body of indirect but convergent evidence implicate rab proteins in vesicular traffic in mammalian cells. Non hydrolyzable GTPgS blocks the fusion of vesicles at several different steps of the secretory pathway(42). Also the diversity of the rab members and the different localization of the rab proteins to various compartments are quite compatible with roles in membrane traffic(43-44-45-46). One might also predict that, very soon, more direct evidence will be obtained by the use of specific antisera against specific rab proteins in semi-intact cells systems.

Ubiquitous or non ubiquitous expression ?

Already in 1987-1988, when it became evident that the *ras* genes constituted a large family, a question was asked: are all these genes expressed evenly everywhere? At that time we didn't have the recombinant protein, nor, a fortiori, the specific antibodies. Therefore, we devised a very coarse experiment where we used cDNAs as probes and polyA RNA from organs in Northern blots. We know that an organ consists of many different cells types, but even coarse as it was, we could see that the G-proteins were not distributed evenly. Some were more ubiquitous than others and a few were not ubiquitous at all. The remarkable case was that of rab3A that was almost exclusively expressed in the brain (47), precisely in neurones (48-49) and even more precisely in certain types of neurones, as can be seen, now by the use of *in situ* hybridization techniques and refined immunocytochemistry with the use of specific antibodies or/and antipeptides.

CONCLUSIONS: ARE THE *ras*-LIKE GENES INVOLVED IN NEOPLASIA?

At present, none of the other members of the large *ras* superfamily, besides the ras proteins proper, have been shown to be activated nor implicated directly in human cancers. The mutated activated cDNAs are unable to transform recipient cells. However, even though there remains much uncertainty as to the exact function of each of these ras-related small G-proteins, one may foretell that they are very probably involved in a wide variety of essential regulatory processes in which they may act as biological switches. It already appears that the expression of some of them are deregulated (either overexpressed or quenched) in some types of human cancers (50). Discrepancies in expression have been observed for rap1 in some lymphosarcomas, for rhoB in certain breast tumors, and rab2 in Sezary syndromes; However, up to now, no correlation has been established with the pathological observations and with other markers. Whether these phenomena are the consequences or the cause of neoplastic transformation remains to be determined. Refined tools, such as very specific antibodies for the characterization of the proteins, identification of the interacting gene products for advanced mammalian cell biology and biochemistry, combined with genetic studies in yeast, should help elucidate the functions and roles of the small guanine nucleotide binding proteins, and/or their interacting molecules, and ascertain their involvement in cancers and other genetic disorders.

REFERENCES

(1) Barbacid, M. (1987), *Ras* genes, Ann. Rev. Biochem., 56, 779-827
(2) Gallwitz, D., Donath, C. and Sander, C. A yeast gene encoding a protein homologous to the c-*ras*/bas proto-oncogene product. Nature, 306: 704-707
(3) Madaule, P., and Axel, R. A novel *ras*-related gene family. Cell, 41: 31-40, 1985
(4) Chardin, P., and Tavitian, A. The *ral* gene: a new *ras* related gene isolated by the use of a synthetic probe. EMBO J., 5: 2203-2208, 1986
(5) Touchot N, Chardin P, Tavitian A: Four additional members of the *ras* gene superfamily isolated by an oligonucleotide strategy: molecular cloning of YPT-related cDNAs from a rat brain library. Proc Natl Acad Sci USA, 84 : 8210, 1987
(6) Lowe, D.G., Capon, D.J., Delwart, E., Sakaguchi, A.Y., Naylor, S.L., and Goeddel, D.V. Structure of the human and murine R-*ras* genes, novel genes closely related to ras proto-oncogenes. Cell, 48: 137-146, 1987
(7) Swanson, M., Elste, A., Greenberg, S., Schartz, J., Aldrich, T. and Furth, M., J. Cell. Biol. 103: 485-492, 1986
(8) Schejter, E.D. and Shilo, B.Z. Characterization of functional domains of p21 ras by use of chimeric genes. EMBO J. 4: 407-412. 1986
(9) Hengst, L., Lehmeier, T. and Gallwitz, D. The *ryh*1 gene in the fission yeast encoding a GTP-binding protein related to *ras*, *rho* and *ypt*: structure, expression and identification of its human homolog. EMBO J. 9: 1949-1955, 1990
(10) Salminen A, Novick, PJ: A ras-like protein is required for a post-Golgi event in yeast secretion. Cell 49: 527, 1987
(11) Chardin, P., and Tavitian, A. Coding sequences of human *ral*A and *ral*B cDNAs. Nucl. Acids. Res., 17: 4380, 1989
(12) Pizon V, Chardin P, Lerosey I, Olofsson B, Tavitian A: Human cDNAs *rap*1 and *rap*2 homologous to the Drosophila gene D*ras*3 encode proteins closely related to ras in the "effector" region. Oncogene 3: 201, 1988
(13) V. Pizon, I. Lerosey, P. Chardin and A. Tavitian. Nucleotide sequence of human cDNA encoding a ras-related protein (rap1B). Nucleic Acids Res., 16, 7719. 1988
(14) Lapetina, E.G., Lacal, J.C., Reep, B.R. and Vedia, L.M. A ras-related protein is phosphorylated and translocated by agonists that increase cAMP levels in human platelets. Proc. Natl. Acad. Sci. U.S. 86: 3131-3135, 1989

(15) Yeramian, P., Chardin, P., Madaule P., and Tavitian, A. Nucleotide sequence of human *rho* cDNA clone 12. Nucl. Acids Res. 4, 1869. 1987
(16) Chardin, P., Madaule P. and A. Tavitian. Coding sequence of human *rho* cDNAs clone 6 and clone 9. Nucl. Acids Res., 16, 2717. 1988
(17) Didsbury J, Weber RF, Bokoch GM, Evans T, Snyderman R: rac, a novel ras-related family of proteins that are botulinum toxin substrates. J Biol Chem 264: 16378, 1989
(18) Polakis, P. G., Snyderman, R. and Evans, T. Biochem. Biophys. Res. Com. 160: 25-32; 1989
(19) Drivas, G.T., Shih, A., Coutavas, E., Rush, M.G. and D'Eustachio, P. Mol. Cell. Biol., 10, 1793-1798
(20) Haubruck, H., Disela, C., Wagner, P., and Gallwitz, D. The Ras-related YPT protein is an ubiquitous eukaryotic protein: isolation and sequence analysis of mouse cDNA clones highly homologous to the yeast YPT1 gene. EMBO J., 6: 4049- 4053, 1987
(21) Matsui, Y., A. Kikuchi, J. Kondo, T. Hishida, Y. Teranishi, and Y. Takai. Nucleotide and deduced amino acid sequences of a GTP-binding protein family with molecular weight of 25.000 from bovine brain. J. Biol. Chem. 263: 11071-11074. 1988
(22) Zahraoui A, Touchot N, Chardin P, Tavitian A. The human *rab* genes encode a family of GTP-binding proteins related to yeast YPT1 and SEC4 products involved in secretion. J Biol Chem 264: 12396, 1989
(23) Bucci, C., R. Frunzio, L., Chiariotti, A.L. Brown, M. M. Rechler, and C.B. Bruni. A new member of the *ras* gene superfamily identified in a rat liver cell line. Nucleic Acids Res. 16: 9979-9993, 1988
(24) Vielh, E., N. Touchot, A. Zahraoui and A. Tavitian. Nucleotide sequence of a rat cDNA: *rab*1B encoding a Rab1-YPT related protein. Nucl. Acid. Res., 17: 1770, 1989
(25) Saxe, S.A., and A.R. Kimmel. Genes encoding novel GTP binding proteins in Dictyostelium. Dev. Genet. 9, 259- 265, 1988
(26) Chavrier, P., Parton, R.G., Hauri, H.P., Simons, K., Zerial. M. Localization of low molecular weight GTP binding proteins to exocytic and endocytic compartments. Cell 62: 317, 1990
(27) Fawel, E., S. Hook, D. Sweet, and J. Armastrong. 1990. Novel YPT1-related genes from *Schizosaccharomyces pombe*. Nucleic Acids Res.. 18: 4264
(28) Miyake, S., and M. Yamamoto. Identification of *ras* related, YPT family genes in *Schizosaccharomyces pombe*. EMBO J. 9: 1417-1422. 1990
(29) Valencia, A., Chardin, P., Wittinghofer, A., and Sander, C. The ras protein family: evolutionary tree and role of conserved amino acids. Biochemistry, 3a: 4637-4648, 1991
(30) Chardin, P. Small GTP-binding proteins of the ras family: a conserved functional mechanism? Cancer cells, 3, 117-126, 1991
(31) Kitayama, H., Sugimoto, Y., Matsuzaki, T., Ikawa, Y. and Noda, M. A *ras*-related gene with transforming suppressor activity; Cell, 56: 77-84; 1989
(32) Beranger, F., Goud, B,. Tavitian, A, and de Gunzburg, J. Association of the Ras-antagonistic Rap1/Krev-1 proteins with the Golgi complex. Proc. Natl. Acad. Sci. USA (1991) 88, 1606-1610
(33) Frech, M., John, J., Pizon, V., Chardin, P., Tavitian, A., Clark, R., McCormick, F. and Wittinghofer, A. Inhibition of GTPase activating protein stimulation of

ras-p21 GTPase by the K-*rev*-1 gene product. Science, 1990, 249, 169-171
(34) Martin, G.A., D. Viskochil, G. Bollag, P.C. McCabe, W.J. Crosier, J.H. Haubruck, L. Conroy, R. Clark, L. O'Gonneli, R.M. Gewthon, M. A. Innis, and F. McCormick. The GAP-related domain of the neurofibromatosis type 1 gene product interacts with ras p21. Cell 63, 843-849. 1990
(35) Kikuchi, A., T. Sasaki, S. Araki, Y. Hata, and Y. Takai. Purification and characterization from bovine brain cytosol of two GTPase activating proteins specific for smg p21, a GTP binding protein having the same effector domain as c-ras p21s. J. Biol. Chem. 264, 9133-9136. 1989
(36) Chardin P., P. Boquet, P. Madaule, M.R. Popoff, E.J. Rubin, and D.M. Gill. The mammalian G protein rhoC is ADP-ribosylated by Clostridium botulinum exoenzyme C3 and affects actin microfilaments in Vero cells. EMBO J. 8: 1087-1092. 1989
(37) Ueda, A. Kikuchi, and Y. Takai. Molecular cloning and characterization of a novel type of regulatory protein (GDI) for the rho proteins, ras p21-like small GTP-binding proteins. Oncogene 5: 1321-1328. 1990
(38) Hall, A., The cellular functions of small GTP-binding proteins; Science, 249: 635-640; 1990
(39) Jahner, D. and T. Hunter. The *ras*-related gene *rho*B is an immediate early gene inducible by v-fps, epidermal growth factor and platelet-derived growth factor in rat fibroblasts. Mol. Cell. Biol. 11: 3682-3690; 1991
(40) Chavrier P., M. Vingron, C. Sander, K. Simons and M. Zerial. Molecular cloning of YPT1/SEC4-related cDNAs from an epithelial cell line. Mol. Cell. Biol. 10: 6578-6585. 1990
(41) Plutner, H., R. Schwaninger, S. Pind, and W.E. Balch. Synthetic peptides of the Rab effector domain inhibit vesicular transport through the secretory pathway. EMBO J., g: 2375-2383, 1990
(42) Melançon, P., B.S. Glick, V. Malhotra, P.J. Weidman, T. Serafini, M.L. Gleason, L. Orci and J.E. Rothman. Involvement of GTP-binding "G" proteins in transport through the Golgi stack. Cell, 51: 1053-1062. 1987
(43) Goud B, Zahraoui A, Tavitian A, Saraste J: Small GTP-binding protein associated with Golgi cisternae. Nature 345: 553, 1990
(44) Darchen, F., Zahraoui, A., Hammel, F., Monteils, M-P., Tavitian, A., and Scherman, D. Association of the GTP-binding protein Rab3A with bovine adrenal chromaffin granules. Proc. Natl. Acad. Sci. USA, 87: 5692-5696, 1990
(45) Fisher v. Mollard, G., Mignery, G.A., Baumert, M., Perin, M.S., Hanson, T.J., Burger, P.M., Jahn, R., and Sudhof, T.C. Rab3 is a small GTP-binding protein exclusively localized to synaptic vesicles. Proc. Natl. Acad. Sci. USA, 87: 1988-1992, 1990
(46) Van der Sluijs P, Hull M, Zahraoui A, Tavitian A, Goud B, Mellman I: The small GTP-binding protein Rab4 is associated with early endosomes. Proc Natl Acad Sci USA 88: 6313, 1991
(47) Olofsson, B., Chardin, P., Touchot, N., Zahraoui, A., and Tavitian, A. Expression of the *ras*-related *ral*A, *rho*12 and *rab* genes in adult mouse tissues. Oncogene, 3: 231-234, 1988
(48) Sano, K., Kikuchi, A., Matsui, Y., Teranishi, Y., and Takai, Y. Tissue-specific expression of a novel GTP-binding protein (smg p25A) mRNA and its increase by nerve growth factor and cyclic AMP in rat pheochromocytoma PC-12 cells. Biochem. Biophys. Res. Commun., 158: 377-385, 1989

(49) Ayala, J., Olofsson, B., Touchot, N., Zahraoui, A., Tavitian, A., and Prochiantz, A. Developmental and regional expression of three new members of the *ras*-gene family in the mouse brain. J. Neuroscience Res., 22: 384-389, 1989
(50) Culine, S., Olofsson, B., Gosselin, S., Honore, N., and Tavitian, A. Expression of the *ras*-related *rap* genes in human tumors. Int. J. Cancer, 44: 990-994, 1989

ONCOGENIC TRANSGENIC MICE IN THE STUDY OF CARCINOGENESIS

Chang-Ho Ahn[1] and Won-Chul Choi[2]

Division of Oncology and Pulmonary Drug Products[1]
and Division of Research and Testing[2]
Center for Drug Evaluation and Research
Food and Drug Administration, Rockville, Maryland 20857, U.S.A.

INTRODUCTION

Many types of chemical carcinogens induce tumors in animals and humans. The study of chemical carcinogenesis consists of two elements: one is to study the mechanism of oncogenesis, and the other to examine the carcinogenic potential of chemicals. The research on carcinogenic mechanisms explores a multistep series of events of carcinogenesis, environmental and endogenous factors, and various genes and transcriptional elements (1,2). Research on the carcinogenic potential of chemicals aims to detect and estimate the risk to humans in a quantitative manner. For these studies, various animal models have been developed and used. Furthermore, rapid development of the field of molecular biology during the last decade has led to the findings of genes responsible for cancers and to the construction of transgenic animals by transferring genes between species (3-5). Transgenic animals which have received oncogenes have been researched for the study of oncogenesis (see reviews 6,7). So far, in spite of their great potential, utilization of transgenic animals for carcinogenic risk assessments has been extremely rare. The purpose of this review is to examine the various aspect of transgenic mice carrying oncogenes and to explore their possible utilization of transgenic mice for the study of the carcinogenic potential of chemicals.

MULTISTEP-CHEMICAL CARCINOGENESIS

Tumors are induced by chemical carcinogens through a multiple step process involving initiation, promotion, and progression (8,9). The initiation is caused through mutations by genotoxic chemicals, steering cells to immortalization, whereas promotion is induced by various agents which release cells from growth control (Fig.1). Progression is a step which changes a benign tumor to a malignant invasive neoplasm, or a highly

MULTISTEP ONCOGENESIS

Normal Cell

↓   INITIATION: mutagenesis by genotoxic chemicals
oncogene activation/suppressor gene deletion

Initiated Cell: immortalization of cells

↓   PROMOTION: tumor promoting chemicals/oncogenes

Tumor Cell: altered growth control of cells

↓   PROGRESSION: progression/angiogenic factors
oncogene/growth factor activation
laminin, fibronectin, collagenase IV

Metastasis: invasive cells

Fig. 1. Simplified model of multistep carcinogenesis: Tumor initiation may be caused by incomplete carcinogen; promotion by chemicals competent to cause the reversible expansion of the initiated cells; and progression by chemical competent to cause the promoted cells to a malignant cells.

---

differentiated cancer to an anaplastic undifferentiated cancer (10). Both initiation and progression are irreversible, but promotion is a reversible stage. Oncogenes are genes important in neoplastic development. Oncogenes are identified in mammalian cells through homology with retroviral oncogenes, through transfection assays, and through their involvement in nonrandom chromosome alterations. In many cases, oncogene products are characterized as tyrosine kinases, growth factors, their receptors, or nucleoproteins. Although at least 60 protooncogenes have been identified in the human genome, the majority of these protooncogenes have not been implicated in human cancers and only about a dozen of them have been clearly demonstrated to be involved in human cancer (e.g., *ras* genes, *myc* genes, c-*erb*B genes, c-*abl* and c-*ret*, etc). A group of nuclear oncogenes (e.g., c-*myc* and N-*myc*) is able to lead to immortalization of normal cells, while a group of cytoplasmic oncogenes (e.g., *sis* and *ras* genes) causes altered control of cell growth (11,12). However, certain type of oncogenes, for example, c-*fos*, although nuclear, are able to produce both cell immortalization and altered growth control (13).

Genetic events important to the development of cancer involve activation of protooncogenes or inactivation/deletion of tumor suppressor genes. Oncogene activation is caused by:

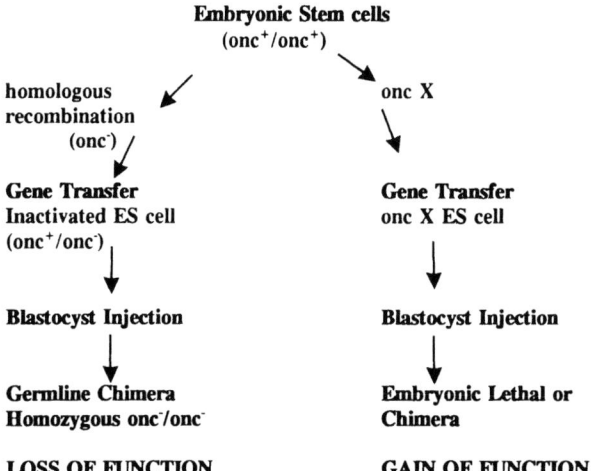

Fig. 2. Embryonic stem cell model for construction of transgenic mice: Expression of oncogenes and tumor suppressor genes may result in gain of function and loss of function, respectively.

1) Insertional/point mutations. The resultant tumors are polyclonal since they comprise multiple independently affected cells. In ras gene family, mutations of all three *ras* genes have been found in many human cancers: mutation of H-*ras* gene in bladder and urinary carcinomas (14,15), of K-*ras* gene in lung, colorectal, and pancreatic carcinomas (16-18), and of N-*ras* gene in melanomas and hematopoietic cancers (19,20). Codon 12 and 61 of ras proteins are two principal regions where point mutation can cause the activation of *ras* oncogene (14,21). In the mouse skin model, in the absence of tumor initiator, retroviral transduction of *ras* oncogene followed by subsequent administration of a tumor promoter (e.g., phorbol esters) causes papillomas, suggesting that the *ras* mutation may be a consequence of the initiation event.;

2) Chromosomal abnormalities. Protooncogenes are activated by their proximity to the breakpoints of chromosomal aberrations except in cases of gene deletions. Translocations with the same chromosomal breakpoints occur in a particular type of malignancy. In Burkitt's lymphoma or chronic myeloid leukemia (CML), the specific translocation is present in most of the tumor samples analyzed and involves the c-*myc* and c-*abl* protooncogenes, respectively (22-24). Chromosome aberrations are examined two ways: by cytogenetic and DNA studies. Cytogenetic studies are crucial in identifying abnormal chromosomes and DNA level studies define the mechanisms of chromosomal abnormalities and of gene function and regulation; and

3) Gene amplification. Some tumors contain amplified copies of genes such as N-*myc* and c-*erb*B2 in neuroblastoma and human breast cancer, respectively (25,26). It is likely that amplification of these genes may provide a greater probability of mutant variant generation.

Inactivation of tumor suppressor genes leads to the loss of function of the involved genes and, similar to oncogene activation, is caused by: 1) insertional mutations; 2) chromosomal loss; 3) homologous recombinations; and 4) gene conversions (27).

GENERATION OF TRANSGENIC MOUSE FOR ONCOGENESIS STUDY

Identification and characterization of oncogenes or protooncogenes has utilized the tissue culture system and are limited by the system itself. However, over the past few years, a new, potent approach has been to construct genetic models of human diseases in animals by a stable integration of exogenous genetic information (human) into the animal genome (28,29). Using these transgenic animal models, many complex issues on carcinogenesis such as tissue-specific expression of oncogenes for tumor growth and identification of oncogenes in specific cell types can be investigated in vivo (30,31). Among many transgenic species, the mouse has been the choice animal and has been extensively used for carcinogenesis studies, since mice can provide a fairly rapid assay system of genetic alterations due to short gestation period, large litter size, easy maintenance, and many mutant strains. A transgenic mouse is defined as a mouse carrying an exogenously introduced gene(s) in some of its tissues and being capable of transmitting the transgene(s) to some proportion of its offspring.

Transgenic mice can be derived by three different methods of introduction of genes into the mouse germline: 1) direct microinjection of DNA into male pronucleus of the one-cell embryo; 2) retroviral infection of preimplantation or postimplantation embryos; and 3) injection of embryonic stem cells into the blastocyst (fig. 2). All methods, except retroviral infection of postimplantation embryos, employ in vitro culture of the embryo prior to introduction into a pseudo-pregnant female mice. In these methods, the one-cell embryos are reintroduced into the fallopian tube, whereas the four- or eight-cell embryos are cultured to blastocyst stage and reintroduced into the uterus. The retroviral infection of postimplantation embryos is performed in utero. Differences in transgenics methodology are described below.

1. Microinjection of DNA into one of male pronuclei of the embryo: The first transgenic mice were produced in 1974 by microinjecting SV40 viral DNA into the blastocoel cavity of early embryos (30). More stable DNA integration into the mouse genome was demonstrated through direct DNA injection into a pronucleus of the fertilized egg and this has become the most widely used method (29). The direct microinjection method is advantageous, since there is no constraint on the size of the DNA to be introduced. The DNA is usually integrated at a random site prior to cleavage. Thus, all cells of the embryo and extraembryonic tissues carry the DNA, resulting in transmission of the DNA to half of the founder's offspring. Usually, with this method, multiple copies integrate into the mouse genome arranged in tandem head-to-tail arrays containing up to several hundred copies, presumably due to homologous recombination between the injected molecules (32). The microinjection

Table 1. Hybrid Oncogenes and Transgenic Tumor Expression

ONCOGENE	PROMOTER/ENHANCER	TUMOR	LATENCY
c-H-ras	WAP	salivary, mammary	long
c-H-ras	WAP	none	
c-H-ras	Ig/Tp	lung	4-wk/10-Mo
c-H-ras	n/a	salivary, mammary	
c-H-ras	suprabasal keratin 10	papillomas	
c-H-ras	albumin promoter-SV40	hepatoma	
c-H-ras	albumin gene	lung	
v-H-ras	MMTV LTR	salivary, mammary, lymphoid	
v-H-ras	ζ-globin gene	papillomas	
v-H-ras	MMTR LTR	mammary, lung	
N-ras	MMTV LTR	salivary, mammary	
c-myc	MMTV LTR	mammary	pregnancy
c-myc	MMTV LTR	testis, breast, lymphoma	
c-myc	WAP	none	
c-myc	Ig/Tp	pre-B cell lymphoma	
c-myc	Eμ IgH	lymphoma	
c-myc	Eμ IgH	B cell lymphoma	
c-myc	Eμ IgH	pre-B cell lymphoma	
c-myc	Eμ IgH	B cell lymphoma	
c-myc	hepato-specific DNA	hapatoma	
c-myc	n/a	pancreas(acinar)	2-7 months
c-myc	albumin gene	hepatic adenoma	15 months
N-myc	Eμ IgH	B cell lymphoma	
c-H-ras/c-myc	WAP	mammary	3-4 months
c-erbB2	MMTV LTR	mammary, B-lymphoma	late
c-erbB2	Ig/Tp	pre-B cell lymphoma	6-10 months
c-neu	MMTV LTR	mammary	5-10 months
c-neu	MMTV LTR	mammary	
pim-1		lymphoma	34-week
E1A/E1B	MMTV LTR	neuroblastoma	6 months
ret	MMTV LTR	salivary	

method is efficient. About 25% of the offspring carry the transgene. However, the method requires somewhat expensive equipment and technical expertise.

2. Retroviral infection of preimplantation embryos. Using the Moloney murine leukemia virus (M-MuLV), the second transgenic mice were produced by infecting preimplantation embryos with the retrovirus in 1976 (31). Although the embryos are not permissive for M-MuLV expression, the virus can integrate into the host germ line. By this method, only single copy of the viral DNA is integrated and integration

occurs at a specific site within the viral DNA with little rearrangement of the flanking sequence. The problem with the proviral long terminal repeat (LTR) as a promoter in embryonic cells can be circumvented by adding an internal promoter to the recombinant retrovirus construct (33,34). The retroviral infection procedure is relatively simple and requires less expensive and advanced equipment than the first method. There are some disadvantages with this method:

i) The viral infection is usually initiated at later embryonic stages, resulting in mosaic founder mice; and

ii) More than one cell may be infected, and thus mice with multiple different integrants are generated, resulting in a difficulty to establish pure lines that are hemizygous for a single insertion site and requirement of outbreeding of the founder mice for the pure line.

In retroviral infection of postimplantation embryos, the transgene can be introduced through retroviral infection at 8.5 days post-coitus into postimplantation embryos which are at a high mitotic rate stage (38,39). The efficiency of this procedure is very low, since only a small number of cases have succeeded to establish the transgenic mouse strains. This method may be used when an oncogene is proven to be toxic if expressed in every cell.

3) Injection of embryonic stem (ES) cells into blastocyst. Embryonic stem cells are derived from the inner cell mass of explanted blastocysts and can grow in culture (35-37). The genes are introduced into these ES cells in culture by microinjection, retroviral vector infection, or transfection. Then, ES cells are transferred back into early embryos by aggregation with eight- or sixteen-cell morula-stage embryos or by microinjection into blastocysts, at 4 days post-coitus, to produce chimeric transgenic mice. This method permits clonal isolation of the transgenic ES cell and analysis of the structure of the transgene prior to generating the transgenic mice, thus, allowing prescreening of the integration site without necessitating the generation of a transgenic mouse.

In contrast to correlation between the expression of an inserted oncogene with aberrant development or tumor in the transgenic mice, genetic cancers such as Wilm's tumor and familial retinoblastoma are caused by a recessive mutations of suppressor genes requiring homozygosity in order for malignancy to occur (40-43). It is extremely difficult to generate a model for the inactivation of suppressor genes. The effect of tumor suppressor genes in vivo is ideally analyzed by functionally deleting the oncogene/proto-oncogene being studied. Two methods have been utilized to develop mutant mouse strains: one using random insertional mutagenesis and the other using site-directed homologous recombination. The former method has successfully introduced mutations into the germline of mice using M-MuLV as an insertional mutagen (44). The generation of transgenic mice with embryonic stem cells by random insertional mutagenesis is more advantageous over microinjection of DNA, since the inactivated suppressor gene can readily be cloned and analyzed using the proviral sequences as a molecular probe (45). However, if the gene is not expressed in the embryonic stem cells, or its function is unknown, then mutant selection on the stem cell line is very difficult. The latter method, homologous recombination into a genomic sequence of cells, has been used to generate specific mutations in embryonic stem cells, thereby permitting the subsequent generation of mutant strains of mice (Fig. 2)(46).

EXPRESSION OF ONCOGENES

Following transgene introduction into the mouse, multiple factors affect the expression of the transgene: DNA methylation, integration site, and the regulatory elements employed.

1) DNA methylation: DNA methylation has been well correlated with the inhibition of transcription of a diverse lot of genes, while demethylation is suggested to be a necessary but not sufficient prerequisite for gene expression (47,48). Infection of ES cells in early studies appeared to be non permissive for the replication of many retroviruses. Treatment of ES cells with 5-azacytidine alters DNA methylation pattern and results in extensive DNA demethylation and transcriptional activation of the unmethylated sequences (49). When the transgenes are transferred by direct microinjection into male pronuclei or by infection of eight- or sixteen-cell embryos with retroviruses, the genes are frequently methylated by enzymes in the embryo (50). Sometimes, the resulting transgenic mice do not express the transgene. Frequently, they are quiescent for the expression of the provirus with a well-defined stochastic activation of the proviral sequences later in life. In contrast to viral promoters, the viral vector with a cellular promoter is frequently expressed, possibly due to the demethylation of the transgene in the tissues where the cellular promoter is expressed (50,51).

2) Site of integration: The site of integration is related to the level of expression of the transgene and may be responsible for the low level of transgene expression (53). The degree of site integration on expression of the transgene appears to depend upon the method of transgene generation and the sequence used. Transgenes introduced by direct DNA microinjection, where multiple tandem genes are integrated and in which only the internal gene sequences dissociated from repressive sites of integration are expressed, appear to be less sensitive to the site of integration than that by proviral DNA following retroviral infection, in which transgenic mice carry inactive transgenes (54). The transgene introduction through ES cell system may reduce this problem, since the method enables the selection of cell clones expressing the transgene prior to production of the founder mice.

3) Regulatory elements: Studies of mechanisms of carcinogenesis as the consequence of oncogene expression in transgenic mice have provided insights into the susceptibility of different cell types to transformation in vivo, and the selectivity of oncoprotein actions. Studies of expression and action of oncogenes in transgenic mice involve three molecular genetic modes: i) oncogene expression using natural regulatory elements; ii) targeted expression via hybrid oncogenes; and iii) deregulated, wide expression of hybrid oncogenes.

   i) Oncogene expression with natural regulatory elements is employed to address the tissue specificity of the regulatory elements and the ability of the oncoproteins to transform the cell types in which they are expressed. As discussed above, although most viral regulatory elements are generally inactive in transgenic mice, the latent viral genomes appear to be susceptible to tissue-specific activation processes, leading to the development of tumors in those

Table 2. In vitro and In vivo Carcinogen Testing

A. *Short-term Tests In Vitro/In Vivo*
   1. Bacterial mutagenesis
   2. Mammalian mutagenesis
   3. Chromosome Aberations
   4. Primary DNA damage
   5. Cell transformation (Balb/c 3T3 cells)
   6. Immunoassay (carcinogen-DNA or -protein adducts)
B. *Limited Bioassys In Vivo*
   1. Altered foci induction in rodent liver
   2. Skin neoplasm induction in mice
   3. Pulmonary neoplasm induction in mice
   4. Breast cancer induction in female rats
C. *Chronic Bioassays In Vivo*
   1. Mice carcinogenesis study (24-month)
   2. Rat carcinogenesis study (24-month)

tissues. For example, when its own early regulatory regions are introduced by direct DNA microinjection, human DNA JC virus with JC T antigens which causes human neural diseases, produces neuroblastomas in founder transgenic mice after a long latency, and a fatal dysmyelination in the central nervous system of the progeny (55). Human DNA BK virus with BK T antigens causes human liver and kidney diseases. It also induces hepatocarcinoma and renal tumors in transgenic mice, when it is transferred with its own promoter regions (e.g., BK virus early region) (56).

ii) Targeted expression through hybrid oncogenes to specific cell types in a transgenic mouse is used to examine the consequences of oncogene expression in that targeted cell type (57). Examples of this category are listed in table 1 and are c-*myc* or N-*myc* gene hybrid with the immunoglobulin heavy chain (IgH) enhancer for B-cell lymphomas (58-62), c-*myc* gene hybrid with the murine mammary tumor virus long terminal repeat (MMTV LTR) for mammary adenocarcinoma (63,64), H-*ras* or N-*ras* hybrid with MMTV LTR for tumors in mammary, salivary gland, and lymphocytes (65-67), and c-*myc* or H-*ras* hybrid with the whey acidic protein gene (WAP) (68,69, and others (table 1) (70-79).

iii) Deregulated, wide expression of hybrid oncogenes is used to assess the consequences of oncoprotein synthesis in a variety of cells and to address the differential activity of oncoprotein action as opposed to the specificity of expression. To achieve wide expression of the transgenes, regulatory regions of cellular genes (e.g., metallothionein and the major histocompatibility complex class I (MHC class I) which are expressed in a variety of cells are utilized (80). When c-*fos* oncogene and FBJ osteosarcoma virus LTR hybrid with the heavy chain of MHC class I (H-2K) or human metallothionein II gene resulted in chondrosarcoma in about 15% of bone hyperplastic transgenic mice, within a year (80).

STUDY OF THE CARCINOGENIC POTENTIAL OF CHEMICALS

In relation to risk assessment, the study of chemical carcinogenesis is traditionally understood to mean detection of chemicals that are genotoxic and/or that act as promoters and possibly progressors. If the study process is susceptible to external control, it is also understood to mean determination of a statistically significant incidence compared with simultaneous controls of given tumors and extrapolation of the results to human risks using mathematical and statistical formulations. There is a battery of tests in vitro for genotoxicity or promoting potential (e.g., bacterial mutagenesis, mammalian mutagenesis, DNA repair, chromosomal aberrations, and cell transformation tests), in vivo limited bioassays (altered foci induction in rodent liver, skin tumor induction in mice, pulmonary tumor induction in mice, and breast cancer induction in female Sprague-Dawley rats, and

Table 3. Protooncogene Activation and Human Cancers

Oncogene	Activation	Human Cancer
abl	translocation(9;22)	chronic myelogenous leukemia
erbB-1	amplification	squamous cell carcinoma
erbB-1	amplification	breast, ovary & stomach cancer
myc	translocation(8;14)	Burkitt's lymphoma
	amplification	lung, breast & cervix cancer
L-myc	amplification	lung carcinoma
N-myc	amplification	neuroblastoma, SCLC
H-ras	point mutation(12,61)	melanoma, colon, lung, pancreas bladder, & breast cancer
K-ras	point mutation(12,59)	ALL, AML, colon, lung & thyroid cancer, melanoma
N-ras	point mutation(13,61)	neuroblastoma, melanoma, thyroid & genitourinary cancer
ret	rearrangement	thyroid cancer
gip	point mutation	ovary & adrenal gland cancer
gsp	point mutation	thyroid & pituitary gland cancer

in vivo long-term bioassay) (Table 2) (see review 81). Limited bioassays and standard chronic bioassays in vivo can be designed to reveal the effect of promoters on their specific target organ. Pretreatment with an appropriate dose of a genotoxic carcinogen for the target organ, followed by a number of dose levels of the promoter chemical provides quantitative information on the effect of the chemical. However, this conventional method for the study of carcinogenesis has many shortcomings: differential susceptibility among species, organ specificity, no-effect dose or threshold, metastasis in human cancer but not in animal tumors, and spontaneous tumors in animals, etc.

Among more than 60 protooncogenes so far identified in the human genome, only a dozen or so have been clearly demonstrated their expression and involvement in human cancers: c-Ha-*ras* 1, c-Ki-*ras*2, N-*ras*, c-*erb*B1, c-*erb*B2, c-*myc*, L-*myc*, N-*myc*, c-*gip*,

Table 4. Tumor Suppressor Genes and Human Cancers

Suppressor Gene	Chromosomal Defect	Deletion	Human Cancer
DCC	del 18q	18q21	colorectal carcinoma (no adenoma)
FAP	del 5q	q21-22	colon carcinoma
MEN-1	del 11q	11q13	adrenal cortex, pancreas, & pituitary cancer
NF1	del 17q	17q11.2	neurofibrosarcomas
p53	del 17marker (heterozygosity)	17q12-13.3	breast, colon, liver & lung cancer, astrocytoma
RB1	del 13q	13q14.13	retinoblastoma, breast, bladder & lung cancer, osteosarcoma
WT1	del 11p	11p13	Wilm's tumor

c-*gsp*, c-*ret*, and c-*abl* (Table 3). At least seven tumor suppressor genes have been implicated in human tumors: Rb1, p53, WT1, DCC, NF1, FAP, and MEN-1 (Table 4). Utilization of transgenic mice with human protooncogene(s) for the study of human carcinogenic potential of chemicals will provide precise information on the mechanisms of expression and activation of certain oncogene(s) by a chemical and potential target organ/tissue of the chemical. The generation of transgenic mice with either targeted or deregulated genes will in the future constitute a most important steps in the study of the carcinogenic potential of chemicals. It will permit: the selection of mouse strains which have low cancer susceptibility; the selection of hybrid strains which have dual potential to develop tumor; observation of formation of tumors possibly throughout the life span of the animals; selection of the method for transgene introduction into mouse; and selection of an oncogene or oncogenes with or without regulatory elements. Appropriate selections will permit determination of the tissue specificity and the latency of tumor generation and comparisons of the size and number of tumors generated in transgenic mice exposed to chemical carcinogens with spontaneous tumors in control animals. Furthermore, molecular, cytogenetic, and cellular analysis of the genes/genomes implicated with cancer will help to assess carcinogenic steps involved and the potential of substances to act as human chemical carcinogen. So will RFLPs in conjunction with classical cytogenetics, NIH 3T3 transfection assays, and examination of karyotypic abnormalities such as translocation, inversion, interstitial deletion, amplification, etc.

CONCLUSION

The transgenic establishment of oncogenes in the germ line of mice is an important new approach to provide information on the oncogene action and the mechanism of carcinogenesis. Studies with transgenic mice have revealed details about the consequences of oncogene expression and about the processes of carcinogenesis. Oncogenes in transgenic mice can induce tumors with tissue or cell specificity in their actions, although the mere presence of an oncogene in transgenic mice is not a sufficient condition for the complete development of tumors. The inheritable nature of transgenic tumors allows the study of different steps in the carcinogenic process.

Transgenic systems will provide new methods for the study chemical carcinogenesis. Critical considerations to study the subject are 1) construction of transgenic mice; 2) oncogene expression through regulatory elements; and 3) assay methods. The selection of the regulatory element which is fused to the oncogene to be transferred is very important because it determines the expression and activation of oncogenes in specific cell types. In addition to the tumor generated (number, size, locality, and latency), detection of expression of oncogenes and genetic changes will provide the additional information about whether a certain chemical has a potential to cause tumors in humans and about the mechanism with which the chemical affects tumor formation.

REFERENCES

1. Bishop, J.M. (1987). Science 235:305-311
2. Knudson, A.G.Jr. (1986). Ann. Rev. Genet. 20:231-251
3. Camper, S.A. (1987). Bio/Techniques 5:638-650
4. Brinster, R.L. and Palmiter, R.D. (1986). Harvey Lect. Ser. 80:1-38
5. Hanahan, D. (1989). Science 246:1265-1275
6. Compere, S.J., Baldacci, P. and Jaenisch, R. (1988). Biochim. Biophys. Acta 948:129-149
7. Hanahan, D. (1988). Ann. Rev. Genet. 22:479-519
8. Weinstein, I.B. (1987). J. Cell. Biochem. 33:213-224
9. Land, H., Parada, L.F. and Weinberg, R.A. (1983). Science 222:771-778
10. Nowell, P.C. (1986). Cancer Res. 46:2203-2207
11. Bos, J.L. (1988). Mutation Res. 195:255-271
12. Escot, C., Thellet, C., Lidereau, R., Spyratos, R. et al. (1986). Proc. Natl. Acad. Sci. USA 83:4834-4838
13. Jenuwein, T. and Muller, R. (1987). Cell 48:647-657
14. Reddy, E.P., Reynolds, R.K., Santos, E. and Barbacid, M. (1982). 300:149-152
15. Fujita, J., Oshida, Ol, Yuasa, Y., Rhim, J.S. et al. (1984). Nature 309:464-467
16. Bos, J.L., Fearon, E.R., Hamilton, S.R., Verlaan-de Vries, M., et al. (1987). Nature 327:293-297
17. Forrester, K., Almoguera, C. Han, K., Grizzle, W.E. and Perucho, M. (1987). Nature 327:298-303
18. Tanaka, T., Slamon, D.J., Battifora, H. and Cline, M. (1986). Cancer Res. 46:1465-1470
19. Eva, A., Tronick, S.R., Gol, R.A., Pierce, J.H. and Aaronson, S.A. (1983). Proc. Natl. Acad. Sci. USA 80:4926-4931
20. Albino, A.P., LeStrange, R., Oliff, A.I., Old, L.J. and Furth, M.E. (1984). Nature 308:69-73
21. Fasano, O., Aldrich, T., Tamanoi, F., Taparowsky, E., Furth, M. and Wigler, M. (1984). Proc. Natl. Acad. Sci. USA 81:4008-4012
22. Lenoir, G.M. and Bornkamm, G.W. (1987). Adv. Vir. Oncol. 7:173-206
23. Klein, G. (1987). Adv. Vir. Oncol. 7:207-211

24. Fialkow, P.J., Martin, P.J., Nayfeld, V., Penfold, G.K., et al. (1981). Blood 58:158-163
25. Fong, C.T., Dracopoli, N.C., White, P.S., Merrill, P.T., et al. (1989). Proc. Natl. Acad. Sci. USA 86:3753-3758
26. Venter, D.J., Tuzi, N.L., Kumar, S. and Gullick, W.J. (1987). Lancet ii:69
27. Marshall, C.J. (1991). Cell 64:313-326
28. Jaenisch, R. (1988). Science 240:1468-1474
29. Brinster, R.L., Chen, H.Y., Trumbauer, M., Senear, A.W., et al. (1981). Cell 27:223-231
30. Jaenisch, R. and Mintz, B. (1974). Proc. Natl. Acad. Sci. USA 71:1250-1254
31. Jaenisch, R. (1976). Proc. Natl. Acad. Sci. USA 73:1250-1254
32. Palmiter, R.D., Chen, H.Y. and Brinster, R.L. (1982). Cell 29:701-710
33. Soriano, P., Cone, R.D., Mulligan, R.C. and Jaenisch, R. (1986) Science 234:1409-1413
34. Stewart, C., Schuetze, S., Vanek, M. and Wagner, E. (1987). EMBO J. 6:383-388
35. Robertson, E., Bradley, A,, Kuehn, M. and Evans, M. (1986). Nature 323:445-447
36. Grossler, A., Doetsch, T., Korn, R., Serfling, E. and Kemler, R. (1986). Proc. Natl. Acad. Sci. USA 83:9065-9069
37. Wagner, E.F. (1990). EMBO J. 9:3025-3032
38. Jaenisch, R. (1980). Cell 19:181-188
39. Soriano, P. and Jaenisch, R. (1986). Cell 46:19-29
40. Gessler, M., Poustka, A., Cavenee, W., Neve, R.L., et al.(1990) Nature 343:774-778
41. Haber, D.A., Buckler, A.J., Glaser, T., Call, K.M., et al. (1990). Cell 61:1257-1269
42. Friend, S.H., Bernards, R., Rogelj, S., Weinberg, R.A., et al. (1986). 323:643-646
43. Cavenee, W.K., Hansen, M.F., Nordenskjold, M., Kock, E., et al. (1985). Science 228:501-503
44. Woychick, R.P., Stewart, T.A., Davis, L.G., D'Eustachine, P. and Leder, P. (1985). Nature 318:36-40
45. Reik, W., Weiher, H. and Jaenisch, R. (1985). Proc. Natl. Acad. Sci. USA 82:1141-1145
46. Smithies, O., Gregg, R.G., Boggs, S.S., Koralewski, M.A. and Kucherlapati, R.S. (1985). Nature 317:230-234
47. Jones, P.A. and Taylor, S.M. (1980). Cell 20:85-93
48. Compere, S.J. and Palmiter, R.D. (1981). Cell 25:233-240
49. Jaenisch, R., Schnieke, A. and Harbers, K. (1985). Proc. Natl. Acad. Sci. USA 82:1451-1455
50. Swain, J.L., Stewart, T.A. and Leder, P. (1987). Cell 50: 719-727
51. Wagner, E.F., Vanek, M. and Vennstrom, B. (1985). EMBO J. 4: 663-666
52. Soriano, P., Cone, R.D., Mulligan, R.C. and Jaenisch, R. (1986) Science 234:1409-1413
53. Chada, K., Magram, J. and Costantini, F. (1986). Nature 319: 685-688
54. Jaenisch, R., Jahner, D., Nobis, P., Simon, I., et al. (1981). Cell 24:519-529
55. Small, J. A., Scangos, G.A., Cork, L., Jay, G. and Khoury, G. (1986). Cell 46:13-18
56. Small., J.A., Khoury, G., Jay, G., Howley, P.M. and Scangos, G.A. (1986). Proc. Natl. Acad. Sci. USA 83:8288-8292

57. Schoenenberger, C.A., Adres, A.C., Groner, b., van der Valk, M., LeMeur, M. and Gerlinger, P. (1988). EMBO J. 7:169-175
58. Ma, A., Smith, R.K., Tesfaye, A., Achacoso, P., et al. (1991). Mol. Cell Biol. 11:440-444
59. Webb, E., Barri, G., Cory, S. and Adams, J.M. (1989). Mol. Biol. Med. 6:475-480
60. Langdon, W.Y., Harris, A.W. and Cory, S. (1988). Oncogene Res. 3:271-279
61. Schmidt, E.V., Pattengale, P.K., Weir, L. and Leder, P. (1988). Proc. Natl. Acad. Sci. USA 85:6047-6051
62. Alexander, W.S., Schrader, J.W. and Adams, J.M. (1987). Mol. Cell Biol. 7:1436-1444
63. Stewart, T.A., Pattengale, P.K. and Leder, P. (1984). Cell 38:627-637
64. Leder, A., Pattengale, P.K., Kuo, A., Stewart, T.A. and Leder, P. (1986). Cell 45:485-495
65. Mangues, R., Seidman, I., Pellicer, A. and Gordon, J.W. (1990). Oncogene 5:1491-1497
66. Tremblay, P.J., Pothier, F., Hoang, T., Tremblay, G. et al. (1989). Mol. Cell Biol. 9:854-859
67. Sinn, E., Muller, W., Pattengale, P., Tepler, I., et al. (1987). Cell 49:465-475
68. Andres, A.C., van der Valk, M.A., Schonenberger, C.A. et al. (1988). Genes Dev. 2:1486-1495
69. Andres, A.C., Schonenberger, C.A., Groner, B., Henninghausen, L., et al. (1987). Proc. Natl. Acad. Sci. USA 84:1299-1303
70. Leder, A., Kuo, A., Cardiff, R.D., Sinn, E. and Leder, P. (1990). Proc. Natl. Acad. Sci. USA 87:9178-9182
71. Bailleul, B., Surani, M.A., White, S., Barton, S.C., et al. (1990). Cell 62:697-708
72. Lee, G.H., Li, H., Ohtake, K., Nomura, K., et al. (1990). Carcinogenesis 11:1145-1148
73. Koike, K., Jay, G., Hartley, J.W., Schrenzel, M.D., et al. (1990). J. Virol. 64:3988-3991
74. Iwamoto, T., Takahashi, M., Ito, M., Hamaguchi, M. et al. (1990). Oncogene 5:535-542
75. Suda, Y., Aizawa, S., Furuta, Y., Yagi, T. et al. (1990). EMBO J. 9:181-190
76. Bouchard, L., Lamarre, L., Tremblay, P.J. and Jolicoeur, P. (1989). Cell 57:931-936
77. Sandgren, E.P., Quaife, C.J., Paulovich, A.G., Palmiter, R.D. and Brinster, R.L. (1989). Oncogene 4:715-724
78. Muller, W.J., Sinn, E., Pattengale, P.K., Wallace, R., Leder, P. (1988). Cell 54:105-115
79. Suda, Y., Aizawa, S., Hirai, S., Inoue, T. et al. (1987). EMBO J. 6:4055-4065
80. Ruther, U., Komitowski, D., Schuber, F.R. and Wagner, E.F. (1989). Oncogene 4:861-865
81. Weisburger, J.H. and Williams, G.M. (1984). In Chemical Carcinogen (ed. by C.E. Searle) Vol. II, pp.1323-1373

LEUKEMIA, LYMPHOMA, EMBRYONIC CARCINOMA AND HEPATOMA INDUCED IN TRANSGENIC RABBITS BY THE c-*myc* ONCOGENE FUSED WITH IMMUNOGLOBULIN ENHANCERS

Helga Spieker-Polet[1], Periannan Sethupathi[1], Herman Polet[2], Pi-Chen Yam[1], and Katherine L. Knight[1]

[1]Department of Microbiology and Immunology
Loyola University Chicago
Stritch School of Medicine, Maywood, IL
[2]Department of Pathology
University of Illinois, Chicago, IL

ABSTRACT

Transgenic rabbits were developed with the rabbit c-*myc* proto oncogene fused to the immunoglobulin heavy or light chain enhancers, E_μ or E_κ. Rabbits transgenic for E_μ-*myc* developed leukemia very early in life, 17-21 days of age, and were terminally ill. The disease was diagnosed as acute lymphoblastic leukemia resembling childhood leukemia. The white blood cell counts were increased as much as 100-fold, the spleens were enlarged and various organs were infiltrated by the neoplastic lymphocytes. Analysis of the DNA showed that immunoglobulin heavy and light chain genes were rearranged, showing the leukemia to be of the B cell type. Since several different rearranged VDJ genes could be cloned from each of three rabbits we concluded that the leukemias were polyclonal. The blood lymphocytes, as well as the cells cloned from various tissues, appear to be lymphoblastic, that is, they are large, have a large nucleus and have very little cytoplasm. These cells do not express surface immunoglobulin, but they have a low concentration of cytoplasmic IgM and they secrete small amounts of light chain.

Rabbits transgenic for E_κ-*myc* also developed tumors, but in contrast to the E_μ-*myc* transgenic rabbits, they developed tumors later in life. The adult onset of tumor development allowed us to establish a colony of rabbits carrying the E_κ-*myc* transgene in the germline. A variety of tumors was observed in the E_κ-*myc* rabbits. Of 16 E_κ-*myc* rabbits four developed lymphoma, one developed embryonic carcinoma and one developed hepatoma. Here we describe the generation of E_μ-*myc* and E_κ-*myc* transgenic rabbits, the tumors that developed in these rabbits and the cell lines that were derived from the tumors. We discuss the interesting observation that non-lymphoid tumors develop in the E_κ-*myc* transgenic rabbits.

INTRODUCTION

The ability to generate transgenic animals provides the opportunity to introduce genes of interest into the germline and to examine their expression in vivo (Gordon and

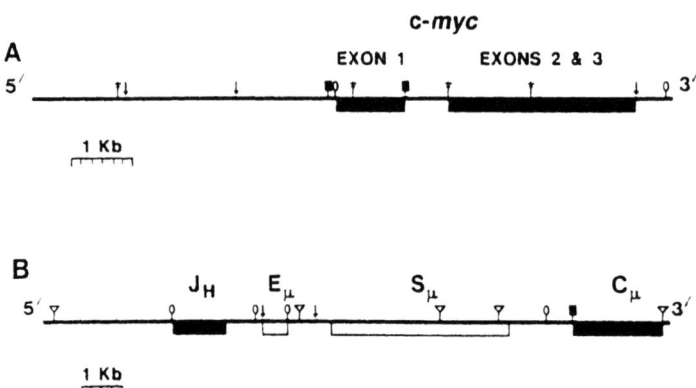

Fig. 1. Restriction maps of rabbit c-*myc* recombinant phage clone (A) and rabbit J_H-C_μ chromosomal region (B). The solid boxes in A represent regions that hybridized with the c-*myc* exon 1 oligomer probe (5' solid box) and the v-*myc* probe (3' solid box). The regions encoding J_H segments and C_μ are designated in B by solid boxes. The E_μ and the switch region (S_μ) are designated by open boxes. The enzyme symbols are as follows: ♀ =BamHI; ↓ =EcoRI; ⊤ =SacII; ▽ =HindIII; and ▮ =XhoI

Ruddle, 1983; Brinster et al., 1985; Hogan et al., 1986). By generating transgenic animals with cellular oncogenes fused with various enhancers and promoters, we can examine the effects of deregulated expression of oncogenes on the development of tumors (for review see Pattengale et al., 1989 , Leder et al., 1986) . Some of the most interesting results of studies with transgenic animals have been obtained when the oncogene was fused with a tissue-specific enhancer or promoter. In these animals, high levels of the oncogene are expressed in the tissue from which the enhancer/promoter was obtained and this in turn leads to increased incidence of tumors of the targeted organ (Pattengale et al., 1989 ; Stewart et al., 1984; Adams et al., 1985; Quaife et al., 1987; Mahon et al., 1987; Schmidt et al., 1988). One proto-oncogene whose expression has been deregulated is the cellular proto-oncogene c-*myc*. The c-*myc* encodes a DNA binding protein which presumably plays an important role in the regulation of the cell cycle and of cellular proliferation (Stewart et al., 1984; Burk et al., 1988; Blackwood and Eisenman, 1991). The c-*myc* is also implicated in the development of lymphocyte tumors because it has been observed that in human Burkitt's lymphoma and many mouse plasmacytomas, *c-myc* is translocated into the immunoglobulin heavy chain gene locus (Adams et al., 1983). The expression of translocated c-*myc* is believed to escape it's normal regulation and is instead under the control of the Ig promotor and/or enhancer. These observations led investigators to suggest that microinjection of c-*myc* linked to an Ig enhancer may lead to B cell transformation. Adams et al. (1985) have developed such transgenic mice with c-*myc* linked to either the heavy chain enhancer, E_μ, or the kappa-light chain enhancer, E_κ and these animals indeed developed lymphomas of the B cell lineage.

We have used the same approach in making transgenic rabbits in the hope of developing another model for human lymphocytic diseases. We generated transgenic rabbits using as the transgene, rabbit c-*myc* coupled to E_μ (Knight et al., 1988) and E_κ. These transgenic rabbits do indeed develop lymphoid tumors. Because rabbits, in

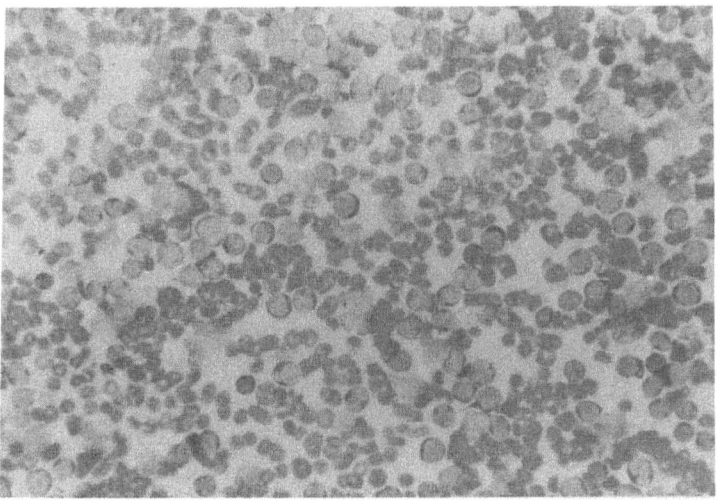

Fig. 2. Blood smear of leukemic rabbit K74-4 stained with hematoxylin/eosin

contrast to inbred mice, are outbred, are large, and have a long lifespan, we think that tumor development in rabbits may more accurately reflect tumor development in humans. Also, because of it's long life span, the rabbit provides the opportunity to follow different stages of tumor development and to study methods of treatment and prevention.

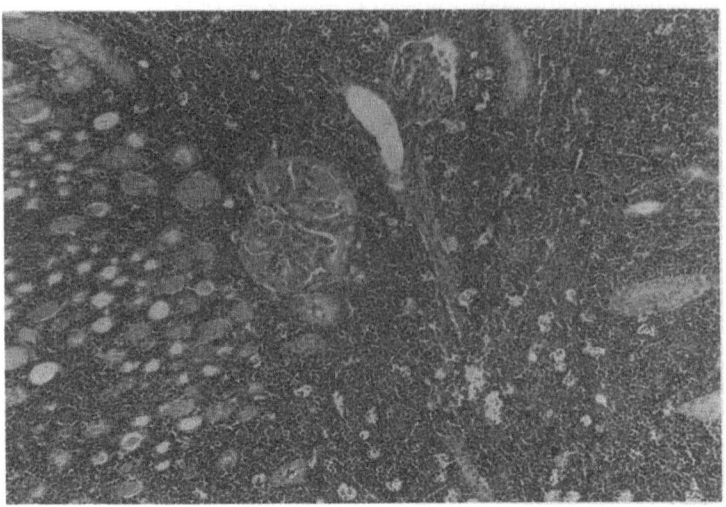

Fig. 3. Section of kidney of leukemic rabbit K75-4 stained with hematoxylin/eosin.

An additional but equally important goal for us in developing transgenic rabbits with *myc* fused to Ig enhancers was to study B cell development and Ig gene rearrangement and diversification. Prior to the development of lymphoid tumors in transgenic animals, only two cases of lymphocytic tumors had been reported in rabbit (Finnie et al., 1980; Cloyd and Johnson, 1978), and we had no ready access to transformed lymphoid cells. We discuss here the development of transgenic rabbits with the E_μ-*myc* and E_κ-*myc* transgenes and the variety of tumors that arise in these rabbits.

LEUKEMIA IN E_μ-*myc* TRANSGENIC RABBITS

In our first studies we used as transgene the rabbit c-*myc* gene coupled to the heavy chain enhancer E_μ (Knight et al., 1988). To construct this transgene rabbit c-*myc* and the heavy chain enhancer E_μ were identified and cloned from a phage library of genomic DNA by hybridization with v-*myc* probe and a murine heavy chain enhancer probe, respectively (Fig. 1). The fragment containing the enhancer region was linked to the 3' end of c-*myc* gene and the resulting construct was used for microinjections. Single cell zygotes from outbred rabbits were microinjected with this construct and were fostered *in utero* in pseudopregnant females. Two to three weeks after birth, the rabbits that had integrated the E_μ-*myc* transgene into their germline developed leukemia. The animals became lethargic and developed splenomegaly. The white blood cell count (WBC) of the transgenic rabbits was elevated to as high as 600,000 cells/μl, while the WBC of normal rabbits is approximately 5000 cells/μl. The peripheral blood smear confirmed the elevated WBC and showed that most of the cells were large atypical lymphocytes, that is, lymphoblasts (Fig. 2). The appearance of the blood resembled closely that of human childhood leukemia. The animals were sacrificed and we found that spleen and liver were enlarged and kidney, bone marrow, liver and spleen were shown to be infiltrated by these neoplastic lymphoblasts (Fig. 3) confirming the

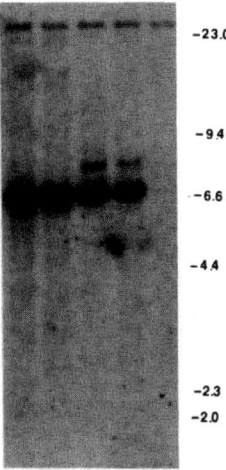

Fig. 4. Southern blot analysis of thymus and PBL DNA restricted with BamHI from rabbits K74-4 and K75-4. The blot was hybridized with the 6.6-kb BamHI rabbit germ-line c-*myc* probe. Lane 1, K74-4 PBL; lane 2, K74-4 liver; lane 3, K75-4 PBL; lane 4, K75-4 liver; lane 5, normal rabbit sperm DNA. The sizes (kb) of *Hind* III-digested λ phage are shown as size standards

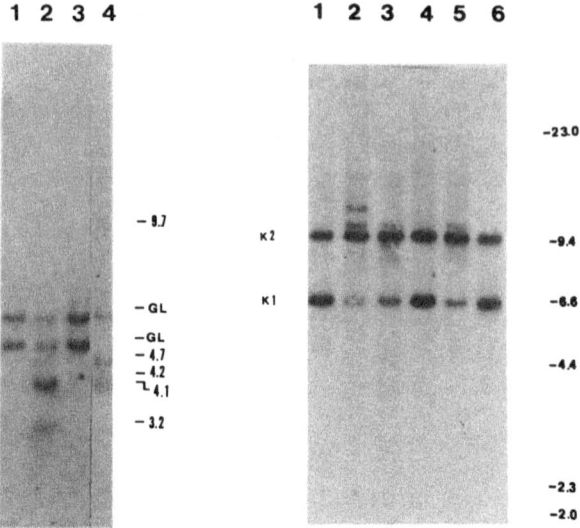

Fig. 5. Southern blot analysis of transgenic rabbit DNA restricted with HindIII. (Left panel): The blot was hybridized with a rabbit J_H probe. Lane 1, K74-4 thymus; lane 2, K74-4 PBL; lane 3, K75-4 thymus; lane 4, K75-4 PBL. The intense bands of thymus DNA are presumably in germ-line configuration. The sizes (kb) of germline (GL)-sized hybridizing fragments are indicated. (Right panel): The blot was hybridized with rabbit C_κ probe. Lane 1, K74-4 thymus; lane 2, K74-4 PBL; lane 3, K75-4 liver; lane 4, K75-4 thymus; lane 5, K75-4 PBL; lane 6, normal rabbit liver. The sizes (kb) of HindIII- digested λ phage are shown as size standards. Germ-line κ1 and κ2 fragments are indicated.

diagnosis of leukemia. In each of four E_μ-myc transgenic rabbits obtained from four different litters, the same pathologic profile was observed. Southern blots of restriction enzyme digested DNA of peripheral blood lymphocytes (PBL) and liver of the transgenic rabbits, hybridized with the myc probe, showed in addition to the endogenous myc gene a hybridizing fragment that was characteristic of the transgene (Fig. 4). The intensity of hybridization to the transgene was several times greater than the intensity of hybridization to the endogenous gene, suggesting that multiple copies of the transgene had been incorporated into the germline.

The leukemic cells of these transgenic rabbits were expected to be of the B-lymphocyte lineage since E_μ is expected to drive the expression of c-myc in B-lineage cells (Adams et al., 1985). Immunoglobulin gene rearrangements in these leukemic cells were examined by Southern blots of PBL DNA using J_H and C_κ probes. Both heavy and κ-light chain gene rearrangements were identified in these cells, indicating that they were indeed of the B-lymphocyte lineage (Knight et al., 1988). While the presence of both heavy and light chain gene rearrangements would suggest that the cells are mature B-lymphocytes, immunofluorescent analysis indicated that these leukemic cells had little or no surface Ig. We conclude that the cells are not mature B-lymphocytes but are either late pre-B-lymphocytes, or early B-lymphocytes. In normal rabbits Gathings et al. (1982) have identified cells containing cytoplasmic heavy and light chains and no surface Ig and these cells were designated "transient" B-lymphocytes.

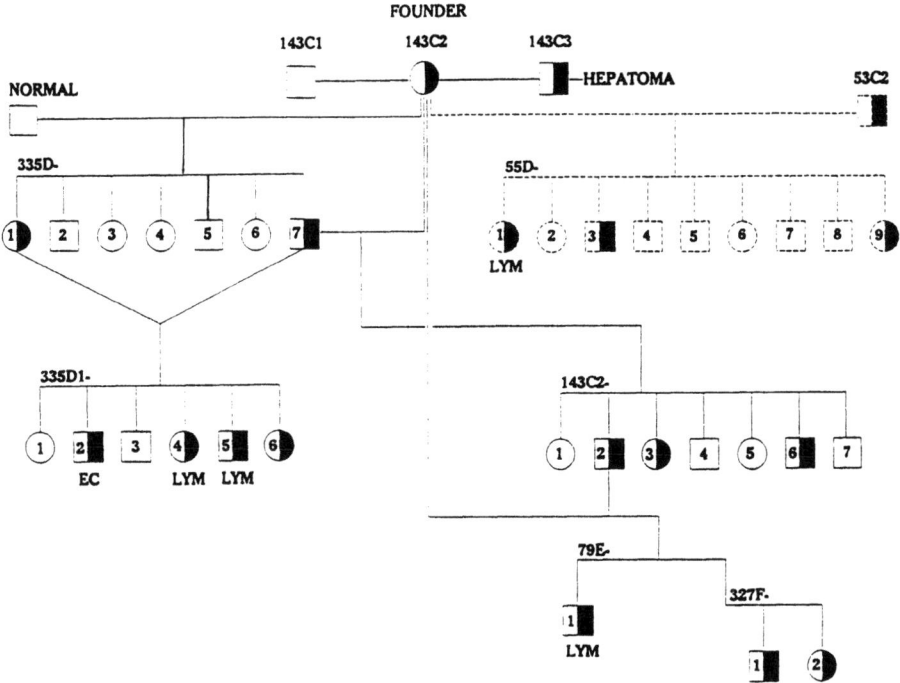

Fig. 6. E$_\kappa$-*myc* transgenic rabbit family. The numbers of the rabbits are indicated. Solid symbols indicate transgenic animals; shaded symbols indicate that the rabbits express the transgene but do not pass it in their germline. Dotted lines indicate animals that are not used for expansion of the transgenic family. Hepatoma indicates that the animal developed hepatoma, Lym = lymphoma and EC = embryonic carcinoma.

Southern analysis (Fig. 5) of DNA of the leukemia cells, probed with J_H and C_κ probes, showed several distinct heavy and light chain gene rearrangements indicating that the leukemias were polyclonal. Indeed two or three different rearranged VDJ gene segments could be cloned from the DNA of each of three rabbits (Becker et al., 1990). These results are different from those obtained for mice for which Adams et al. (1985) had reported monoclonal tumors. The development of polyclonal tumors in rabbits suggests that the second transforming signal is not a rare event but instead may be a normal growth factor.

Cell lines were established from PBL and bone marrow of one of the four leukemic rabbits. These lines were similar to each other and to the PBL leukemic cells described above as they had no surface Ig but had both heavy and κ-light chain genes rearranged. In addition, cytoplasmic IgM was found in these cell lines. Electron microscopic analysis of the cloned cells (data not shown) demonstrated moderately large mononuclear cells (at least twice the size of a mature lymphocyte) with irregularly shaped nuclei, minimal peripheral chromatin condensation and distinct nucleoli. The cytoplasm was moderately abundant and was almost devoid of rough endoplasmic reticulum. These data are consistent with our conclusion that the leukemic cells are late pre-B or early-B-lymphocytes.

Since all E_μ-myc transgenic rabbits became leukemic and died at an early age, we were unable to establish a family of rabbits that carried the transgene in their germline.

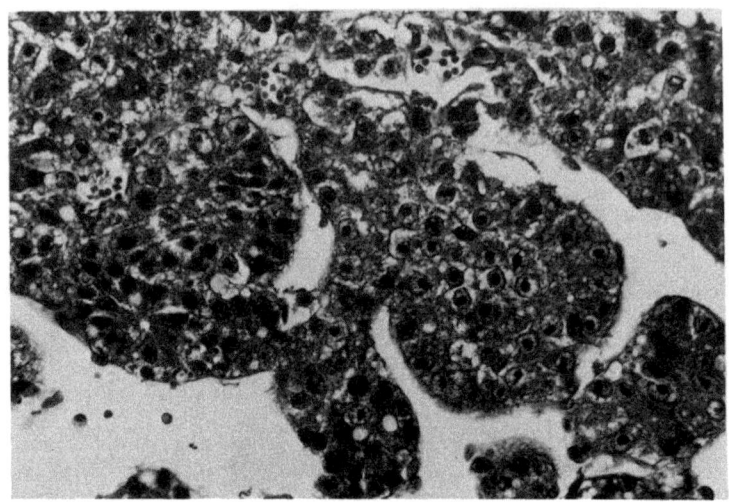

Fig. 7. Hepatoma. Section of rabbit 143C3 liver stained with hematoxylin/eosin: x100

Adams et al. (1985) had developed transgenic mice with c-*myc* coupled to the κ chain enhancer instead of the heavy chain enhancer and they found that these mice developed tumors less frequently and later in life than Eμ-*myc* transgenic mice. The tumors were mostly of the B cell type rather than of the pre-B cell type that developed in the E_μ-*myc* transgenic mice. We reasoned that if the same holds true in rabbits, then an E_κ-*myc* construct could lead to a rabbit model of lymphocytic diseases of more developed B lymphocytes, i.e. lymphomas and plasmacytomas. For that reason we developed rabbits with the E_κ-*myc* transgene.

Development of a Family of E_κ-*myc* Transgenic Rabbits

The E_κ-*myc* transgene was constructed by coupling rabbit c-*myc* to the κ1 (Emorine et al., 1983) and the proposed κ2 enhancers (Emorine et al., 1983; Karlin et al., 1985)from the J-C$_\kappa$1 region of the genomic kappa chain locus.

A total of 251 embryos microinjected with the E_κ-*myc* transgene were transferred into 22 foster mothers and 12 live pups were born to 8 of these females. Southern blot analysis of PBL DNA from these 12 rabbits showed that four, 143C2, 143C3, 53C1 and

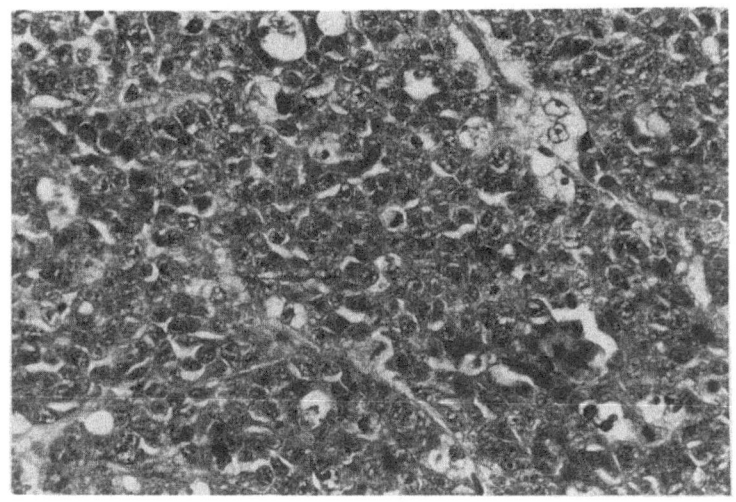

Fig. 8a

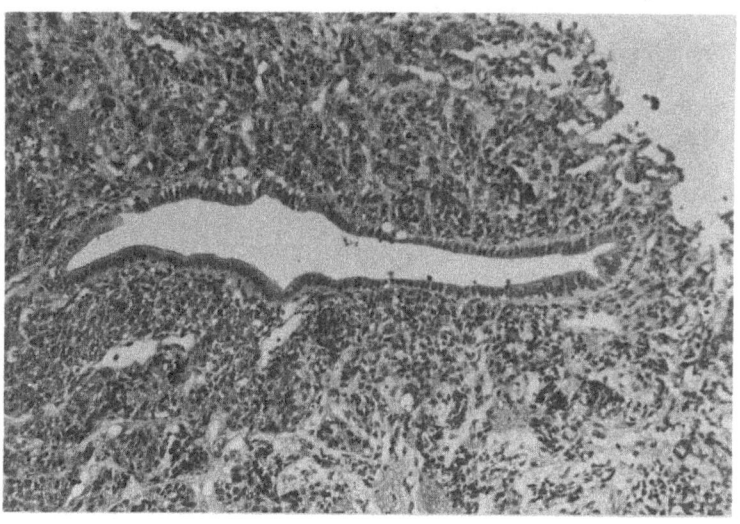

Fig. 8b

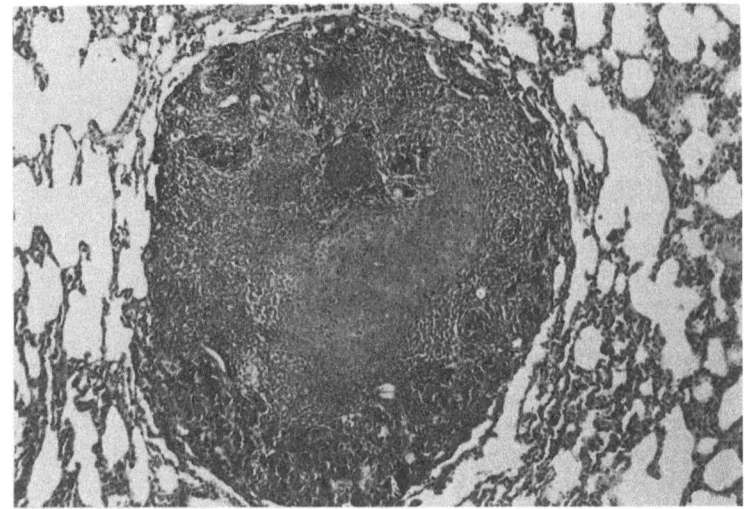

Fig. 8c

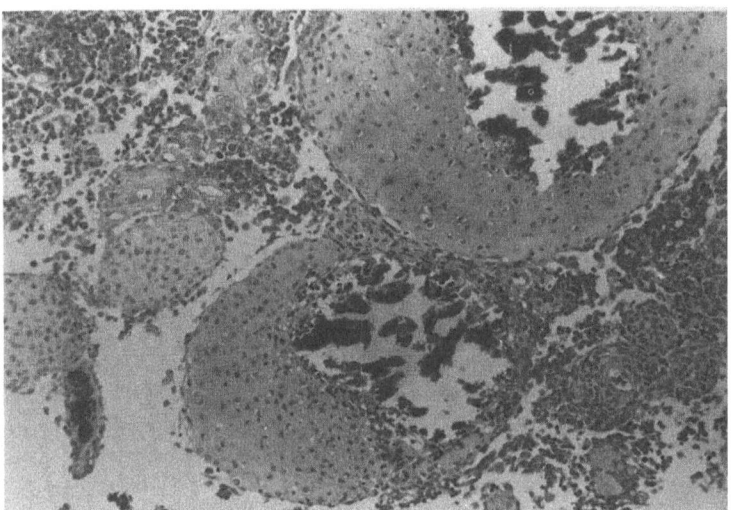

Fig. 8d

Fig. 8. Embryonic carcinoma. a-c, tumor sections of rabbit 335D1-2 stained with hematoxylin/eosin. a. Tumor section showing undifferentiated cells: x400. b. Section in which differentiation to duct structures can be seen: x100. c. Section in which differentiation to cartilage is apparent: x100. d. Section of a tumor that developed in the peritoneum of a newborn rabbit injected i.p. with *in vitro* cultured EC cells: x100. Differentiation to cartilage is apparent.

53C2, carried the E_κ-*myc* transgene. The founder transgenic rabbit, 143C2, was bred with a normal, non-transgenic rabbit; two of the seven offspring carried the transgene, indicating that the transgene was in the germline of the founder rabbit. These two transgenic offspring, 335D1 and 335D7 were used, along with the founder, to develop a family of E_κ-*myc* transgenic rabbits (Figure 6). Two other transgenic rabbits 53C1 and 53C2 were also mated with non-transgenic rabbits but none of 35 offspring of these matings were transgenic suggesting that the transgene was not carried in the germline of these two rabbits. Therefore it appears that the transgene of each offspring of the transgenic family (Figure 6) is derived from the founder 143C2.

Malignancies of E_κ-*myc* Transgenic Rabbits

Hepatoma. The transgenic rabbit 143C3 developed swelling in the lower jaw at the age of four months. The blood hemoglobin was elevated to approximately 20 g/dl and the hematocrit was approximately 60%. X-rays showed bony destruction of the mandible and multiple radio-opaque nodules of both lung parenchyma, suggesting the presence of a remote primary tumor. The rabbit was sacrificed and a 5 x 4 cm solitary tumor along with a few micronodules were found in the liver. Multiple nodular deposits were also present in both lungs, pleura, and the lateral thoracic wall.

This tumor was diagnosed by histologic analysis as a primary hepatoma with secondary deposits in lung and bone. The tumor tissue in all sites was composed of compact masses of typical liver cells with large nucleoli forming nodules of variable size, or cords of variable width (Figure 7).

Embryonic Carcinoma. At four weeks of age, rabbit 335D1-2 (Figure 6) developed a swelling in the right foot and thigh. The soft tissue swelling was firm, nodular, and was fixed to the deeper structure, whereas the thigh swelling was mobile. The rabbit was sacrificed and the autopsy showed enlarged kidneys with nodules that were confined to the outer margin. Similar nodules were found in the liver, spleen, mesenteric lymph nodes, and lung; all other abdominal organs appeared normal.

This tumor was diagnosed by histologic examination as embryonic carcinoma, a tumor arising from primitive cells. (Figure 8). These cells were large and had indistinct cytoplasm, and vesicular nuclei with one or more variable sized nucleoli (Figure 8a). The tumor tissue showed areas of mesenchymal differentiation and duct (Figure 8b) formation. In addition cartilage formation (Figure 8c) was observed in a lung metastasis. These observations indicate that the transformed cells are pluripotential. To obtain a cell line from this embryonic carcinoma, cells from blood and from the tumors of liver and lung were cultured *in vitro* and after three weeks of *in vitro* growth the adherent cells were injected s.c. into nude mice. Large s.c. tumors developed in these mice after two weeks. The cells harvested from the tumors could be passed either *in vitro* or *in vivo* in nude mice indicating that the cells were tumorigenic. Histologic analysis of the tumors obtained after several passages in nude mice showed that the cells remained undifferentiated. These undifferentiated cells were also injected into the heart of newborn rabbits less than 5 hours after birth. After four weeks growth of tumors composed of undifferentiated cells was observed in several locations in each of five rabbits. In one case the tumor cells were injected into the peritoneum of a newborn rabbit and several tumors developed within the peritoneal cavity. In one of these tumors differentiation to cartilage was observed (Figure 8d). These results suggest that the established cell line consists of undifferentiated embryonic carcinoma cells which have the ability to differentiate.

B-Lymphoma Three of the transgenic rabbits of the E_κ-*myc* family developed lymphomas. At five to six months of age, rabbits 335D1-4 and 335D1-5, became ill with dehydration and weight loss and were sacrificed. Both rabbits had an appendicular mass and enlarged mesenteric lymph nodes. Rabbit 335D1-4 had nodular deposits on the ovaries; other abdominal organs appeared normal. In rabbit 335D1-5 both kidneys were enlarged and contained multiple nodules; other abdominal organs appeared normal. The third rabbit, 79E, became emaciated at 11 months of age and at autopsy a grossly enlarged mesenteric lymph node (approximately 6 cm in diameter) was found. In addition we found multiple nodular swellings throughout the small intestine and ulcerated circular growths over the gastric mucosa; the other abdominal organs appeared normal.

Histologic analysis of tissues from the three lymphomatous rabbits showed the normal lymph node architecture to be replaced by cells of monotonous morphology arranged in sheets or follicles (Figure 9a). The cells were large polygonal and rounded cells with large vesicular nuclei, prominent nucleoli and little cytoplasm (Figure 9b). Cells infiltrating other organs like lymph node, kidney, gut (Figure 9b) had a similar appearance. We concluded that the tumors in these three animals were lymphomas. To determine if the lymphomas were of B lineage DNA was examined by Southern analysis for Ig gene rearrangements using J_H and C_κ probes. Southern analysis of *Hind* III digested DNA from germline (liver) and tumor DNA from rabbits 335D1-4, 335D1-5 and 79E identified J_H-hybridizing fragments, 4.1 and 3.0 kb in size, that were absent in germline DNA (Figure 10). Similarly, the C_κ probe identified a C_κ-hybridizing fragment in tumor DNA that was absent in the germline DNA (data not shown). These J_H and C_κ-hybridizing fragments represent rearranged J_H and C_κ genes, respectively, demonstrating that the lymphomas are of the B-lymphocyte lineage.

Undifferentiated B-Lineage Tumor. The transgenic rabbit 55D1 developed difficulty in breathing at 2 months of age and a proptosis of the left eye. The animal was sacrificed. We found a large maxillary tumor and the spleen and a cervical lymph node enlarged; all other abdominal organs appeared normal. A swelling over the left maxilla extended to the nasal cavity, causing obstruction, and to the orbital fossa causing protrusion of the left eye.

Histologic analysis of tissues of rabbit 55D1 showed that tumor cells were dispersed throughout the spleen and the cervical lymph node and that the maxillary tumor was composed of sheets of compactly arranged large round and polygonal cells with a thin rim of cytoplasm and vesicular nuclei. Histologically this tumor appears undifferentiated and cannot be unambiguously classified.

To determine if this tumor was of B-lymphoid lineage, Southern analysis was performed with DNA from liver, from the maxillary tumor, the cervical lymph node and from tumorous spleen to identify Ig gene rearrangements. Germline JH fragments were found on two bands 5.8 and 6.0 kb in size in liver DNA. In tumor, lymph node and splenic DNA the 6 kb JH hybridizing fragment was reduced in intensity, and a new

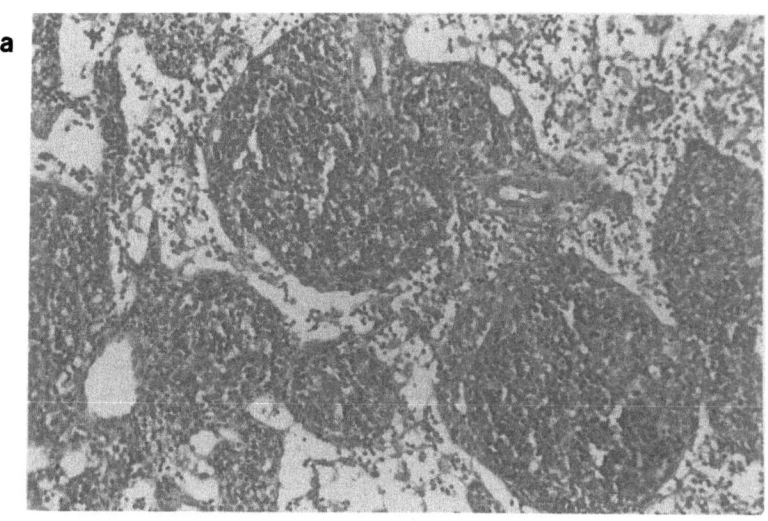

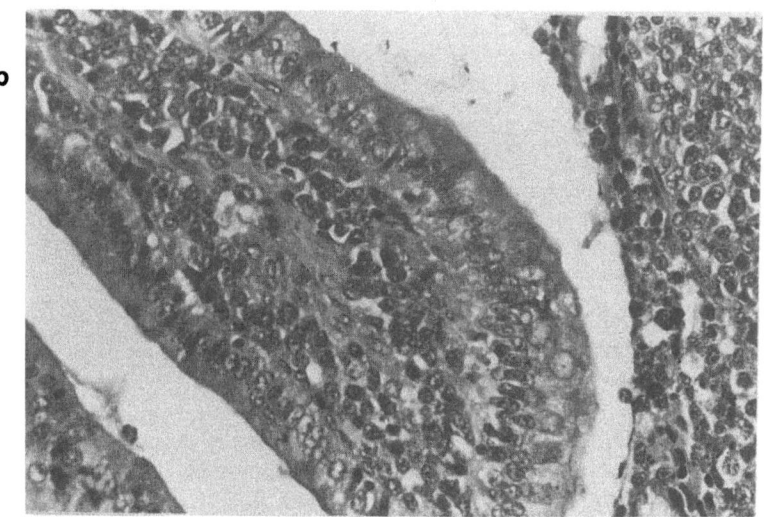

Fig. 9. Lymphoma. Tissue sections stained with hematoxylin/eosin. a. Lymph node of rabbit 335D1-4 showing follicles of monomorphic lymphoid cells: x100. b. Section of gut tissue of rabbit 79E, infiltrated by lymphoid cells: x100.

band, 4.0 kb in size, was observed (Figure 11). Similarly, with a C_κ-probe a new band, not present in liver DNA, was observed in DNA of tumorous tissues (data not shown).

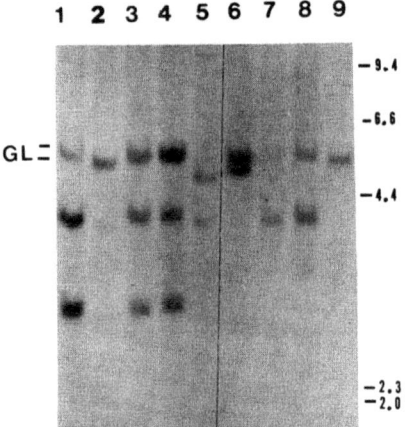

Fig. 10. Southern blot of *Hind*III restricted DNA of three lymphomatous rabbits probed with J_H. Lanes 1-4, rabbit 335D1-4: 1, appendix; 2, liver; 3, kidney; 4, MLN. Lanes 5 and 6 gut and liver respectively, of rabbit 79E. Lanes 7-9, rabbit 335D1-5: 7, appendix; 8, kidney; 9, liver. GL = germ line. The sizes of *Hind*III-digested λ phage are shown as size standards

These data indicate that cells of the lymph nodes as well as cells in the spleen had undergone J_H and C_κ rearrangements and therefore the 55D1 tumor is of B-lymphoid lineage.

Cells from the 55D1 tumor were placed into culture and after four weeks viable cells began to grow. These cells were cloned and they have continued to divide in culture for more than 2 years. Staining of the cultured cells with FITC-labeled anti Ig antibodies revealed small amounts of surface IgM indicating that the cells are B cells. Southern analysis of DNA of the cloned cell line revealed two rearranged J_H hybridizing fragments at 3.5-4.0 kb indicating that J_H genes of both chromosomes were rearranged (Fig. 11). Since only one rearranged J_H fragment was observed in 55D1 tumor DNA, the tumor cells that grew *in vitro* may represent a small subpopulation of the original tumor. We cannot however rule out the possibility that the original tumor cells had J_H rearrangement on only one chromosome and that these cells rearranged their J_H genes on the other chromosome while growing *in vitro* The significance of the two rearranged J_H fragments is not yet clear.

DISCUSSION

B-ALL Model in E_μ-myc Transgenic Rabbits

The transgenic rabbits obtained with the E_μ-myc transgene developed acute B-lymphoblastic leukemia that closely resembled childhood leukemia. In human, B-ALL affects primarily children and has a very rapid onset. Almost all B-ALL patients are anemic and their PBL levels vary widely. Similar to human ALL, our leukemic rabbits were anemic and they developed the disease early in life; by 17 to 20 days of age, the rabbits had from 10 to 100 times the normal number of PBL and they had enlarged spleen and liver due to infiltration by leukemic cells. The PBL of the leukemic rabbits, like PBL of B-ALL patients, were mainly immature blastic-like cells. The characteristic difference between human B-ALL and the rabbit model lies in the clonality of the leukemia. Human B-ALL is characteristically monoclonal in origin whereas the rabbit leukemias are oligo- or polyclonal in origin. The polyclonality suggests that the oncogenic "events" are not random in these leukemic rabbits but that similar oncogenic "signals" must be present in most, if not all, early B-lymphocytes (Langdon et al., 1986).

This animal model can be very useful for studying the course and development of B-ALL. It provides an opportunity to study the effect of drugs on the course of the disease and to develop other much needed therapies that may lead to the prolongation of the life of patients and eventually to the cure of the disease. One major positive aspect of this model is that all transgenic animals (four of four) developed the disease at approximately the same age. The disadvantage of this model is that the leukemic rabbits are terminally ill before they reach sexual maturity and therefore, a colony of rabbits carrying the transgene is impossible to establish. Each leukemic rabbit must be independently generated as a transgenic animal. In our experience, however, 10 to 20% of rabbits born from microinjection of zygotes with the E_μ-myc DNA construct are transgenic, making it feasible to generate these leukemic rabbits on a regular basis.

Another potential use of this model is to identify cellular factors that may serve as second signals for transformation. The leukemias are striking in that they occur early in the life of the animals and are polyclonal in nature. We believe that c-myc, driven by E_μ provides the first signal for transformation and the second signal may be a factor normally expressed in lymphoid cells, for example, a growth factor. Cells from the leukemic rabbits could provide a source of the "second signal" and thereby the means to investigate the molecular mechanisms of transformation.

Tumor Development in E_κ-myc Transgenic Rabbits

The rabbits with the E_κ-myc transgene were different in several aspects from rabbits with the E_μ-myc transgene. First, the E_κ-myc transgenic rabbits developed tumors later in life, at ages varying from 1 to 11 months while all E_μ-myc transgenic rabbits had leukemia at an age of 17-21 days. Second, the tumors in the E_κ-myc rabbits were monoclonal whereas the tumors in E_μ-myc were polyclonal. Third and most striking, the E_κ-myc rabbits developed different types of tumors, lymphoma, embryonic carcinoma and hepatoma, whereas all the E_μ-myc rabbits developed leukemia.

Transformation generally requires the activation or deregulation of at least two oncogenes. For the E_μ-myc and E_κ-myc transgenic rabbits, deregulation of c-myc presumably represents one signal. Because all E_μ-myc transgenic rabbits develop pre-B cell leukemia and do so at a young age, we suggest that the second signal in these rabbits may be a growth or differentiation factor (oncogene) that is normally expressed at an early stage in B cell development. Simultaneous expression of myc and such a growth or differentiation factor in these early B-lymphoid cells presumably leads to transformation of multiple B cells at an early stage of differentiation and consequently to polyclonal tumors. In the E_κ-myc transgenic rabbits on the other hand, the second event appears to occur randomly at any stage of B-cell development, transforming a single cell, leading to development of monoclonal tumors at any age of the rabbit.

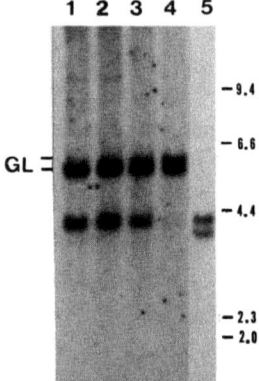

Fig. 11. Southern analysis of HindIII restricted DNA of rabbit 55D1 probed with J_H. Lane 1, tumor of the cervical lymph node; lane 2, maxillary tumor; lane 3, spleen; lane 4, liver; lane 5, cloned 55D1 cell line. GL = germ line. The sizes of HindIII-digested λ phage are shown as size standards.

The most surprising result from the E_κ-myc transgenic rabbits was the development of tumors of different cell types. Using an Ig-enhancer, we had expected the transgenic rabbits to develop only tumors of B-lymphoid origin. Since two of the tumors, the embryonic carcinoma and the hepatoma, were not of B cell lineage we propose that the κ-chain enhancer is not strictly lymphoid-specific but may be active in other tissues as well.

In our established family of E_κ-myc transgenic rabbits derived from the founder rabbit, 143C2, 16 of 34 rabbits were transgenic. Of these 16 transgenic rabbits 4 developed lymphoma and one developed embryonic carcinoma. It is possible that the incidence of tumor development could be increased if the animals were bred to homozygosity. If we could generate a family in which all or most transgenic rabbits developed tumors, these rabbits would provide an excellent model to study the progression of tumor development and growth. And, as discussed before for the leukemic rabbits, this animal model would also offer an opportunity to study the effects of drugs and to develop new ways of therapies which may lead to prolongation of life or to an eventual cure of disease. The family of E_κ-*myc* transgenic rabbits may also be useful to identify the oncogenes which deliver the second signal for transformation. One might speculate that the different types of tumors are caused by different oncogenes delivering the second signal. But it is also possible that the same oncogene is activated in all cases and different tumors arise because the κ-chain enhancer can be active in different tissues.

Ig-gene Rearrangement

In addition to disease models, the E_μ-*myc* and the E_κ-*myc* transgenic rabbits provide us with the opportunity to study Ig-gene rearrangement and VH gene usage. We have already shown that in eight different VDJ genes which were cloned from three different E_μ-*myc* transgenic rabbits only one VH gene, the D-proximal gene, VH1, was utilized (Becker et al., 1990). The V-region sequences of the VDJ genes were identical to the germline VH1 indicating that no somatic hypermutation had taken place in these early B cells. The E_κ-*myc* rabbits which develop tumors of later stage B cells will be useful to study VH gene usage and diversification in more mature B cells.

REFERENCES

Adams, J. M., Gerondakis, S., Webb, E., Cocoran, L. M., and Cory, S., 1983. Cellular *myc* oncogene is altered by chromosome translocation to an immunoglobulin locus in murine plasmacytomas and is rearranged similarly in human Burkitt lymphomas. Proc. Natl. Acad. Sci. USA 80:1982.

Adams, J. M., Harris, A. W., Pinkert, C. A., Corcoran, L. M., Alexander, W. S., Cory, S., Palmiter, R. D., and Brinster, R. L., 1985. The c-*myc* oncogene driven by immunoglobulin enhancers induces lymphoid malignancy in transgenic mice. Nature 318:533.

Becker, R. S., Suter, M., and Knight, K. L., 1990. Restricted utilization of VH and DH genes in leukemic rabbit B cells. Eur. J. Immunol. 20:397.

Blackwood, E. M., and Eisenman, R. N., 1991. Max: A Helix-Loop-Helix Zipper protein that forms a sequence-specific DNA-binding complex with *myc*., Science 251:1211.

Brinster, R. L., Chen, H. Y., Trumbauer, M. E., Yagle, M. K., Palmiter, R. D., 1985. Factors affecting the efficiency of introducing foreign DNA into mice by microinjecting eggs. Proc. Natl. Acad. Sci. USA 82:4438-4442.

Burck, K. B., Liu, E. T., Larrick, J. W., 1988. Oncogenes: An introduction to the concept of cancer genes. New York, SpringerVerlag.

Cloyd, G. G., and Johnson, G. R., 1978. Lymphosarcoma with lymphoblastic leukemia in a New Zealand white rabbit. Lab. Anim. Sci. 28:66.

Emorine, L., Kuehl, M., Weir, L., Leder, P., and Max, E. E., 1983. A conserved sequence in the immunoglobulin J_κ-C_κ intron: possible enhancer element. Nature 304:447.

Finnie, J. W., Bostock, D. E., and Walden, N. B., 1980. Lymphoblastic leukemia in a rabbit: a case report. Lab. Anim. 14:49.

Gathings, W. E., Mage, R. G., Cooper, M. D., and Young, G. O., 1982. A subpopulation of small pre-B cells in rabbit bone marrow expresses K light chains and exhibits allelic exclusion of b locus allotypes. Eur. J. Immunol. 12:76.

Gordon, J. W., Ruddle, F. H., 1983. Gene transfer into mouse embryos: production of transgenic mice by pronuclear injection. Methods Enzymol. 101:411-433.

Hogan, B., Constantini, F., Lacy, E., 1986. Manipulating the mouse embryo: a laboratory manual. Cold Spring Harbor Laboratory Manual. New York, Cold Spring Harbor. (D. Hanahan, ed).

Karlin, S., Ghandour, G., 1985. Alignment maps and homology analysis of the J-C intron in human, mouse and rabbit immunoglobulin kappa gene. Mol. Biol. Evol. 2:53.

Knight, K. L., Spieker-Polet, H., Kazdin, D. S., and Oi, V. T., 1988. Transgenic rabbits with lymphocytic leukemia induced by the c-*myc* oncogene fused with the immunoglobulin heavy chain enhancer. Proc. Natl. Acad. Sci. USA 85:3130.

Langdon, W. Y., Harris, A. W., Cory, S., and Adams, J. M., 1986. The c-*myc* oncogene perturbs B lymphocyte development in E_μ-*myc* transgenic mice. Cell 47:11.

Leder, A., Pattengale, P. K., Kuo, A., Stewart, T. A., and Leder, P., 1986. Consequences of wide-spread deregulation of the c-*myc* gene in transgenic mice: multiple neoplasms and normal development. Cell, 45:485.

Mahon, K. A., Chepelinsky, A. B., Khillan, J. S., Overbeek, P. A., Piatigorsky, J., Westphal, H., 1987. Oncogenesis of the lens in transgenic mice. Science 235:1622-1625.

Pattengale, P. K., Stewart, T. A., Leder, A., Sinn, E., Muller, W., Tepler, I., Schmidt, E. and Leder, P., 1989. Animal Models of Human Disease. Pathology and Molecular Biology of Spontaneous Neoplasms Occurring in Transgenic Mice Carrying and Expressing Activated Cellular Oncogenes. Am. J. Path. 135:39.

Quaife, C. J., Pinkert, C. A., Ornitz, D. M., Palmiter, R. D., and Brinster, R. L., 1987. Pancreatic neoplasia induced by ras expression in acinar cells of transgenic mice. Cell 48:1023-1031.

Schmidt, E. V., Pattengale, P. K., Weir, L. and Leder, P., 1988. Transgenic mice bearing the human n-*myc* gene activated by an immunoglobulin enhancer: A pre-B cell lymphoma model. Proc. Natl. Acad. Sci. USA 85:6047-6051.

Stewart, T. A., Pattengale, P. K. and Leder, P., 1984. Spontaneous mammary adenocarcinomas in transgenic mice that carry and express MTV/*myc* fusion genes. Cell 38:627-637.

AN OVERVIEW OF THE OUTSTANDING ISSUES IN THE RISK ASSESSMENT OF METHYLENE CHLORIDE

Rory B. Conolly, Kannan Krishnan and Melvin E. Andersen

Chemical Industry Institute of Toxicology
Post Office Box 12137
6 Davis Dr.
Research Triangle Park, NC 27709

ABSTRACT

Methylene chloride is a liver and lung carcinogen in male and female B6C3F1 mice. Accurate assessment of human cancer risk for this chemical requires mechanistic knowledge of both target tissue dosimetry (pharmacokinetics) and of the tissue responses culminating in tumor development (pharmacodynamics). A major step in this direction was taken by Andersen et al. (Toxicol. Appl. Pharmacol., 87:185, 1987) who used a physiologically based pharmacokinetic (PBPK) model to define target tissue doses of methylene chloride and its metabolites, produced by either cytochrome P-450 or by glutathione (GSH) conjugation. With this approach they found that methylene chloride metabolism by GSH conjugation correlated better with observed tumor outcome than did metabolism via cytochrome P-450. The PBPK model was scaled to humans and used to predict human metabolism of methylene chloride by the GSH pathway. When these calculated dose surrogates were used as input to a linearized multistage (LMS) model, a human cancer risk of 3.7×10^{-8}, for lifetime exposure at 1 mg/m^3 was predicted, more than two orders of magnitude lower than the risk estimated when bioassay exposure concentrations were used as input to the LMS for high to low dose extrapolation, and a body surface area correction factor was used for scaling from mice to humans. This report reviews the essential features of the Andersen et al. PBPK assisted risk assessment and considers how on-going developments in biologically based risk assessment should lead to further refinement of the human health risk assessment for methylene chloride.

INTRODUCTION

An overall goal of biologically-based chemical risk assessment is to predict accurately the shape of the dose-response curve for people, and also for non-human species, exposed to carcinogens and other classes of toxic chemicals. Knowledge of the shape of the dose-response curve allows a direct translation from exposure to risk of adverse effect. To approach this goal, mechanistic data must be obtained in laboratory animals and the utility of such data for predicting human risk evaluated. Two main classes of data, (1) tissue dosimetry and, (2) tissue response, are required.

Biologically based (or equivalently, physiologically based) descriptions of the mechanistic factors controlling target tissue dosimetry (PBPK models) have been developed in recent years for a number of chemicals, including volatile hydrocarbons[1], organophosphates[2], drugs[3,4], and non volatile chemicals such as 2,3,7,8-tetrachlorodibenzo-p-dioxin[5,6] and polychlorinated biphenyls[7]. Much less work has been done on biologically based descriptions of the responses which ensue once the active chemical reaches the target tissue. Some PBPK models include descriptions of initial tissue responses which have feedback effects on pharmacokinetic behavior. These include descriptions of enzyme induction[5], enzyme inhibition[3,4,8,9], and GSH depletion[10]. Extension of a PBPK model for chloroform to describe its cytolethal effects in mice was recently described[11]. In general, however, biologically based modeling of tissue responses is not as well developed as dosimetry modeling.

Biologically based descriptions of long term sequelae of chemical exposure such as cancer have also received attention in recent years. In particular, Moolgavkar and colleagues[12,13] have developed the mathematical theory of the two-stage cancer model (MVK model) where cells in each stage undergo stochastic birth, death and mutation processes. This model is consistent with incidence curves for a wide variety of cancers, such as childhood retinoblastoma and breast cancer, which are not well described by simpler models[12,13,14]. Of particular advantage for cancer risk assessment, the MVK model allows explicit definition of the cellular effects of different classes of carcinogens, including DNA reactive chemicals, mitogens, and cytolethal agents[15,16]. Conolly et al.[15,16] have proposed the development of comprehensive exposure response models for chemical carcinogens consisting of biologically based (sub)models for (1) target tissue dosimetry, (2) initial tissue response, and (3) cancer. Initial definition of such comprehensive models, well before the data necessary for their development and validation are available, serves to identify data gaps to provide an explicit framework guiding experimental design and data interpretation.

In the following, the biologically based (PBPK) model for tissue dosimetry of methylene chloride and its use for cancer risk assessment are reviewed. On-going developments which should lead to further refinement of the human health risk assessment for methylene chloride are then discussed.

PBPK MODEL FOR METHYLENE CHLORIDE

Methylene chloride caused significant increases in the incidence of liver and lung tumors in male and female B6C3F1 mice after chronic inhalation of 2000 or 4000

Table 1. Tumor Incidence and Calculated Dose of Methylene Chloride and Its Metabolites Following Inhalation Exposure of 2,000 or 4,000 ppm in Female Mice

			Tissue Dose (mg/L Tissue/day)	
CH_2Cl_2 in ppm	Tumor	GSH Pathway	Oxidative Pathway	CH_2Cl_2
Liver				
0	6	-	-	-
2,000	33	851	3,575	362
4,000	83	1,800	3,701	771
Lung				
0	6	-	-	-
2.000	63	123	1,531	381
4,000	85	256	1,583	794

ppm[17]. The toxic moiety responsible for methylene chloride tumorigenicity has not been identified; however, it is known that potentially reactive intermediates are produced by two major metabolic pathways[18,19,20]. Methylene chloride is metabolized in both target organs, by a cytochrome P-450 mediated oxidative pathway that yields formyl chloride, and by conjugation with GSH yielding chloromethyl GSH.

Using the PBPK modeling approach, information on tissue dosimetry of parent chemical and its metabolites in the most sensitive test species (the mouse) was obtained[21]. A target tissue dose surrogate which correlated well with the tumor levels seen in the NTP bioassay was used to estimate an acceptable exposure concentration for humans[22]. These predictions were then compared to those obtained with the conventional risk assessment approach adopted by the U.S. Environmental Protection Agency[22] (U.S. EPA).

Model Development

The PBPK model for methylene chloride consisted of the following tissue compartments: liver, lung, fat, slowly perfused tissues and richly perfused tissues[21]. The rate of change in the amount of methylene chloride in the tissue compartments (dA_i/dt) was described by a series of mass balance differential equations of the following form:

$$\frac{dA_i}{dt} = Q_i \cdot (C_a - C_{vi}) - \frac{dA_{met}}{dt}$$

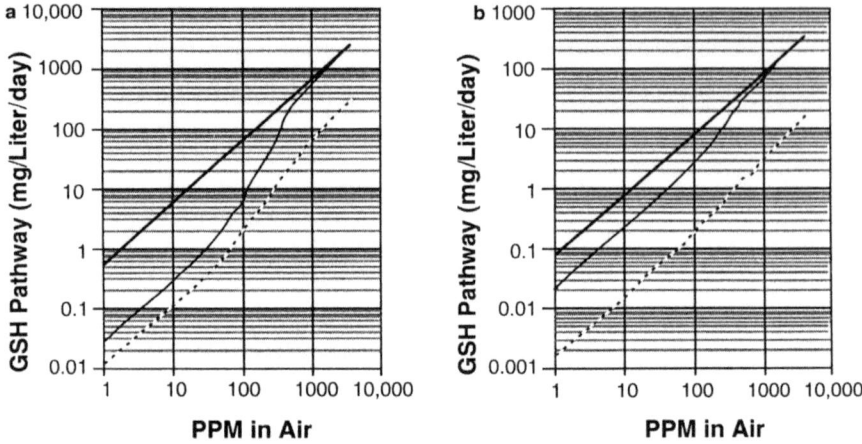

Fig. 1. Relationship between tissue dose and external exposure concentration for humans and mice for the glutathione pathway. The physiological modeling approach was used to determine the target tissue dose from the GSH pathway in both liver (a) and lung (b) at various exposure concentrations. Mice (---) and humans (- - -). The lighter solid line depicts linear back extrapolation from 1 ppm (from Andersen et al.[21])

where Q_i is the rate of blood flow to tissue "i" (L/hr), C_a is the methylene chloride concentration in arterial blood (mg/L), C_{vi} is the methylene chloride concentration in venous blood leaving tissue "i" (mg/L), and dA_{met}/dt is the rate of methylene chloride metabolism (mg/hr).

The rate of methylene chloride metabolism per unit time in the liver and lung was described by accounting for both microsomal oxidation, a saturable process, and GSH conjugation, an effectively first order process at all exposure concentrations used in the NTP cancer bioassay:

$$\frac{dA_{met}}{dt} = \frac{V_{max} \cdot C_{vi}}{K_m + C_{vi}} + K_f \cdot C_{vi} \cdot V_i$$

where V_{max} is the maximum enzymatic reaction rate (mg/hr), K_m is the Michaelis constant (mg/L), K_f is the first order rate constant for GSH conjugation (1/hr), and V_i is the volume of tissue "i" (L).

The physiological parameters required for the PBPK model (i.e., alveolar ventilation rate, blood flow rates, tissue volumes) were obtained from the literature[23,24]. The blood:air and tissue:air partition coefficients for methylene chloride were determined by vial equilibration[25,26]. The tissue:blood partition coefficients required for the model were determined by dividing tissue:air values by the blood:air value. The rate constants for methylene chloride metabolism were determined by apportioning the whole body metabolic capacity between lung and liver by assuming that the distribution of enzyme activities metabolizing methylene chloride was the same as the distribution of enzyme activities acting upon two model substrates, 7-ethoxycoumarin for microsomal oxidation, and 2,5-dinitrochlorobenzene for GSH conjugation[27].

The mouse PBPK description, once validated by comparing model predictions with observed pharmacokinetic data, was scaled to predict the tissue dosimetry of methylene chloride and its metabolites in humans. This was accomplished by scaling the physiological parameters of the model, and determining chemical-specific parameters for humans. Thus, the tissue:blood partition coefficients for humans were calculated by dividing mouse tissue:air partition coefficients by the human blood:air partition coefficient. Further, the metabolic rate constants for humans were estimated from volunteer human exposure studies, in which levels of methylene chloride and carboxyhemoglobin in blood were determined during and following a 6-hr exposure to 100 and 350 ppm methylene chloride[28]. The level of GSH S-transferase activity in humans was set equal to the highest activity reported in rodents.

Choice of Dose Surrogate to Correlate With Tumor Formation

The mouse PBPK model for methylene chloride, formulated by integrating information on mouse physiology, methylene chloride solubility characteristics, and metabolic rate constants, was successfully utilized to describe the disposition of methylene chloride[21]. The mouse PBPK model was then used to calculate the tissue dose of metabolites and parent chemical arising from exposure scenarios comparable to those of the NTP bioassay studies and their relationship to the observed tumor incidence was examined. Since methylene chloride is unreactive, it is unlikely to be directly involved in its tumorigenicity. Hence the relationship between the tissue exposure to its metabolites and tumor incidence was examined (Table 1). While the dose surrogate based on the oxidative pathway did not vary between methylene chloride exposure concentrations of 2000 and 4000 ppm, the flux through the GSH conjugation pathway and tumor incidence both increased in rough proportion with exposure concentration. These observations are consistent with a role for the metabolite(s) arising from the GSH conjugation pathway in methylene chloride-induced lung and liver cancer. The conjugation of methylene chloride with GSH, reported to be mediated by a new class of GSH S-transferase enzymes[29], yields formaldehyde as a metabolite. Very recently, Casanova et al.[30] have reported DNA-protein crosslinks in mouse liver after methylene chloride exposure. These cross-links are due to formaldehyde, a metabolite of methylene chloride. This provides an opportunity to compare kinetics of methylene chloride metabolism by the GSH-dependent and oxidative pathways with the formation of a DNA-reactive metabolite.

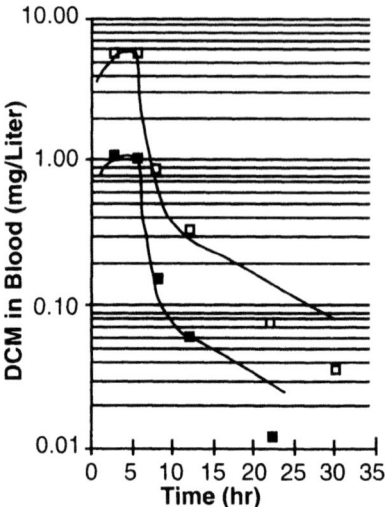

Fig. 2. Comparison of model simulations (solid lines) with data (symbols) on blood methylene chloride levels in humans during and following six-hour inhalation exposure of 100 or 350 ppm. (From Andersen et al.[21])

High Dose - Low Dose Extrapolation

The model prediction of the target tissue dose of the methylene chloride-GSH conjugate resulting from 6-hr inhalation exposures of 1 to 4000 ppm of methylene chloride is presented in Figure 1. The estimation of target tissue dose of methylene chloride-GSH conjugate by linear back-extrapolation gives rise to 21-fold higher estimate than that obtained by the PBPK modeling approach. This discrepancy arises from the nonlinear behavior of methylene chloride metabolism at high exposure concentrations. At exposure concentrations exceeding 300 ppm, the cytochrome P-450 mediated oxidation pathway is saturated giving rise to a corresponding disproportionate increase in the flux through GSH conjugation pathway.

Interspecies Extrapolation

Interspecies extrapolation of the model describing methylene chloride disposition was

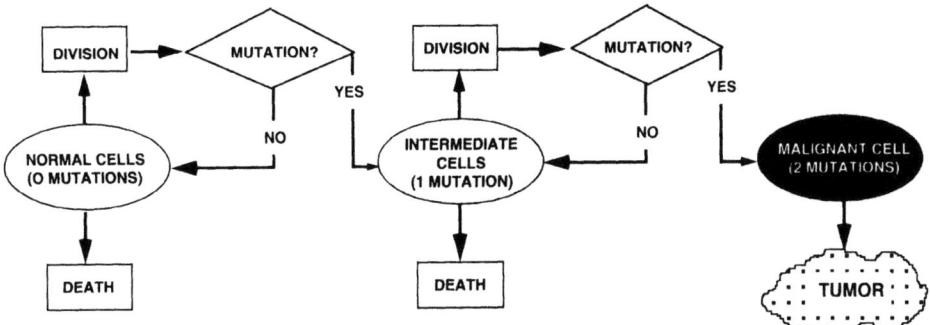

Fig. 3. Two-stage cell growth and mutation model for cancer. Cells are either normal, intermediate, or malignant. Normal and intermediate cells divide and die. Critical, pro-carcinogenic mutations may occur spontaneously, or due to effects of carcinogenic chemicals, during cell division. Accumulation of two such mutations produces a malignant cell which expands clonally into a tumor.

possible because the critical determinants of this disposition were known for the mouse. Thus, the physiological parameters were scaled allometrically, the metabolic parameters were determined experimentally, and the tissue:air partitioning of methylene chloride was assumed to be species-invariant. The PBPK model adequately simulated the blood levels of methylene chloride observed in humans after a 6-hr inhalation exposure to 100 or 350 ppm methylene chloride (Figure 2). The target tissue dose of methylene chloride to humans was estimated to be some 2.7 times lower per unit of tissue volume than that for the mouse. The human tissue dose of methylene chloride-GSH conjugate for a 6-hr exposure to 1 ppm methylene chloride was predicted by the PBPK model to be some 57 times lower than that expected by linear extrapolation of its behavior at high doses, such as the doses used in the mouse bioassay[21].

Risk Assessment

The cancer risk assessment for methylene chloride was conducted using the LMS model to relate tissue dose of the methylene chloride-GSH conjugate (instead of exposure concentration) to the observed tumor incidence rates in the mouse. In assessing the tumorigenic risks associated with human exposure to this chemical, it was assumed that humans are as sensitive as the mouse, the most sensitive target species. Therefore, equal target tissue doses are expected to produce similar tumor incidence regardless of the species. This conclusion is in contrast to that obtained by U.S. EPA, which estimates that people are more sensitive than mice based on the use of a surface area scaling approach[31]. In the human methylene chloride risk assessment based on the PBPK model using the GSH S-transferase pathway dose, the predicted human low dose cancer risk was about 100 to 200 fold less than that estimated by U.S. EPA using their standard default assumptions[21]. With refinement of the model, by estimation of the metabolic rate constants for humans in vitro, Reitz et al.[22], using the delivered dose calculated with the PBPK model, predicted a cancer risk of 3.7×10^{-8} for a lifetime inhalation exposure of 1 mg/m^3. This risk estimate is lower, by more than two orders of magnitude, than that calculated by U.S. EPA (4.1×10^{-6}) using default assumptions and the methylene chloride exposure concentration. The U.S. EPA has amended its original risk assessment for methylene chloride[32] and incorporated some but not all of the concepts used in the physiological pharmacokinetics-based risk assessment approach outlined here.

The use of PBPK models in quantitative risk assessment does not necessarily result in the estimation of lower risk than the conventional approach adopted by U.S. EPA. For example, if the test chemical is a directly-acting agent, the PBPK approach could actually predict more risk to humans than rodents because the rate of enzyme-mediated metabolic clearance (detoxification) is expected to be slower in the larger species. Similarly, if the toxicity of a chemical is mediated by reactive intermediate(s) resulting from a saturable metabolic process, then the high dose to low dose extrapolation conducted with the PBPK modeling approach would predict a greater risk at low doses than that predicted by the linear extrapolation procedure.

OUTSTANDING ISSUES: WHERE DO WE GO FROM HERE?

Mechanistic Studies to Assess Methylene Chloride-GSH Dose Surrogate

Use of the PBPK model to identify a correlation between the flux of metabolites via the GSH pathway and tumor incidence was a significant advance. However, there are not yet any unambiguous data showing a direct mechanistic link of components of this pathway and tumors. Recent work showing DNA-protein cross-link formation in mice after methylene chloride exposure[30] via the formaldehyde metabolite is promising. Formaldehdye is carcinogenic to the nasal respiratory epithelium of the F344 rat[33]. Formaldehyde-stimulated tumors only arise under exposure conditions causing increased cell replication, presumably secondary to cytolethality[34,35,36]. Methylene chloride has not been reported to stimulate cell division in the mouse liver, though it apparently is toxic to selected cell types in mouse lung[37]. Evaluation of the role of formaldehyde in the carcinogenicity of methylene chloride will require careful evaluation of potential methylene chloride induced changes in death and division rates of target cell populations in liver and lung.

Exposure to dibromoethane leads to formation of sulfur-containing DNA adducts[38] by a mechanism, involving GSH conjugation, which is similar to that for methylene

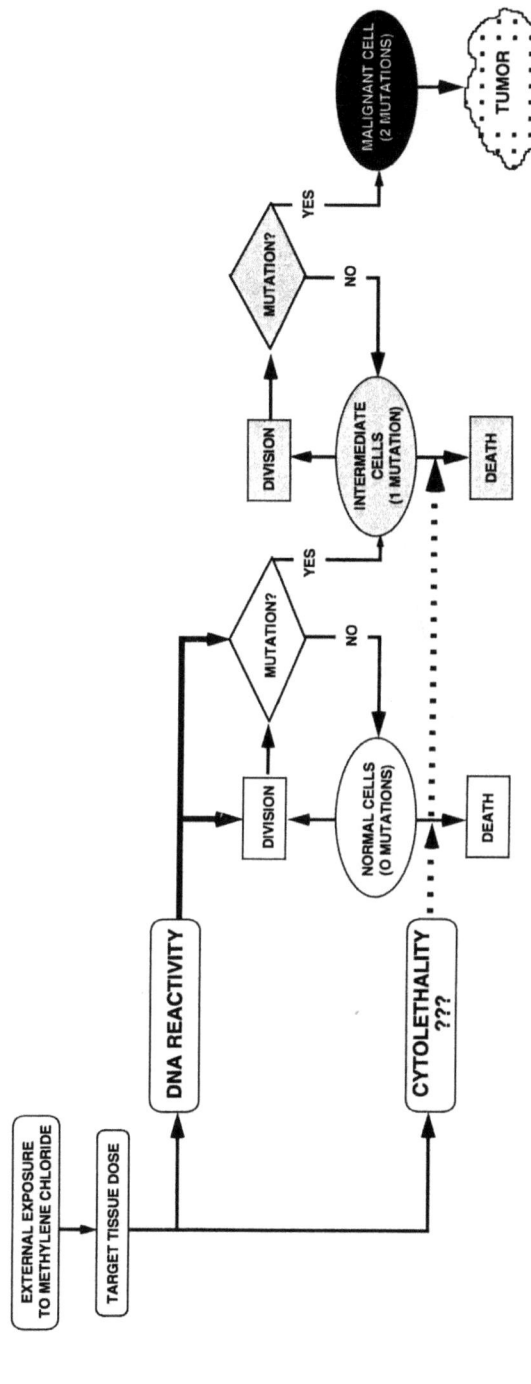

Fig. 4. Possible mechanisms of methylene chloride carcinogenicity. The physiologically-based pharmacokinetic model converts exposure concentration into doses of relevant metabolites in target tissues, i.e., liver and lung. Tumor formation may be stimulated by (1) DNA reactivity of these metabolites, leading to increases in the probability of mutation per cell division or, (2) by cytolethality, leading initially to increases in the death rates of cells. The affected tissue would then undergo compensatory cellular regeneration to replace killed cells, leading to an increased mutation rate, even under conditions where the probability of mutation is not affected. It is quite possible that both mechanisms operate simultaneously, though there is little evidence for cytolethal effects of methylene chloride.

chloride. Although there have been no reports of sulfur containing DNA adducts due to methylene chloride, this possibility should be considered in mechanistic investigations of methylene chloride mutagenicity and carcinogenicity.

Use of Human Tissues for Parameter Estimation

Reitz et al.[22] refined the PBPK-based risk assessment for methylene chloride by measuring its GSH-dependent metabolism in human tissue samples. An *in vitro* to *in vivo* extrapolation was conducted by applying the *in vivo/in vitro* ratio for the mouse to the human in vitro value to predict the human *in vivo* constant. There is an almost total lack of mechanistic data on differences between *in vitro* and *in vivo* rates of xenobiotic metabolism. More work is needed in this area to provide a sound understanding of conditions under which *in vitro* constants are predictive of *in vivo* behavior. This should include mechanistic studies of *in vivo/in vitro* differences and expansion of the data base correlating *in vivo* and *in vitro* rates[39].

Biologically Based Models for Carcinogenic Response

The PBPK-based risk assessment described by Andersen et al.[21] uses dose surrogate predictions from the PBPK model as input to LMS model. In this case the LMS specifies tissue response, i.e., the quantitative relationship between the dose surrogate and tumor formation. This is quite different from the traditional use of the LMS with external measures of exposure such as airborne or drinking water concentration. In these latter cases the LMS is used, albeit implicitly, as a comprehensive model, accounting for both target tissue dosimetry and tissue response. The LMS is not an accurate specification of tissue response in tumor development. For example, there is no good way to incorporate data on cell replication and on intermediate, "preneoplastic" lesions, which play basic roles in the process by which normal cells become malignant.

A logical extension of the current PBPK-based risk assessment would be to combine it with a cancer model, such as the Moolgavkar-Venzon-Knudson model[15,16] (Figure 3), which provides a more realistic description of tumor development. Conolly et al.[15,16] have described generic approaches to the construction of such comprehensive, biologically based models. For the MVK model, time course data on (1) target tissue replication rates for normal, preneoplastic, and neoplastic cells and, (2) number and size of preneoplastic lesions and tumors are required. A critical aspect is specification of the relationship between target tissue dosimetry and the parameters of the cancer model. Consider, for example, how the methylene chloride-GSH dose surrogate might be linked to an MVK model. It is important to realize that the surrogate is only a dosimeter. At present we have no hard data on its mechanistic significance, if any. The surrogate may be linked to the cell turnover parameters of the model, or to the probabilities of mutation per cell division, or to both the cell turnover and mutation parameters (Figure 4). A proportionality constant defines increases in the values of the parameters as a function of amount of dose surrogate. Clearly, insight into the mechanism of carcinogenic action of methylene chloride is needed to guide the specification of this linkage. A biologically-based risk assessment strategy of this type has recently been proposed for formaldehdye[40] where there is a much richer data base on tissue effects and tumor progression.

Variability of Risk Estimates

The use of point estimates of model parameter values, rather than distributions of values for each parameter, is an important issue in both dose-response and interspecies extrapolation. Using ranges of parameter values leads to ranges of risk estimates[41], which is reasonable, given the heterogenous human population. The issue deserves serious attention from a generic point-of-view, not solely as it applies to PBPK model-based assessments. Presently, default appraoches to interspecies scaling take little explicit account of variability in population characteristics to derive ranges of risk.

Many of the parameters in PBPK models can be measured in the laboratory. Thus, the variability of parameter estimates can be addressed experimentally. Some work has recently been done in this area[41,42,43]. These studies illustrate the need for data sets sufficient to allow realistic inferences about parameter variability and for refined statistical approaches using assumptions consistent with complex, interacting biological processes.

REFERENCES

1. J. C. Ramsey and M. E. Andersen, 1984, A physiologically based description of the inhalation pharmacokinetics of styrene in rats and humans, Toxicol. Appl. Pharmacol., 73:159.
2. J. M. Gearhart, G. W. Jepson, H. J. Clewell, III, M. E. Andersen and R. B. Conolly, 1990, A physiologically-based model for in vivo inhibition of acetylcholinesterase by diisopropylfluorophosphate, Toxicol. Appl. Pharmacol., 106:295.
3. K. B. Bischoff, R. L. Dedrick, D. S. Zaharko and J. A. Longstreth, Methotrexate pharmacokinetics, J. Pharm. Sci., 60:1128.
4. P. F. Morrison, R. L. Dedrick and R. J. Lutz, 1971, Methotrexate: Pharmacokinetics and assessment of toxicity, in: "Pharmacokinetics and Risk Assessment," 8:410, Natl. Academy Press, Washington, D.C.
5. H. W. Leung, D. J. Paustenbach, F. J. Murray and M. E. Andersen, 1990, A physiological pharmacokinetic description of the tissue disposition and enzyme inducing properties of 2,3,7,8,-tetrachlorodibenzo-p-dioxin in the rat, Toxicol. Appl. Pharmacol., 103:399.
6. H. W. Leung, A. P. Poland, D. J. Paustenbach and M. E. Andersen, 1990, Dose-dependent pharmacokinetics of [^{125}I]-2-iodo-3,7,8-trichlorodibenzo-p-dioxin in mice: Analysis with a physiological modeling approach, Toxicol. Appl. Pharmacol., 103:411.
7. H. B. Matthews and R. L. Dedrick, 1984, Pharmacokinetics of PCBs, Annu. Rev. Pharmacol. Toxicol., 24:85.
8. H. J. Clewell, III and M. E. Andersen, 1987, Dose, species, and route extrapolations with a physiologically based model, in: "Drinking Water and Health," 8:159-182.
9. M. L. Gargas and M. E. Andersen, 1988, Physiologically based approaches for examining the pharmacokinetics of inhaled vapors, in: "Toxicology of the Lung," J. Crapo, E. Masaro and D. Carpenter, eds., Raven Press, pp. 449-476.
10. R. W. D'Souza and M. E. Andersen, 1988, Physiologically-based pharmacokinetic model for vinylidene chloride, Toxicol. Appl. Pharmacol., 95:230.

11. R. H. Reitz, A. L. Mendrala, R. A. Corley, J. F. Quast, M. L. Gargas, M. E. Andersen, D. A. Staats and R. B. Conolly, 1990, Estimating the risk of liver cancer associated with human exposures to chloroform using physiologically based pharmacokinetic modeling, Toxicol. Appl. Pharmacol., 105:443.
12. S. H. Moolgavkar and D. J. Venzon, 1979, Two-event models for carcinogenesis: Incidence curves for childhood and adult tumors, Math. Biosci., 47:55.
13. S. H. Moolgavkar and A. G. Knudson, 1981, Mutation and cancer: A model for human carcinogenesis, J. Natl. Cancer Inst., 66:1037.
14. S. H. Moolgavkar, N. E. Day and R. G. Stevens, 1980, Two-stage model for carcinogenesis: Epidemiology of breast cancer in females, J. Natl. Cancer Inst., 65:559.
15. R. B. Conolly, R. H. Reitz, H. J. Clewell, III and M. E. Andersen, 1988, Pharmacokinetics, biochemical mechanism and mutation accumulation: A comprehensive model of chemical carcinogenesis. Toxicol. Lett. 43:189.
16. R. B. Conolly, R. H. Reitz, H. J. Clewell, III and Andersen, M.E., 1988, Biologically-structured models and computer simulation: Application to chemical carcinogenesis, Comments on Toxicology 2:305.
17. National Toxicology Program (1985) NTP Technical Report on the toxicology and carcinogenesis studies of dichloromethane in F-344 rats and B6C3F1 mice (inhalation studies). NTP TR No. 306.
18. V. L. Kubic, M. W. Anders, R. R. Engel, C. H. Barlow and W. S. Caughey, 1974, Metabolism of dihalomethanes to carbon monoxide, Drug Metab. Dispos., 2:53.
19. M. L. Gargas, M. E. Andersen and H. J. Clewell, III, 1986, Metabolism of inhaled dihalomethanes: Differentiation of kinetic constants for two independent pathways. Toxicol. Appl. Pharmacol., 87:211.
20. R. H. Reitz, F. A. Smith and M. E. Andersen, 1986, In vivo metabolism of ^{14}C-methylene chloride, Toxicologist, 6:260.
21. M. E. Andersen, H. J. Clewell, III, M. L. Gargas, F. A. Smith and R. H. Reitz, 1987, Physiologically based pharmacokinetics and the risk assessment process for methylene chloride, Toxicol. Appl. Pharmacol., 87:185.
22. R. H. Reitz, A. L. Mendrala, C. N. Park, M. E. Andersen and F. P. Guengerich, 1988, Incorporation of in vitro enzyme data into the physiologically-based pharmacokinetic (PB-PK) model for methylene chloride: Implications for risk assessment, Toxicol. Lett., 43:97.
23. W. O. Caster, J. Poncelet, A. B. Simon and W. D. Armstrong, 1956, Tissue weights of the rat. I. Normal values determined by dissection and chemical methods, Proc. Soc. Exp. Biol. Med., 91:122.
24. International Commission on Radiation Protection, 1975, Report of the task group on reference man. ICRP Publication No 23, Pergamon, NY.
25. A. Sato and T. Nakajima, 1979, Partition coefficients of some aromatic hydrocarbons and ketones in water, blood and oil, Br. J. Ind. Med., 36:31.
26. M. L. Gargas, R. J. Burgess, D. E. Voisard, G. H. Cason and M. E. Andersen, 1989, Partition coefficients of low molecular weight volatile liquids and tissues, Toxicol. Appl. Pharmacol., 98:87.
27. J. Lorenz, H. R. Glatt, R. Fleischmann, R. Ferlinz and F. Oesch, 1984, Drug metabolism in man and its relationship to that in three rodent species. Monooxygenase, epoxide hydrolase and GSH S-transferase activities in subcellular fractions of lung and liver. Biochem. Med., 32:43.

28. M. E. Andersen, H. J. Clewell, III, M. L. Gargas, M. J. MacNaughton, R. H. Reitz, R. Nolan and M. McKenna, 1991, Physiologically based pharmacokinetic modeling with dichloromethane, its metabolite carbon monoxide and blood carboxyhaemoglobin in rats and humans, Toxicol. Appl. Pharmacol., 108:14.
29. Meyer, D.J., Coles, B., Pemble, S.E., Gilmore, K.S., Fraser, G.M., and Ketterer, B. 1991. Theta, a new class of glutathione transferases purified from rat and man, Biochem. J., 274:409.
30. M. Casanova, H. d'Heck and D. F. Deyo, 1991, Dichloromethane: Metabolism to formaldehyde and formation of DNA-protein crosslinks in mice and hamsters, Toxicologist 11:180.
31. D. V. Singh, H. L. Spitzer and P. D. White, 1985, Addendum to the health risk assessment for dichloromethane. Updated carcinogenicity assessment for dichloromethane, EPA/600/8-82/004F.
32. United States Environmental Protection Agency, (1987), Update to the health assessment document and addendum for dichloromethane: pharmacokientics, mechanism of action and epidemiology. EPA 600/8-87/030A.
33. W. D. Kerns, K. L. Pavkov, D. J. Donofrio, E. J. Gralla and J. A. Swenberg, 1983, Carcinogenicity of formaldehyde in rats and mice after long-term inhalation exposure, Cancer Res., 43:4382.
34. T. M. Monticello, 1990, Formaldehyde-induced pathology and cell proliferation, Ph.D. Dissertation, Duke University, Durham, NC.
35. T. M. Monticello, F. J. Miller and K. T. Morgan, 1991, Regional increases in rat nasal respiratory epithelial cell proliferation following acute and subacute inhalation of formaldehyde, Toxicol. Appl. Pharmacol., 111:409.
36. K. T. Morgan and T. M. Monticello, 1990, Formaldehyde toxicity: Respiratory epithelial cell injury and repair, in: "Biology, Toxicology and Carcinogenesis of Respiratory Epithelium," D. G. Thomassen and P. Nettesheim, eds., Hemisphere Publishing Inc., New York, pp. 155-171.
37. T. Green, 1989, A biological data base for methylene chloride risk assessment, in: "Biologically Based Models for Cancer Risk Assessment," C. C. Travis, ed., NATO ASI Series, Series A: Life Sciences, V. 159, Plenum Press, pp. 289-300.
38. D. H. Kim and F. P. Guengerich, 1990, Formation of the DNA adduct S-[2-(N^7-guanyl)ethyl]glutathione from ethylene dibromide: Effects of modulation of glutathione and glutathione S-transferase levels and lack of a role for sulfation, Carcinogenesis, 11:419.
39. M. L. Gargas and M. E. Andersen, 1992, Kinetic constants for biotransformation reactions of volatile organic chemicals (VOCs): In vivo/in vitro comparisons, The Toxicologist, 12:000.
40. R. B. Conolly, T. M. Monticello, K. T. Morgan, H. J. Clewell, III, and M. E. Andersen, 1991, Strategy for biologically-based risk assessment of inhaled formaldehyde, Comments on Toxicology, 4:269.
41. C. J. Portier and N. L. Kaplan, 1989, Variability of safe dose estimates when using complicated models of the carcinogenic process, Fundam. Appl. Toxicol., 13:533.
42. F. Y. Bois, L. Zeise and T. N. Tozer, 1990, Precision and sensitivity of pharmacokinetic models for cancer risk assessment: Tetrachloroethylene in mice, rats, and humans, Toxicol. Appl. Pharmacol., 102:300.
43. F. Y. Bois, T. J. Woodruff and R. C. Spear, 1991, Comparison of three physiologically based pharmacokinetic models of benzene disposition, Toxicol. Appl. Pharmacol., 110:79.

CANCER DOSE-RESPONSE MODELING AND METHYLENE CHLORIDE

Gail Charnley

Consultant in Toxicology
Arlington, V.A.

ABSTRACT

The two-stage dynamic model for cancer dose-response modeling has advantages over standard empirical procedures because it reflects the biologic mechanisms of carcinogenesis. The biologic mechanisms of methylene chloride-induced rodent tumorigenesis are still controversial, however. As a result, it is impossible to apply a biologically-based modeling procedure until its activity is clarified. Several possible models are proposed based on different assumptions regarding its mechanisms, along with the experiments that would be required to distinguish among them. Study of the roles of oncogenes in methylene chloride-induced rodent carcinogenesis may both clarify the relevance of observations in rodents to human carcinogenesis, as well as permit an appropriate choice of dose-response model for estimating low-dose risks.

INTRODUCTION

Dose-response modeling of putative carcinogens for regulatory purposes has been an evolutionary process. Earlier in this symposium, Drs. Anderson and Moolgavkar discussed this evolution, from models based on radiation biology and a multistage, genotoxic theory of carcinogenesis to a two-stage dynamic model that incorporates information on rates of cell proliferation as well as on genotoxicity. As data availability permits, use of a dose-response model that reflects the biologic mechanisms of action of a carcinogen is more likely to produce an accurate estimate of cancer risk.

The biologic mechanisms of action of methylene chloride are still controversial. In general, it is weakly genotoxic in some microorganisms and not genotoxic in mammalian systems. There is no conclusive evidence that it alkylates DNA. It enhances tumor rates in organs that already have significant spontaneous tumor rates, such as the B6C3F$_1$ mouse liver. Tumors occur at doses that also produce toxicity. These observations are consistent with a non genotoxic mechanism of cancer induction.

Regulatory agencies in the United States are not yet comfortable with the widespread application of biologically-based dose-response models to non genotoxic agents. Such applications are made using the assumption that these agents increase rates of cell proliferation and as a consequence, rates of spontaneous mutation, to produce tumors. In the absence of conclusive evidence that a substance is non genotoxic and in the absence of sufficient dose-response data for phenomena such as increased rates of cell proliferation, the conservative and health-protective course of action is to regulate on the basis of the results of the linearized multistage model. Following is a discussion of some different kinds of dose-response models for methylene chloride that could be developed using different assumptions and the data that would be required to develop them.

CANCER DOSE-RESPONSE MODELING

The cancer potency factors that the U.S. Environmental Protection Agency uses to estimate potential human cancer risks associated with exposure to environmental contaminants are usually based on the results of bioassays conducted using laboratory animals and doses of agents that are several orders of magnitude greater than those to which humans are exposed environmentally. Extrapolation from the dose-response relationship obtained at high doses to lower environmental doses is generally performed using the linearized multistage model (Anderson et al. 1983). This model is based on the assumption that all carcinogens are tumor initiators and exert their effects exclusively by virtue of their ability to interact directly with DNA and cause mutations, although it contains enough arbitrary constants to be able to fit almost any monotonically increasing dose-response data. As a result, it is an empirical curve-fitting procedure with questionable biological relevance.

A more plausible general alternative is the two-stage model described by Moolgavkar, Venzon, and Knudson (1979, 1981), which has advantages over standard empirical modeling techniques because it permits incorporation of information on the biological mechanisms of action of an agent into the dose-response equations, when such information is available. For example, genotoxic carcinogens can be modeled on the basis of their effects on the transition rates between cell types (normal → initiated → cancerous). In addition, many putative carcinogens are not mutagens and may be classified as tumor promoters. One of the ways tumor promoters are thought to increase rates of tumor formation is by increasing rates of cell proliferation. Using the two-stage model, dose-dependent effects on rates of cell proliferation can be modeled as well.

The disadvantage of a biologically-based model is that it requires knowledge of an agent's mechanisms of action. Information other than a rodent tumor response may be unavailable for an agent, permitting little insight into its mechanism of carcinogenesis. In such a case, use of an empirical dose-response modeling procedure is the sole alternative. Where mechanistic data are available, however, their use in estimating cancer risk is likely to produce a more accurate and scientifically defensible estimate.

METHYLENE CHLORIDE TUMORIGENESIS

There have been two studies of the potential carcinogenicity of methylene chloride in humans exposed occupationally. These were cohort mortality studies performed on Kodak employees exposed to 30-120 ppm methylene chloride (Friedlander et al. 1978, Friedlander et al. 1985, Hearne and Friedlander 1981) and on fiber production workers exposed to 0-1,700 ppm (Ott et al. 1983a,b). The former study reported a suggestive relationship between methylene chloride exposure and an increased risk of pancreatic cancer, as well as potentially reduced risks of colon, prostate, and bladder cancer, although none of the differences was statistically significant or dose-related. The latter study reported a possible elevation of liver and biliary tract cancers (SMR = 20). Neither study was controlled for smoking.

Of the five inhalation bioassays in laboratory rodents, increases in salivary gland sarcomas in male rats, increases in benign mammary gland tumors in female rats, and increases in alveolar/bronchiolar adenomas and carcinomas as well as hepatocellular adenomas and carcinomas in mice were reported (Burek et al. 1984, Dow 1980, NTP 1986). No toxicity was reported in hamsters (Burek et al. 1984). The drinking water bioassay performed in rats was negative but that in mice reported an increased incidence of liver tumors when adenomas and carcinomas were combined (Serota 1986a,b; National Coffee Association 1982, 1983). All of the studies in rats and mice also reported liver toxicity, including hepatocellular vacuolization and multinucleation, hemosiderosis, bile duct fibrosis, fatty degeneration, necrosis, and the appearance of foci of cellular alteration.

Studies of methylene chloride's potential carcinogenicity thus do not demonstrate a clear relationship between exposure and cancer incidence in humans, within the constraints of the studies performed, while animal bioassay data support an association between lung and liver tumor incidences in mice, although in only one study were liver tumors statistically elevated. The U.S. EPA classifies methylene chloride as a B2 or probable human carcinogen, based on inadequate human data and sufficient evidence of carcinogenicity in animals. Its cancer potency factor for oral exposure was based on combined hepatocellular adenomas and carcinomas from the NTP study and combined hepatocellular carcinomas and "neoplastic nodules" from the National Coffee Association study, while the potency factor for inhalation exposure was based on combined adenomas and carcinomas in the mouse liver and lung from the NTP study. Both were estimated using the linearized multistage procedure.

Mechanistic data relating to methylene chloride tumorigenesis are controversial. While bacterial assays for point mutation such as *Salmonella typhimurium* have yielded positive results, *in vitro* mammalian assays for genotoxicity have been equivocal and *in vivo* mammalian assays have been negative. No binding to DNA has been detected within the limits of the assays performed. While chronic bioassays produced liver toxicity and proliferative lesions, only small, variable increases in hepatocyte proliferation have been demonstrated. Enhanced liver tumorigenesis in the $B6C3F_1$ mouse is suggestive, but not at all conclusive, of a non genotoxic mechanism of action.

Thus there is no clear evidence to support either a genotoxic or non genotoxic mechanism for methylene chloride's tumorigenesis in rodents. Applying a biologically based dose-response model to estimate its cancer risk is consequently risky, if not impossible, although applying models based on different assumptions about its mechanism might, through goodness-of-fit comparisons, indicate whether one mechanism is more likely than another.

SPECULATIONS ON CANCER DOSE-RESPONSE MODELING OF LIVER LESIONS

Developing a dose-response model for methylene chloride using the two-stage paradigm requires assumptions about the stage in the model at which it might exert its activity. It may be a mutagen and increase the transition rates from normal to initiated to cancerous cells, or it may be non genotoxic and act by increasing cell proliferation rates. If the latter, it could increase either the proliferation rates of normal stem cells or of initiated cells.

It is apparent that the target organ for methylene chloride toxicity in most species is the liver. It causes histopathologic changes that are indicative of alterations in cells' abilities to control normal proliferation and differentiation, especially foci of cellular alteration, adenomas, and carcinomas. These changes could arise by either genotoxic or non genotoxic mechanisms. A number of assumptions must be made to classify the adenomas and other proliferative hepatic lesions within the context of the model. While hepatocellular carcinomas are clearly malignant, the status of the other lesions is uncertain.

Hepatic foci of cellular alteration can be distinguished morphologically from surrounding tissue by a number of phenotypic and enzymatic markers. They do not show any disruption of normal hepatic lobular architecture although they often exhibit elevated rates of proliferation as compared to surrounding tissue. Adenomas develop later than, and are generally larger than, altered foci. They have lost some of the characteristics of normal cellular architecture, tend to compress surrounding tissue, and are clearly distinguished from it. In the absence of continued chemical exposure, foci and adenomas have been shown to regress. Within the context of the two-stage model, altered foci may be considered to have one of three roles: as initiated cells, as normal cells, or as being irrelevant. These three possibilities are depicted in Figure 1.

EPA (1986) has proposed a hypothetical scheme for neoplastic development in the rat liver:

$$\text{normal hepatocyte} \rightarrow \text{initiated cell} \rightarrow \text{focus} \rightarrow \rightarrow \rightarrow \text{carcinoma}$$
$$\searrow \qquad \nearrow$$
$$\text{adenoma}$$

The general acceptance of the characterization of hepatocellular foci as initiated cells and their progression to adenoma and possibly carcinoma is based primarily on indirect, morphological evidence. The observations that some carcinomas have arisen within foci (Shulte-Hermann et al. 1982) and carcinomas within adenomas (Goldfarb 1973, Farber 1982) are consistent with this characterization. Adenomas would thus also be initiated

Initiated Foci
   Approach

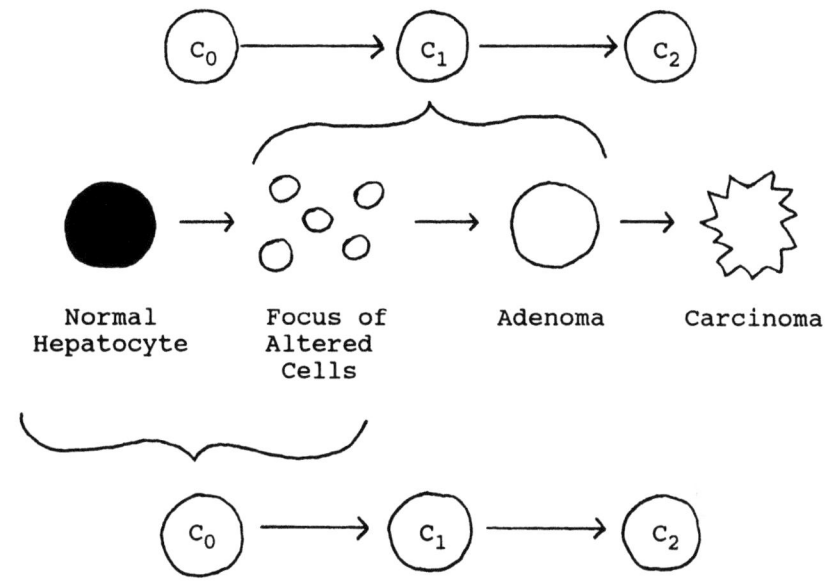

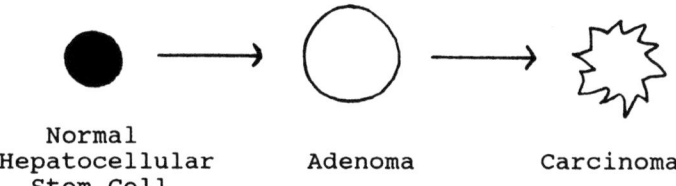

Irrelevant Foci
   Approach

Fig. 1. Classification scenarios for hepatocellular lesions within the two-stage model framework. C_0, C_1, and C_2 are normal, initiated, and cancerous cells, respectively.

cells, so for the purpose of the two-stage model, altered foci and adenomas would be considered together to represent the initiated cell population.

Alternatively, altered foci could be considered to be genotypically normal, although phenotypically abnormal, hepatocytes that are proliferating more rapidly than usual. Foci would thus represent areas of rapidly proliferating normal cells that are susceptible to an enhanced spontaneous mutation rate, which could lead to an increase in the number of initiated, or adenoma, cells that are formed. This possibility is supported by the fact that altered foci are described on the basis of morphologic and phenotypic changes in very labile cytoplasmic constituents, in the absence of any direct evidence to demonstrate that irreversible genetic alterations have occurred. Morphologic and phenotypic changes can occur in the absence of mutation. While it is recognized that foci are areas of focal hyperplasia, hyperplasia in other organs has not been considered a positive end point for determining carcinogenicity. Thus there may be no more compelling evidence that rodent liver foci are necessarily preneoplastic than that other forms of focal epithelial hyperplasia are necessarily preneoplastic.

Finally, foci may be irrelevant to hepatocarcinogenesis. They may arise independently as a result of exposure to hepatocarcinogens and represent one of several pleiotropic effects of exposure. Investigations of the comparative developmental and phenotypic properties of altered foci and hepatic tumors in rats following initiation by diethylnitrosamine and promotion by phenobarbital have provided evidence that these lesions are developmentally independent (Peraino et al. 1988). Comparisons of the dose-response and growth characteristics of enzyme-altered foci, morphologically altered foci, and tumors indicate that these lesions appear to develop independently, with the developmental characteristics of each lesion class determined at the time of initiation. Using the detection of histochemical markers as the criterion for focus identification and phenotypic characterization, Peraino et al. (1988) concluded that foci are phenotypically stable and do not evolve through progressively more deviated forms into tumors. Foci and tumors were considered to be separate manifestations of carcinogen activity. In this case, adenoma would be considered initiated cells for the purpose of the two-stage model.

The role of foci in hepatocarcinogenesis remains controversial. Studies of patterns of oncogene activation during hepatocarcinogenesis may help clarify their role. Until that time, dose-response models should consider all three possible roles.

SPECULATIONS ON CANCER DOSE-RESPONSE MODELING OF METHYLENE CHLORIDE

There are several roles that methylene chloride may play in the process of rodent hepatocarcinogenesis. Possible roles and the dose-response models that would reflect those roles are described below.

Methylene chloride increases stem cell proliferation

As mentioned above, observations of the continuum of morphological changes that precede liver cancer (altered foci, adenoma) have led many investigators to conclude that hepatocellular carcinoma arises by dedifferentiation of adult liver cells. It is likely that there is a subpopulation of mature hepatocytes that is capable of dedifferentiation and proliferation, as is observed following partial hepatectomy. Other evidence has shown, however, that large numbers of small, nondescript "oval cells" appear and proliferate in the livers of rodents exposed to carcinogens (Sell and Dunsford 1989). These cells arise early after carcinogen exposure, develop prior to the appearance of other morphological alterations such as altered foci, may differentiate into hepatocytes or ductal cells, and have been found to proliferate in most, although not all, models of carcinogenesis (Sell 1990). In contrast to hepatocytes, oval cells are resistant to the toxic effects of carcinogens (Sells et al. 1981). These observations suggest that the oval cell or a precursor may be the susceptible stem cell for hepatocarcinogenesis. In either case, it is plausible that stem cells would proliferate in response to methylene chloride-induced hepatotoxicity. There are no data to support this hypothesis at present.

Under the assumption that increased stem cell proliferation leads to an increased spontaneous rate of mutation and ultimately, to an increased hepatocellular tumor rate, the probability that a tumor will develop by time t under conditions of constant exposure to methylene chloride at dose x is proportional to the dose-related increase in the number of susceptible stem cells:

$$P(x,t) = 1 - e^{-\frac{M e^{G(x)t} - 1 - G(x)t}{G(x)^2}}$$

where $M = M_0 M_1$. Using this model, the growth rate of susceptible stem cells, $G(x)$, may be expressed in the form:

$$G(x) = G(0) + (G(\infty) - G(0)) \cdot R(x)$$

where $G(0)$ is the normal stem cell growth rate, $G(\infty)$ is the upper bound on the methylene chloride-induced stem cell growth rate, and $R(x)$ is the fraction of the maximal increase in the stem cell growth rate that is induced by a constant exposure to methylene chloride at level x. This increase could be described by a bounded log-logistic-like function on the basis of the following logic: at low doses, the function is bounded by the background number of stem cells while at high doses, it is reasonable to assume that proliferation and therefore the number of stem cells reaches a plateau because it cannot increase indefinitely. This logic has not been examined carefully or validated due to the difficulty inherent in identifying stem cells and their kinetics, although in the case of hepatic oval cells, such an experiment is feasible.

Development of a dose-response model for methylene chloride using the assumption that it increases the number of susceptible stem cells in the liver thus requires not only the identification of the appropriate stem cell population but the dose-response data for this phenomenon, in order to characterize the function $R(x)$.

Methylene chloride is a tumor promoter

One hypothesis regarding the mechanism of action of non genotoxic carcinogens is that they selectively induce the proliferation and clonal expansion of previously initiated cells in a target tissue. Within the context of the two-stage model, methylene chloride's tumorigenicity could thus be proposed to result from its ability to increase the birth rate of intermediate cells while having no direct effect on the transition rates between cell stages. There are no data to support this hypothesis, either.

The probability that a tumor will develop by time t following constant exposure to methylene chloride at level x under this assumption can be expressed as it was above. Using this model, the growth rate of preneoplastic cells would be described as it was above for stem cells and would still require characterization of the functional form of R(x). Experiments to characterize this relationship would be more difficult to perform than those for stem cells, however, due to the greater difficulty inherent in defining and identifying preneoplastic cells. In addition, there is no particular reason to think that methylene chloride selectively increases the proliferation of initiated cells; it appears to be a simple hepatotoxicant without any direct mitogenic or hormonal capabilities such as those of TCDD, for example.

Methylene chloride is genotoxic and affects two loci

Despite a lack of evidence for genotoxicity *in vivo*, there is some evidence that methylene chloride is a weak mutagen and clastogen *in vitro*. In addition, although it has not been demonstrated to alkylate DNA, it is possible that alkylation occurs below the detection limits of the assays that have been performed. Thus there is a possibility that methylene chloride initiates hepatocarcinogenesis, although there is no direct evidence supporting this hypothesis.

Using the two-stage model and assuming that both transition rates are affected by methylene chloride exposure, the probability of tumor development may be expressed as:

$$P(x) = 1 - e^{A\{1+(S+I)x\}\{1+Sx\}}$$

where $1 - e^{-A} > 0$ and is the background, spontaneous tumor rate, $S > 0$ and is the smaller of the relative methylene chloride exposure-dependent transition rates, and $I \geq 0$ and is the incremental change between the largest and smallest relative transition rates. Verification of this model would require some evidence that methylene chloride is genotoxic in the target organ for carcinogenesis; studies of oncogene expression following methylene chloride exposure might provide some basis for such a mechanism in the absence of other evidence.

DISCUSSION

It is clear that much work remains before it is appropriate to model methylene chloride-induced tumorigenesis using the two-stage model. In addition, the dose-response models for methylene chloride that have been speculated upon above are proposed in the absence of consideration of tumors at sites other than the liver as well as of judgments regarding the relevance of methylene chloride-induced rodent tumors to human carcinogenesis. The regulatory agencies of many countries have rejected the U.S. EPA's assessment that methylene chloride is a probable human carcinogen and have chosen not to regulate it is as such. The basis for this decision is the difference in its metabolism between mice and humans and the assertion that the particular mechanism that causes tumors in mice exposed to high doses of methylene chloride does not exist in humans. Further study of the oncogenetic mechanisms of mouse liver tumor induction could confirm whether the tumors seen following methylene chloride exposure are induced by the same mechanism as spontaneous tumors, whether they are induced by an independent mechanism, and whether either of these mechanisms is likely to be operative in humans. Study of the role of oncogenes in methylene chloride-induced tumorigenesis could thus clarify the relevance of the observations in rodents to humans and if found to be relevant, could help elucidate the underlying biologic mechanism that would then permit an appropriate choice of dose-response model for estimating low-dose risks.

REFERENCES

Anderson, E.L. and the Carcinogen Assessment Group of the U.S. Environmental Protection Agency, 1983, Quantitative approaches in use to assess cancer risk, Risk Anal., 3:27.

Burek, J.D., Nitschke, K.D., Bell, T.J., Wadkerle, D.L., Childs, R.C., Beyer, J.E. Dittenber, D.A., Rampy, L.W., and McKenna, M.J., 1984, Methylene chloride: A two year inhalation toxicity and oncogenicity study in rats and hamsters, Fund. Appl. Tox., 4:30.

Dow Chemical Company, 1980, Methylene Chloride: A Two-Year Inhalation Toxicity and Oncogenicity Study in Rats and Hamsters, FYI-OTS-0281-0097, Office of Toxic Substances, U.S. Environmental Protection Agency, Washington, DC.

Environmental Protection Agency (EPA), 1986, Proliferative hepatocellular lesions of the rat: Review and future use in risk assessment, EPA/625/3-86/011, Washington, DC.

Farber, E., 1982, The biology of carcinogen-induced hepatocyte nodules and related liver lesions in the rats, Toxicol. Pathol., 10:197.

Friedlander, B.R., Hearne, T., and Hall, S., 1978, Epidemiologic investigation of employees chronically exposed to methylene chloride, Mortality analysis, J. Occup. Med., 20:657.

Friedlander, B.R., Pifer, J.W., and Hearne, F.T., 1985, 1964 Methylene Chloride Cohort Mortality Study, Update Through 1984, Eastman Kodak Company, Rochester, NY.

Goldfarb, S., 1973, A morphological and histochemical study of carcinogenesis of the liver in rats fed 3'-methyl-4-dimethylaminoazobenzene, Cancer Res., 33:1110.

Hearne, F.T. and Friedlander, B.R., 1981, Follow-up of methylene chloride study, J. Occup. Med., 23:660.

Moolgavkar, S.H. and Knudson, A.G., 1981, Mutation and cancer: A model for human carcinogenesis, J. Natl. Cancer Inst., 66:1037.

Moolgavkar, S.H. and Venzon, D.J., 1979, Two-event models for carcinogenesis:

incidence curves for childhood and adult tumors, Math Biosci., 47:55.

National Coffee Association, 1982, 24-month chronic toxicity and oncogenicity study of methylene chloride in rats, Final report, Prepared by Hazleton Laboratories, Inc., Vienna, VA.

National Coffee Association, 1983, 24-month chronic toxicity and oncogenicity study of methylene chloride in mice, Final report, Prepared by Hazleton Laboratories, Inc., Vienna, VA.

National Toxicology Program (NTP), 1986, Toxicology and Carcinogenesis Studies of Dichloromethane (Methylene Chloride) (CAS No. 75-09-2) in F344/N Rats and B6C3Fa Mice (Inhalation Studies), NIH Publication No. 86-2562.

Ott, M.G., Skory, L.K., Holder, B.B., Bronson, J.M., and Williams, P.R., 1983a, Health evaluation of employees occupationally exposed to methylene chloride: General study design and environmental considerations, Scand. J. Work Environ. Health, 9:1.

Ott, M.G., Skory, L.K., Holder, B.B., Bronson, J.M., and Williams, P.R., 1983b, Health evaluation of employees occupationally exposed to methylene chloride: Mortality, Scand. J. Work Environ. Health, 9:8.

Peraino, C., Carnes, B.A., Stevens, F.J., Staffeldt, E.F., Russell, J.J., Prapuolenis, A., Blomquist, J.A., Vesselinovitch, S.D., and Maronpot, R.R., 1988, Comparative developmental and phenotypic properties of altered hepatocyte foci and hepatic tumors in rats, Cancer Res., 48:4171.

Sell, S., 1990, Is there a liver stem cell?, Cancer Res., 50: 3811.

Sell, S. and Dunsford, H.A., 1989, Evidence for the stem cell origin of hepatocellular carcinoma and cholangiocarcinoma, Am. J. Pathol., 134:1347.

Sells, M.A., Katyal, S.L., Shinozuka, H., Estes, L.W., Sell, S., and Lombardi, B.L., 1981, Isolation of oval cells and transitional cells from the livers of rats fed the carcinogen DL-ethionine, J. Natl. Cancer Inst., 66:355.

Serota, D.G., Thakar, A.K., Ulland, B.M., Kirschman, J.C., Brown, N.M., Coots, R.H., and Morganeidge, K., 1986a, A two year drinking water study of dichloromethane on rodents. I. Rats, Food Chem. Toxicol., 24:951.

Serota, D.G., Thakar, A.K., Ulland, B.M., Kirschman, J.CC., Brown, N.M., Coots, R.H., and Morganeidge, K., 1986b, A two year drinking water study of dichloromethane on rodents. II. Mice, Food Chem. Toxicol., 24:951.

Shulte-Hermann, R., Timmermann-Trosiener, I., and Schuppler, J., 1982, Response of liver foci in rats to hepatic tumor promoters, Toxicol. Pathol., 10:63.

RISK ASSESSMENTS FOR BENZENE-INDUCED LEUKEMIA -- A REVIEW

Kenny S. Crump

Clement International Corporation
Ruston, Louisiana USA 71270

INTRODUCTION

Benzene has a long history of extensive use and is still an important industrial chemical. Exposure to benzene has been associated with blood disorders for more than 70 years (Selling 1916). More recently, increased incidences of leukemia have been related to occupational exposure to benzene.

Increasing knowledge of the health effects of benzene has resulted in progressive lower industrial standards (Table 1). The adoption of the current U.S. Occupational Safety and Health Administration (OSHA) standard of 1 ppm benzene was an important milestone in the use of risk assessment in standard setting. In 1978 OSHA promulgated a new occupational standard of 1 ppm (8-hour TWA) to replace the then existing standard of 10 ppm. The U.S. Court of Appeals vacated this standard in 1978 and the U.S. Supreme Court affirmed that decision in 1980. OSHA did not conduct a risk assessment to support the new standard, but instead argued that since there was strong evidence that benzene was a human carcinogen, exposures should be reduced to the lowest feasible level. The Supreme Court ruled, however, that OSHA must provide evidence that a significant risk to workers exists at the old standard and that this risk would be significantly reduced by implementation of the new standard.

OSHA subsequently undertook to satisfy these requirements for benzene using quantitative risk assessment (OSHA 1985, 1987). Based on this approach, OSHA promulgated a final benzene standard of 1 ppm in 1987. In the meantime, OSHA had used quantitative risk assessment in successfully lowering occupational standards for a number of other putative human carcinogens (*e.g.*, arsenic, ethylene oxide, asbestos).

Adherence to the current 1 ppm standard will not result in any exposures as high as those at which health effects have been documented. Nevertheless, there is still concern that current industrial exposures may be capable of causing benzene-related disease.

These concerns focus primarily on leukemia, both because of the serious nature of the disease and because of the theory that there is no level of exposure to benzene that is totally without risk of leukemia. This theory is not unique to benzene-induced leukemia, but has been widely used as the basis for the regulation of carcinogens in the U.S., particularly for genotoxic chemicals that are thought to increase the risk of cancer by inducing one or more inheritable changes in a single cell (Anderson et al., 1983).

Since the late 1970s, benzene has been the subject of numerous risk assessments involving increasingly more extensive and more reliable data and increased sophistication and detail. Despite these advances, many problems and uncertainties remain. This paper will review the epidemiological studies on benzene with particular emphasis on those that have been used in quantitative risk assessments. Risk assessments for benzene based on epidemiological data that have been conducted up to the present time will be described. Limitations of current approaches and additional work that is underway will also be discussed.

Although a number of risk calculations have been based upon animal data, only those based upon human epidemiological data will be reviewed here. Generally, when adequate human data are available, as is the case with benzene, these are given the most emphasis when developing quantitative estimates.

REVIEW OF EPIDEMIOLOGICAL STUDIES

Case Control Studies

Case control studies conducted in France (Girard and Revol, 1970), Japan (Isimaru et al., 1971), the U.S. (Linos et al., 1980; Arp et al., 1983; Checkoway et al., 1984) and the United Kingdom (Rushton and Alderson, 1981) (all reviewed by Austin et al., 1988) each found an elevated risk for leukemia among groups with potential exposure to benzene or other potential hematologic toxins. The elevated risk was statistically significant in the studies conducted by Girard and Revol and by Isimaru et al. Collectively, these studies strongly suggest a link between benzene and leukemia.

However, in none of these studies were there any direct measures of exposure to benzene that could be used to develop a quantitative risk assessment. For example, Girard and Revol relied on questioning of subjects to document exposure and did not distinguish between benzene or toluene exposures, whereas Isimaru et al. used occupation as a surrogate for exposure and grouped occupations with potential exposure to benzene and medical x-rays.

These shortcomings are common to case control studies. Generally, these studies rely on interviews or questionnaire responses of subjects or acquaintances to estimate the potential for exposure, or else use occupation as a surrogate for exposure. Rarely are these sources able to furnish reliable quantitative estimates of exposure to environmental toxins. Moreover, in the case of interview or questionnaire data, there is often a question of recall bias. Because of such shortcomings, case control studies often do not furnish exposure information necessary for a quantitative risk assessment.

Table 1. Various Occupational Exposure Limits for Benzene, 1940-1987

Year	Recommended Standard	Source
1941	100 ppm	USDOL
1947	50 ppm 8-hour TWA	ACGIH
1948	35 ppm 8-hour TWA	ACGIH
1957	25 ppm 8-hour TWA	ACGIH
1963	25 ppm ceiling	ACGIH
1969	10 ppm 8-hour TWA	ACGIH
1971	10 ppm 8-hour TWA	OSHA
1987	1 ppm 8-hour TWA	OSHA

SOURCE: Rinsky *et al.* (1981).
NOTE: USDOL = U.S. Department of Labor
ACGIH = American Conference of Governmental Industrial Hygienists
OSHA = Occupational Safety and Health Administration

Study of Istanbul Shoe Workers

Aksoy and coworkers reported on leukemia incidence of Istanbul shoe workers in a series of studies. Aksoy (1985) reported 51 cases of leukemia between 1967 and 1983 among men occupationally exposed to benzene (mainly shoe workers). He estimated the crude rate of leukemia was 13 per 100,000 person-years among shoe workers compared with six per 100,000 in the general population. The rate of six per 100,000 was based on general mortality data from Western nations. Aksoy (1977) indicated that the leukemia rate for Turkey was 2.5 to 3 per 100,000, but that vital statistics reporting in Turkey could not be relied upon to draw scientific conclusions. Some very limited and not well characterized exposure data were described by Aksoy (1977).

This study is not based upon follow-up of a well-defined cohort. Also, exposures of the cohort are poorly characterized. Because of these shortcomings, this cohort has not been used in most risk assessments for benzene.

Study of Workers at DOW Chemical Company in Michigan

Ott *et al.* (1978) studied the mortality experience of 594 men who were exposed to benzene at the Michigan division of DOW Chemical. The cohort consisted of individuals who had been employed during 1940 or later and follow-up was through 1973. Benzene exposures were estimated in different job categories as time-weighted average (TWA) concentrations. Job histories were combined with these TWA concentrations to construct a complete exposure profile for each cohort member.

Ott *et al.* found a non-significant increase in total malignancies (30 versus 22.8 expected) and two leukemia deaths where 1.0 was expected. Leukemia was listed under "other significant conditions" in a third death. One leukemia case had 20 to 80 ppm-years of benzene exposure, whereas the other case had only 0 to 5 ppm-years of

Table 2.   Extent of Benzene Air Sampling Data from Pliofilm Plants

St. Marys (operated 1939-1976)
1946-1950   15 samples collected by the company at various locations
1956   Summaries of measurements collected in six areas by the Ohio Department of Health on January 17
1963-1974   Results of "112 surveys" performed routinely by the company between 1963 and 1974
1974   22 samples collected by the University of North Carolina during an industrial survey
1976   20 samples collected by NIOSH during an industrial hygiene survey
1973-1976   Charcoal tube measurements collected by the company between 1973 and the time the company ceased operations in 1976

Akron Facility 1 (operated 1937-1949)
1948   Brief summary of report of industrial hygiene survey by Ohio Department of Health on April 27

Akron Facility 2 (operated 1949-1965)
Unknown time A summary of average concentrations in 16 locations based on 414 samples collected some unknown time, but believed to be around 1957

exposure. No cases occurred in either the 80-200 or the 200+ ppm-year categories (0.43 expected). The authors concluded that "no mortalities directly attributable to benzene exposure occurred."

An update of this study included an additional 362 employees and follow-up was through 1982 (Bond *et al.*, 1986). Four leukemia deaths were observed in this study versus 2.1 expected, which was a non-significant increase.

Studies of Pliofilm Workers in Ohio

Rinsky *et al.* (1981) reported on the mortality of a group of Goodyear workers exposed to benzene in the manufacture of rubber hydrochloride (Pliofilm) in two facilities in Akron, Ohio and one in St. Marys, Ohio. The cohort consisted of 1,006 non-salaried white males who had at least one day of exposure to benzene between 1940 and 1960. Follow-up, which was 98% complete, began in 1950 and continued until midway through 1975. There were seven deaths from leukemia among workers first exposed before 1950 versus 1.25 expected, for an standard mortality ratio (SMR) of 560 ($p > 0.001$). Among those first exposed after 1950 there was one leukemia death versus 0.46 expected, for an SMR of 217 (not significant). The mean duration of exposure was brief, with 58% of those first exposed before 1950 being exposed for less than one year. For those in this group who attained at least five years of exposure, there were five leukemia deaths with only 0.23 expected, for an SMR of 2100. Although Rinsky *et al.* described the facilities at the two locations and presented detailed data on exposures measured in the various locations within the facilities, no attempt was made in this paper to relate benzene exposure level to leukemia risk.

Rinsky et al. (1987) extended follow-up in this study through 1981 (an additional 6.5 years). The cohort was also expanded to include all non-salaried white males (1,165 in all) who had at least 1 ppm-day of cumulative exposure to benzene in a pliofilm department between 1940 and 1965. In this population there were 69 deaths from all malignant neoplasms versus 66.8 expected. However, 15 lymphatic and hematopoietic cancers were observed with only 6.6 expected, for an SMR of 227 (95% CI = 127-376). The excess resulted from an excess of leukemia (nine observed versus 2.7 expected for an SMR of 337 [95% CI = 154-641]) and multiple myeloma (four observed versus 1.0 expected, for an SMR of 409 [95% CI = 110-1047]).

Rinsky et al. found a strong trend of increasing SMRs for leukemia with increasing cumulative exposure (SMR = 109 for 0-40 ppm-yr; SMR = 332 for 40-200 ppm-yr; SMR = 1186 for 200-400 ppm-yr.; SMR = 6637 for > 400 ppm-yr.). SMRs in the highest two exposure categories were statistically significantly increased. SMRs for multiple myeloma did not increase with increasing exposure.

The data and methods used by Rinsky et al. to quantify exposures will be discussed later in this review. Similarly, a case control analysis will be described later in the section dealing with risk assessments.

Study of U.S. Chemical Workers

Wong et al. (1983) studied the mortality experience of 4,062 male chemical workers from seven chemical plants who were occupationally exposed to benzene for at least six months between 1946 and 1975. Also considered was a comparison group of 3,074 male chemical workers from the same plants who were employed for at least six months during the same time period, but who were never occupationally exposed to benzene. Follow-up continued through December, 1977. All jobs were classified as to whether they involved "peak" exposures to benzene. Jobs with continuous exposure to benzene (potential exposure at least three days per week) were further classified on the basis of 8-hour TWA exposures. A cumulative exposure index in ppm-months was calculated for each member of the continuous exposure group (3,636 workers).

Seven deaths from leukemia were observed in the group with continuous exposure compared to 6.0 expected deaths based on rates for U.S. males. No leukemia deaths were observed in the control group versus 3.4 expected. Among workers with continuous exposures, no strong evidence of higher mortality from leukemia among workers with higher cumulative exposures was noted (SMR = 97 for < 180 ppm-months; SMR = 78 for 180-719 ppm-months; SMR = 276 for 720+ ppm-months). None of these SMRs were statistically significant.

Study of Benzene Workers in China

Yin et al. (1987, 1989) conducted a retrospective cohort study during 1982 and 1983 among 28,460 benzene-exposed workers from 233 factories and 28,257 unexposed workers from 83 factories located in 12 large cities in China. An exposed subject had to have worked in an operation involving benzene exposure for at least six months between January 1, 1972 and December 31, 1981. Controls had to be employed at least six months during this same period. Follow-up also covered this period.

Benzene exposures were estimated from factory records of benzene exposures, benzene concentrations in various materials and products, and records on environmental measures and protective devices. Control factories were selected from among those with no evidence of benzene, radiation or other carcinogenic exposures.

Mortality from all forms of cancer was significantly higher among benzene-exposed male workers; however, there was a (non-significant) deficit of cancer among exposed female workers. SMRs for leukemia were significantly elevated among both males (SMR = 501) and females (SMR = 830). There was a total of 25 leukemia deaths among benzene-exposed workers, but only four among controls. Lung cancer was significantly elevated in males (SMR = 231) but not in females. Elevated but non-significant SMRs were noted among males for a number of other cancers.

The crude lung cancer rate among benzene-exposed male workers was nearly as high among non-smokers (235/100,000) as among smokers (283/100,000). Further, the rate for non-smokers was significantly higher than the rate among smoking workers not exposed to benzene (122/100,000). This suggests that lung cancer was more strongly related to benzene exposure than to smoking. This finding is surprising, since other cohort studies have not found an association between lung cancer and benzene, and additional evaluation is needed.

So far as we know, this study has not been used in a quantitative risk assessment. However, the relatively large number of leukemias[1] and the availability of exposure information suggest that a risk assessment based on this study could provide useful information. If such a study is attempted, the following points should be considered: 1) the basis for exposure estimates should be laid out in more detail; 2) since exposures to some members of the cohort possibly began less than 10 years prior to the end of follow-up, additional follow-up of this cohort is suggested; 3) information should be obtained on benzene exposures prior to the beginning of follow-up; 4) the suggested relationship between lung cancer and benzene should be explored further; and 5) in order to have the flexibility to develop the most definitive risk assessment, the analysts should have access to the raw data from this study.

REVIEW OF RISK ASSESSMENTS

A number of risk assessments for leukemia following benzene exposure have been conducted based on one or more of the epidemiological studies reviewed above. Several

---

[1] The 25 leukemias found in this cohort exceed the total number (17) found in the Wong et al. (1983), Bond et al. (1986) and Rinsky et al. cohorts combined.

Table 3. Additional Leukemia Deaths per 1,000 Workers from 40 Years of Occupational Exposure Beginning at Age 20

	1 p	10 ppm
Relative Risk Model (Crump and Allen, 1984)		
Cumulative Exposure		
Rinsky et al., Ott et al., and Wong et al. data	9.5	88
Rinsky et al. and Ott et al. data	3.7	35
Rinsky et al. data	6.6	61
Weighted Cumulative Exposure		
Rinsky et al. and Ott et al. data	3.0	29
Rinsky et al. data	4.9	48
Window Exposure		
Rinsky et al. and Ott et al. data	1.2	12
Rinsky et al. data	1.7	17
Absolute Risk Model (Crump and Allen, 1984)		
Cumulative Exposure		
Rinsky et al. and Ott et al. data	2.0	19
Rinsky et al. data	2.1	20
Weighted Cumulative Exposure		
Rinsky et al. and Ott et al. data	1.5	15
Rinsky et al. data	1.9	19
Window Exposure		
Rinsky et al. and Ott et al. data	1.1	11
Rinsky et al. data	1.3	13
Multistage Model, Three Stages Rinsky et al. Data (Crump et al., 1984)		
First Stage Dose-Related	3.3	32
Second Stage Dose-Related	3.0	30

of these have had access to the raw data from the epidemiological studies, whereas others have not. Conductors of epidemiological studies may not have risk assessment needs in mind when summarizing the results of their study and consequently may not report data in a format that is most suitable for risk assessment. In some instances, a piece of information that is critical to risk assessment is not provided. Even when this is not the case, having access to the raw data greatly expands the type of analyses that can be conducted in a risk assessment based upon epidemiological data.

Consequently, risk assessments based upon raw epidemiological data often analyze the data in more sophisticated and varied ways than risk assessments based only upon reports. Several risk assessments of both types have been conducted for benzene. We will review only those that have access to the raw data. Other risk assessments for benzene based upon human data have been performed by Albert (1979), White et al. (1982), Austin et al. (1988), and Swaen and Meijers (1989).

Table 4. Summary of Brett *et al.* Risk Calculations

Exposure Assumptions	Control Set[a]	Additional Lifetime Leukemia Deaths per 1000 Workers Due to Benzene Exposure	
		45 ppm-years	450 ppm-years
Rinsky *et al.* (1987)	1	5.1 (0.83 - 11.7)[b]	635 (15.6 - 986)
	2	6.4 (1.2 - 14.7)	819 (26.4 - 991)
	3	4.2 (1.0 - 8.7)	449 (21.3 - 953)
Crump and Allen (1984) I	1	0.5 (0.13 - 1.0)	8.3 (1.4 - 20.4)
	2	0.7 (0.1 - 1.3)	11.0 (1.4 - 30.9)
	3	0.5 (0.1 - 1.0)	7.9 (1.1 - 20.4)
Crump and Allen (1984) II	1	1.3 (0.3 - 2.3)	29.8 (4.2 - 106)
	2	1.6 (0.3 - 3.1)	47.0 (4.0 - 218)
	3	1.2 (0.3 - 2.3)	27.8 (3.3 - 103)

SOURCE: Brett *et al.* (1989).
[a] Control sets: 1) Rinsky *et al.* controls matched on date of birth, date entering pliofilm work; 2) Brett *et al.* controls matched on date of birth, date entering pliofilm work; 3) Brett *et al.* controls matched on date of birth, date entering pliofilm work, and plant.
[b] Numbers in parentheses are 95% confidence intervals.

Crump and Allen Risk Assessment

Crump and Allen (1984) conducted a risk assessment of benzene for OSHA after OSHA's 1 ppm standard had been vacated by the courts. Crump and Allen based their assessment on the studies of Rinsky *et al.* (1981), Ott *et al.* (1978) and Wong *et al.* (1983). They had access to the tapes containing the raw data from the former two studies. Since only 17 leukemias were observed in all three studies combined, in order to increase statistical power, Crump and Allen combined the data from the three studies in some analyses.

Crump and Allen determined an exposure pattern for each worker in the Rinsky *et al.* cohort using: 1) a job code and time period for each job held by each worker from a data tape provided by Rinsky, 2) a personal communication from Rinsky indicating the areas of work for each job code, and 3) data from Rinsky *et al.* (1981) on exposures in each work area.

The sampling data used by Crump and Allen to estimate exposures are summarized in Table 2. As this table indicates, although there were a total of over 2,000 samples available, there were a very limited number available prior to 1960 and none prior to 1946. Unfortunately, exposures were likely to have been highest during the earlier years and consequently exposures during these years are the most critical for risk assessment. To overcome the limitations in the data base, Crump and Allen were forced to make a number of assumptions.

Average exposures were estimated for eight different work areas and seven different time periods. The time periods selected were based on the times at which occupational benzene standards changed (see Table 1), and the year (1946) at which a major improvement to the exhaust system was made at St. Marys. An assumption underlying this categorization was that the company's industrial hygiene procedures would remain approximately constant unless standards were revised. At that point, the company may have been prompted to make some improvements. Measurements available within each of these time periods were averaged. It was also assumed that average concentrations at all locations were non-decreasing over time. Thus, if measurements in an area were higher in one time period than in the subsequent time period, the common average of measurements over both time periods were assigned to the two time periods. Since measurements were more numerous in the later time periods, this approach was applied sequentially from later time periods to earlier ones. If no measurements were available for a location in a particular time period, the estimated concentration in the following time period was multiplied by the ratio of the applicable standard in the particular time period to the standard for the following time period.

It was generally assumed, unless monitoring data indicated otherwise, that data collected from a given location at one facility were applicable to the comparable location at the other facility. Based on these assumptions, year-by-year estimates of average exposure were estimated for each worker.

The data tape furnished to Crump and Allen contained an additional three years of follow-up (to 1978) over that reported by Rinsky *et al.* (1981). The cohort definition and follow-up used by Crump and Allen differed from that of Rinsky *et al.* in two additional ways. Whereas Rinsky *et al.* began follow-up in 1950, Crump and Allen included follow-up between 1940 and 1950,[2] and whereas Rinsky *et al.* omitted 177 workers in the pliofilm department assumed to have minimal exposures to benzene, Crump and Allen included all workers in their analysis, and assigned those thought to be minimally exposed TWA exposures of 5 ppm prior to 1946 and 1 ppm thereafter.

A data tape from the study of Ott *et al.* (1978) was also made available to Crump and Allen. TWA estimates of the exposures of each worker for various time periods were provided on the tape. Since the Ott *et al.* study only included two leukemia deaths and did not indicate a dose response relationship, Crump and Allen did not use data from this study independently to make risk estimates, but only in conjunction with the data from the Rinsky *et al.* cohort.

Crump and Allen also used data from the Wong *et al.* (1983) study in their risk

---

[2] However, Crump and Allen found little difference in results whether or not follow-up began in 1940 or in 1950.

assessment. However, since a data tape was not available from this study, analyses involving data from this study were more limited.

The general method of analysis used by Crump and Allen involved categorizing the person-years in the studies by exposure to benzene. The models applied to the categorized data were the relative risk model

$$E(O_i) = aE_i(1 + bd_i),$$

and the absolute risk model

$$E(O_i) = E_i + (a + bd_i)Y_i,$$

where

$E(O_i)$ = the expected number of leukemia deaths in the ith dose category,
E_i = the expected number of deaths in the ith dose group based on age-, sex-, and race-specific U.S. rates,
a = a parameter that allows for the possibility that the background rates in the cohort differ from general U.S. rates (a = 1, in the relative risk model, and a = 0, in the absolute risk model, corresponds to U.S. rates being appropriate),
b = the potency of benzene for causing leukemia,
d_i = the average benzene dose in the ith dose category,
Y_i = the number of person-years observed in the ith dose category.

The relative risk model assumes the increased mortality from benzene is proportional to the background mortality, and the absolute risk model assumes this increased mortality is independent of the background mortality. With either model, the observed number of leukemia deaths in each exposure category were assumed to be distributed as a Poisson variable. Consequently, this approach is often referred to as "Poisson regression" (NAS, 1990).

Crump and Allen used three different dose measures in applying this model. The first was cumulative dose in ppm-years ignoring the most recent 2.5 years. The second measure was "weighted cumulative dose," defined by assigning exposures in the most recent five years zero weight, those in the next earlier five years full weight, those in the next earlier period one-third weight, and still earlier exposures one-sixth weight. This measure of dose was based upon the latency pattern of leukemia in Japanese atomic bomb survivors. The third measure of exposure was "window dose" and defined by assigning exposures 5 to 15 years in the past full weight and ignoring all other exposures. This exposure measure assumes that previous exposures outside this "window" have no effect upon current leukemia mortality.

In more recent work, Crump et al. (1984) applied the multistage model of cancer (Armitage and Doll, 1961; Crump and Howe, 1984) to the data on the pliofilm cohort. In this model it is assumed that a cell line becomes cancerous after progressing through k stages in succession. The rate at which a cell line goes through the jth stage at time t is given by

$$\tau_j = a_j + b_j d(t),$$

where d(t) is the instantaneous dose rate at time t. Once a cell line arrives at the kth cancerous stage, it is assumed that after a fixed latency, t_0, the cancer will be fatal. Since this model utilizes the instantaneous dose rate, unlike the relative risk and absolute risk models, it does not require a summary dose measure. The multistage model also differs conceptually from these models in that, whereas the relative risk and absolute risk models are statistical models only, the multistage model is derived from assumptions regarding the underlying biological processes.

Crump *et al.* (1984) developed a computer program for applying the multistage model to data from a cohort study. The program requires that only a single stage be affected by benzene (*i.e.*, $b_j = 0$ for all j not equal to r, where r is the stage affected by benzene) and that the number of stages, k, the affected stage, r, and the latency, t_0, be provided.

Outputs from models were converted into estimates of lifetime risk of cancer from occupational exposure to benzene at various levels and durations. This was done through a life table approach using mortality rates for U.S. white males as background mortality rates for leukemia and total deaths. The only model parameter needed for this approach was the benzene potency parameter (designated as "b" in each of the models).

Table 3 lists risk estimates made from various models, various combinations of data sets, and applying various summary dose measures to the relative risk and absolute risk models. This table permits one to consider separately the effects of different models and different data sets upon the risk estimates.

Since the Rinsky *et al.* data were applied to all of the models, results from these data allow a comparison among different models. Because the raw data were not available for the Wong *et al.* study, only the relative risk model using cumulative dose could be applied to data from this study.

The highest risks were obtained by applying cumulative dose to the relative risk model. This combination will predict high risks in elderly persons because background rates are high in this group, and the cumulative dose measure allows any exposure to contribute to the increased risk, no matter how early in life the exposure occurred. Inclusion of the Wong *et al.* data with this model caused the risks to be still higher. However, one reason for the large dose response slope predicted from the Wong *et al.* data is the deficit of leukemias observed in the internal control population from that study. The cause of this deficit was not determined by Wong *et al.*

The lowest risks were estimated using window exposure as the summary exposure measure, and the estimates based upon weighted cumulative exposure were intermediate between those based upon (unweighted) cumulative exposure and window exposure. The predictions of the multistage model were not highly dependent upon whether the first or second stage in the carcinogenic process was assumed to be affected by benzene. Both assumptions provided estimates of risk that were intermediate between those obtained from the relative risk and absolute risk models using weighted cumulated dose.

A reviewer of a draft of the Crump and Allen report suggested that their exposure estimates may have been too high and that acute benzene poisoning would have occurred at some of the higher levels estimated by Crump and Allen. As a result, Crump and Allen developed alternative exposure estimates. Risk estimates based on these alternative estimates were about 25% higher than those presented in Table 3.

All of the models studied by Crump and Allen fit the data adequately. Although the models using cumulative exposure fit somewhat worse than those based on weighted cumulative exposure and window exposure, this difference was not statistically significant. Similarly, the fits of the relative risk and absolute risk models could not be distinguished.

All of the models used by Crump and Allen (1984) and Crump et al. (1984) assumed a linear relationship between risk of leukemia and a measure of benzene exposure. Because of the small number of leukemias observed in the studies used by Crump and Allen, linear models cannot be distinguished from a variety of non-linear models. To explore the effect upon the risk estimates of using a non-linear model, Crump and Allen applied a relative risk quadratic model (defined by replacing d_i in the relative risk model described above by d_i^2) to the same data that were used to obtain the estimates in the first row of Table 3. This model also described the data adequately and it predicted an additional leukemia risk of 24 per 1,000 for occupational exposure to 10 ppm for 40 years beginning at age 20. This value is about one-fourth to one-third that obtained from the linear model. As lower exposures are considered, this model will predict lower risks than those predicted by a linear model by increasingly greater amounts.

U.S. EPA Risk Assessment

The United States Environmental Protection Agency (EPA, 1985) conducted a risk assessment for benzene to take account of new information that had been developed since EPA's earlier risk assessment (EPA, 1979). This risk assessment was termed "interim," and was derived from the risk assessment conducted by Crump and Allen. EPA reviewed the work of Crump and Allen and considered the analyses based on the relative risk and absolute risk models using either cumulative dose or weighted cumulative dose to be the most reliable. Since they were unable to differentiate among these analyses, EPA took the geometric mean of the "unit risks"[3] obtained from these four models applied to the Ott et al. and Rinsky et al. data sets. To make their risk assessment also reflect the Wong et al. data set, EPA then multiplied this geometric mean by the ratio of the unit risk obtained from applying the cumulative dose relative risk model to the combined Ott et al., Rinsky et al. and Wong et al. data to the unit risk obtained from applying the same model to only the Ott et al. and Rinsky et al. data sets. (This was necessary since Crump and Allen did not have access to the raw data from the Wong et al. study and consequently could only apply the relative risk, cumulative dose model to these data.)

---

[3] The "unit risk" is the additional risk of leukemia from continuous lifetime exposure to 1 ppm benzene.

Rinsky et al. Risk Assessment

In addition to the cohort analysis reported earlier, Rinsky et al. (1987) explored the shape of the dose response for leukemia using a case control methodology. Each of the nine leukemia cases were matched to ten controls by year of birth and year first employed. Conditional logistic regression analysis was used to estimate the dose response for the odds ratio.

This analysis used independent estimates of the exposures of each worker. Although these estimates were based on the same data as that used by Crump and Allen, there were important differences in the two sets of estimates. These differences were due mainly to different assumptions made concerning exposures in early years when exposure measurements were very limited. Whereas Crump and Allen assumed exposures decreased steadily as the threshold limit value (TLV) decreased, Rinsky et al. assumed that no changes in workplace concentrations occurred unless there was specific information to indicate otherwise. In each location, samples collected in a given year were averaged to estimate exposures for that year. For a year in which no samples were available, the value closest in time, either before or after, was used to estimate exposure for that year. This means that for years prior to the earliest measurement, the value of the earliest year was used. Whenever measurements did not exist at Akron, measurements from St. Marys were used as substitutes.

The logistic regression analysis considered odds ratios in the form

$$OR = e^{B_1 X_1 + \cdots + B_n X_n}$$

where X_i represents exposure variables, potential confounders, or effect modifiers, and the B_i are coefficients estimated by fitting the model to the data. Cumulative exposure, duration of exposure, and average exposure rate were all applied in the model, both individually and jointly. Cumulative exposure (incorporated by setting X_i = cumulative dose) was found to be the strongest predictor of leukemia, and the corresponding B_i was significantly elevated ($p = 0.011$). With cumulative exposure in the model, the remaining exposure variables did not contribute materially to the prediction.

The shape of the dose response function was evaluated by considering several alternative ways of incorporating cumulative dose into the model. A logarithmic transform of cumulative exposure was found not to describe the data as well as the untransformed variable. A quadratic term for cumulative exposure was added, but did not significantly improve the fit. Redefining cumulative exposure by disregarding exposures in the most recent five years improved the statistical significance of the coefficient slightly. (See footnote 2 for additional evaluation of these alternative analyses.)

Rinsky et al. expressed their findings only in terms of the equation for the odds ratio

$$OR = e^{0.0126\, ppm\text{-}yrs}$$

for cumulative exposure, or

Table 5. <u>Summary of Risk Calculations by Thorslund *et al.*</u>

Modification	Excess Risk at 1 ppm Continuous Lifetime Exposure
None (Combined EPA (1985) estimate)	2.60×10^{-2}
Model restricted to absolute risk and weighted cumulative dose form; data restricted to Rinsky cohort	1.76×10^{-2}
Three years of follow-up added	1.71×10^{-2}
Job code errors corrected	1.84×10^{-2}
New weighted cumulative dose form from epidemiological latency data; new statistical method for estimating transition rate parameters	3.19×10^{-3}
New definition of diseases induced by benzene; new estimates of background rates in U.S. population	3.48×10^{-3}
Quadratic model: two molecules of benzene metabolite required to induce initial event producing malignant cell	1.43×10^{-4}
Linear-quadratic model: upper bound on quadratic model	1.00×10^{-3}

SOURCE: Thorslund *et al.* (1988).

$$OR = e^{0.0169 \cdot ppm\text{-}yrs}$$

for cumulative exposure lagged five years. It is difficult to compare results in this form with applicable standards or with results of the Crump and Allen risk assessment. However, in addition to making independent risk assessment calculations, Brett *et al.* (1989) translated the results of the Rinsky *et al.* analysis into a form that is comparable to that used by Crump and Allen. Therefore, further discussion of the results of Rinsky *et al.* will be delayed until after the risk assessment of Brett *et al.* is discussed.

<u>Brett *et al.* Risk Assessment</u>

Brett *et al.* (1989) performed a number of additional analyses based on the Rinsky *et al.* (1986) data set using the same case control format and applying the same logistic regression methodology as used by Rinsky *et al.* Brett *et al.* utilized the exposure measures developed by Rinsky *et al.*, as well as both of the exposure measures developed by Crump and Allen for this cohort. In addition to using controls selected by Rinsky *et al.* in their case control analysis, they also developed two additional sets of controls. One group was selected using the same matching criteria as Rinsky *et al.* In addition to using these criteria, the other group was also matched according to plant location, because, as explained by Brett *et al.*, "the two plants were located more than 100 miles apart and because of the differences in non-pliofilm benzene exposure, which may have influenced plant-specific mortality patterns."

The results of the analyses by Brett *et al.* are summarized in Table 4. These are expressed in terms of additional lifetime leukemia deaths per 1,000 workers exposed to either 45 or 450 ppm-years of benzene. The results of the risk assessment developed by Rinsky *et al.* (1987) are also expressed in this same format. Details on how odds ratios were converted to measures of additional lifetime risk were not presented in Brett *et al.* (1989).

Comparison of Results Obtained by Crump and Allen, Rinsky *et al.*, and Brett *et al.*

The differences between the analyses of Rinsky *et al.* and Brett *et al.* stemming from the use of three different sets of controls in the case control study and in use of different exposure estimates may be evaluated from Table 4. Use of different control groups has a relatively minor effect upon the risk estimates; estimates of the risk from 45 years of occupational exposure at 1 ppm using different controls but the same exposure estimates differ by no more than 50%. However, estimates for 1 ppm exposure using the same controls but different exposure estimates differ by as much as an order of magnitude. For exposures of 10 ppm, the risk estimates differ by as much as a factor of 80. In all cases, estimates of risk based upon the Rinsky *et al.* exposure estimates are higher.

The primary reason for these differences is the differences in estimates of exposure made by Crump and Allen and by Rinsky *et al.* In this regard Kipen *et al.* (1986 and 1987; see also Kipen *et al.*, 1988) found that the estimates made by Crump and Allen correlated more closely with hematological data on St. Marys workers from the 1940s than the estimates made by Rinsky *et al.*

Differences associated with the use of different statistical methods and dose response models can be determined by comparing estimates in Table 3 with estimates labelled Crump and Allen I in Table 4, all of which are based on the same exposure estimates. In particular, the estimates in Table 3 based upon the Rinsky *et al.* cohort and cumulative exposure and the estimates labelled as Crump and Allen I in Table 4 all involve the same epidemiological data (except that those in Table 4 are based upon a few additional years of follow-up) and all utilize cumulative exposure as the exposure summary variable. The estimates of Crump and Allen range from 2.1 to 6.6 per 1,000 workers for exposure to 1 ppm and from 20 to 61 per 1,000 for exposures to 10 ppm. On the other hand, the corresponding estimates in Table 4 are smaller, ranging from 0.5 to 0.7 per 1,000 for 1 ppm exposures, and from 7.9 to 11 per 1,000 for 10 ppm exposures.

These differences appear to be due largely to differences in the dose response curves. Whereas Crump and Allen assumed that relative risk varies linearly with exposure, Rinsky *et al.* and Brett *et al.* assumed exponential variation. Although some scientific arguments in support of linear models have been presented (Crump *et al.*, 1976; NAS, 1977), we are not aware of any scientific arguments supporting an exponential dose response model. This model was apparently used by Rinsky *et al.* because software for its implementation was conveniently available. Such models simplify certain statistical procedures and therefore have been used by statisticians in testing statistical hypotheses. However, these applications do not involve extrapolation out of the range of the data. Thus, the conclusion by Rinsky *et al.* that, "from this model, it can be calculated that protection from benzene-induced leukemia would increase *exponentially* [emphasis added] with any reduction in the permissible exposure limit," actually was a direct result of the dose response function they chose to use.[4]

Risk Assessment of Thorslund

Thorslund *et al.* (1988) conducted a risk assessment of benzene based upon the Rinsky *et al.* (1986) cohort and the Crump and Allen exposure estimates. The analysis of Thorslund *et al.* was similar to that of Crump and Allen in several respects. However, there were some significant differences. Whereas Crump and Allen considered both relative risk and absolute risk models, Thorslund *et al.* used only an absolute risk model on the basis that a relative risk model predicts that the risk of benzene-induced leukemia increases with age, which does not agree with epidemiological data on this question. Thorslund *et al.* adopted a weighted dose measure similar to that used by Crump and Allen for the same reason that an absolute risk model was adopted. However, the weights were based on leukemia response in patients treated for ankylosing spondylitis with radiation therapy. Thorslund *et al.* considered all types of acute leukemias and myelodyplastic syndromes to be related to benzene exposure, but not chronic leukemia. This led to deletion of one chronic leukemia in the Rinsky *et al.* cohort and adding one case of aplastic anemia. Finally, Thorslund *et al.*'s statistical analysis involved treating the experience of each individual in the cohort separately, rather than aggregating the data in exposure levels as was done by Crump and Allen. Thorslund *et al.*'s method of data analysis is more precise than the method used by Crump and Allen.

Thorslund *et al.* expressed their results in terms of risks from continuous lifetime

---

[4] Rinsky *et al.* did consider two modifications to their model. One used the logarithm of cumulative dose rather than cumulative dose itself. However, in this model the odds ratio is given by: $\exp[b \ \text{Ln}(dose)] = (dose)^b$. Whereas the odds ratio should be 1.0 with no exposure, in this model it is 0.0, which is nonsensical. This model also predicts an odds ratio of 1.0 whenever the exposure is 1.0, independent of the units used to express exposure; this also is clearly nonsensical. The other dose response model considered by Rinsky *et al.* involved adding a quadratic term to their original model, resulting in the following expression for the odds ratio: $\exp\{b_1 \ \text{Ln}(dose) + b_2 \ [\text{Ln}(dose)]^2\}$. In this model a positive b_2 will exacerbate the exponential nature of the model, whereas a negative b_2 will force the odds ratio to approach zero at high doses. Neither Rinsky *et al.* nor Brett *et al.* investigated how well any of the models they used fit the observed data. Obviously, they also did not compare the fits of their models with those of other more plausible models.

exposure to 1 ppm benzene, which are not strictly comparable to the estimates made by Crump and Allen. Thorslund *et al.*'s results are summarized in Table 5.

DETAILED INVESTIGATION OF EXPOSURES IN THE PLIOFILM COHORT

The U.S. cohort exposed to benzene in the manufacture of pliofilm (Rinsky *et al.*, 1981, 1986) appears to be the one best suited for developing quantitative estimates of risk. However, there is still considerable uncertainty regarding the exposures experienced by this cohort. Recently, a reevaluation of exposures in that cohort was conducted that was much more detailed than any conducted in the past (Paustenbach *et al.*, 1991). This review involved recent information obtained from worker interviews and historical records. The following factors were accounted for that were ignored or given less weight in earlier exposure assessments for this cohort: uptake of benzene due to short-term, high-level exposures; uptake due to dermal contact; morbidity and mortality data on workers indicative of extremely high exposures in some workers; installation of engineering controls for reducing exposures; extraordinarily long work weeks during the 1940s; and information on biassed readings from devices used to measure exposures of this cohort. Based upon this review, Paustenbach *et al.* developed independent estimates of exposures to the pliofilm cohort. Certain key features of this review will be summarized.

Paustenbach *et al.* reviewed several assessments of the deficiencies in the methods used to measure benzene concentrations in air, and concluded that the two methods used prior to 1963 (detector tube and the Combustible Gas Indicator) underestimated benzene concentrations by at least a factor of 1.5. Paustenbach *et al.* adjusted early exposures by this factor.

Paustenbach *et al.* accounted for evidence of extended hours of work that began during World War II and extended through the 1940s. They also accounted for greatly reduced production at St. Marys during World War II due to shortage of rubber.

Paustenbach *et al.* reported on several sources of information indicating that pliofilm workers were exposed to high levels of benzene during the early years of operation. A 1942 U.S. Department of Labor report indicated that 25 workers (including those from the pliofilm department) had marked blood changes; nine of these were sufficiently affected to require hospitalization, of whom three died (one of whom worked in the pliofilm department). Another report from 1942 indicated that exposures could occasionally reach 500 to 1,000 ppm (estimated by Paustenbach *et al.* to be 750 to 1,500 ppm after adjusting for bias in the measuring devices), with "average concentrations being about 100 ppm" (150 ppm after bias adjustment). Kipen *et al.* (1989) analyzed blood monitoring data available from St. Marys workers that indicated that white cell counts were depressed between 1940 and 1949, and particularly so in the earlier years.

Paustenbach *et al.* identified four groups of workers that had opportunity for significant dermal exposure to benzene (*e.g.*, by having their hands and forearms immersed in benzene for 30 minutes per day while cleaning a press). The dermal uptake was estimated for each of these groups of workers and converted to equivalent ppm air exposures.

Paustenbach *et al.* also reviewed in detail the opportunity for exposure in specific jobs based on worker records and information on the physical layout of the plants. They also reviewed information on changes in engineering controls for airborne levels and reports on the efficacy of these controls.

Based on this investigation, Paustenbach *et al.* made independent estimates of airborne concentrations at different work stations in the different plants. In making these estimates, they generally used the assumption made by Crump and Allen that, during time periods in which no exposure measurements were available, exposures decreased in proportion to decreased TLVs. However, there were some instances in which they thought such an assumption was not justified. An effort was made to incorporate information on peak exposures as well as information on the use of respirators and the effectiveness of respirators. Job assignments to exposure categories were reviewed and some were revised. Separate estimates of exposures were made for St. Marys and Akron.

Figure 1 compares estimates made by Paustenbach *et al.* with earlier estimates made by Crump and Allen and Rinsky *et al.* for selected job categories. As this figure illustrates, there are substantial differences among the three sets of estimates. In general, Paustenbach *et al.*'s estimates follow those of Crump and Allen more closely than those of Rinsky *et al.*, and are about five-fold higher than those of Rinsky *et al.* for most job categories. However, exposures in some jobs were estimated to be significantly lower than estimated by Crump and Allen.

FUTURE WORK

The exposure estimates developed by Paustenbach *et al.* are based on a thorough review and take into account new information not available to either Crump and Allen or Rinsky *et al.* These new estimates appear to be more definitive than those developed by these earlier authors. There are major differences between these new estimates and the two older sets, and it is not clear at this time what effect these new estimates will have on the risk estimates.

The next step is to apply these new estimates developed by Paustenbach *et al.* in a quantitative assessment of risk. Such a project is currently underway in our office. These analyses will use models that are special cases of the general model discussed by Thorslund *et al.* (1988) in which the probability of benzene-induced death from leukemia at age t from exposure to an airborne concentration of $X(v)$ ppm benzene at age v is given by: $h(v,t) = G[X(v)]w(t-v)$. In this equation, $G[X(v)]$ represents the instantaneous dose-dependent probability that exposure to a specified level of benzene at age v, $X(v)$, will induce an event that alters the expression of DNA, which leads to a cellular transformation that results in leukemia at some time in the future, and $w(t-v)$ is the weighing function that defines the conditional probabilities of death due to leukemia at age t given that the cellular transformation occurred at age v. The data will be analyzed using a "full likelihood" approach in which each individual in the cohort contributes a separate term likelihood expression. This approach should be somewhat more efficient than the Poisson regression approach used by Crump and Allen, which involves grouping person-years of observation data into exposure categories.

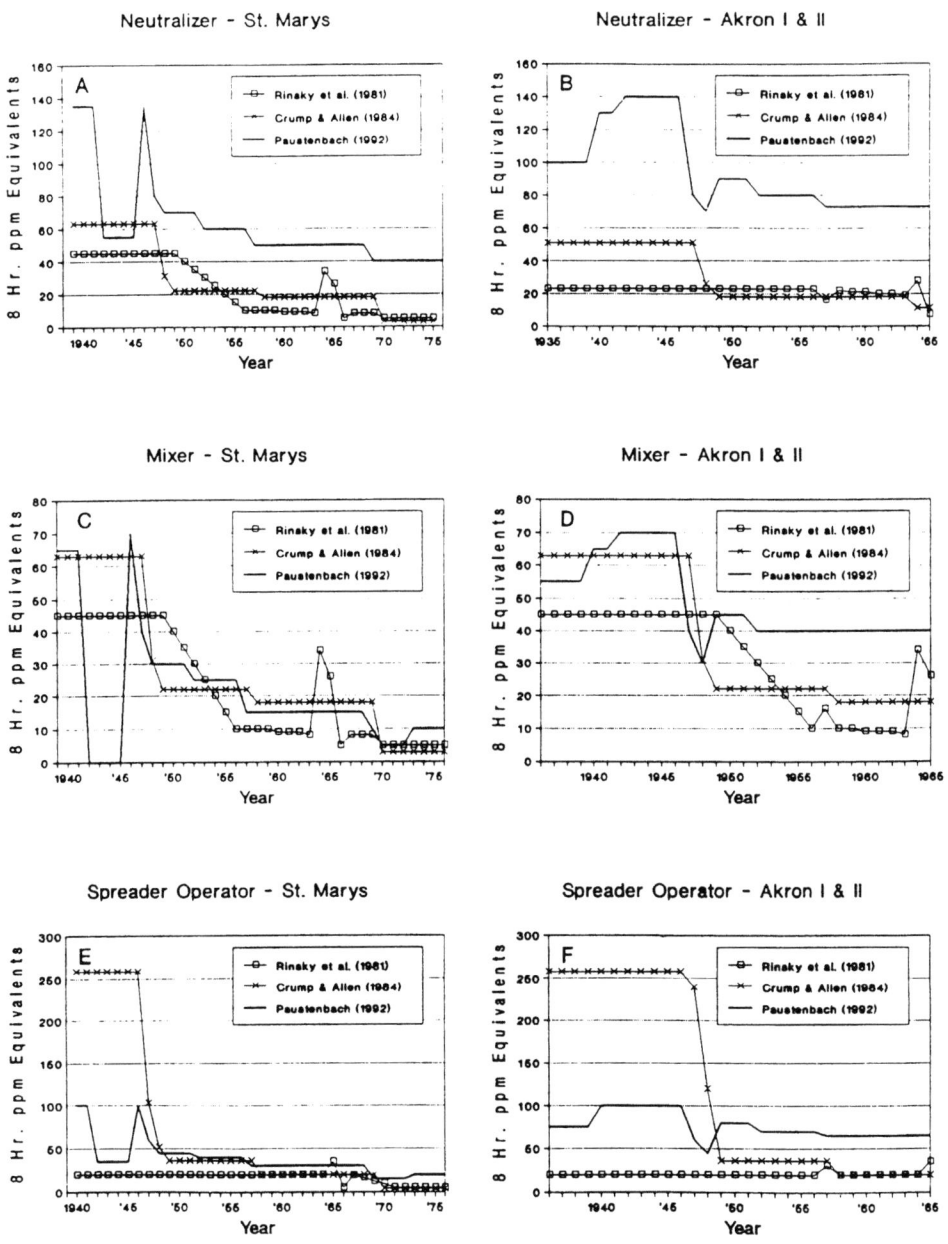

Source: Paustenbach et al. 1992.

Fig. 1. Comparison of Three Sets of Exposure Estimates for Pliofilm Cohort

This approach will be used to calculate new estimates of potency based on the latest update of the pliofilm cohort. A sensitivity analysis will also be performed to evaluate the effect of choice of latency distribution, additive or relative risk model, and leukemia endpoint to model (*e.g.*, whether to include chronic leukemias). A linear dose response model will be the basic model used ($G[X(v)] = \beta*X(v)$), and statistical tests for departure from linearity will be conducted. It is hoped that this new risk assessment will provide a sounder basis for evaluating the potential leukemogenic effect of low exposures to benzene vapors upon human populations.

REFERENCES

Aksoy, M., 1985, Malignancies due to occupational exposure to benzene, *Amer. J. Ind. Med.*, 7:395-402.

Aksoy, M., 1977, Leukemia in workers due to occupational exposure to benzene. *New Istanbul Contrib. Clin. Sci.*, 12:3-14.

Aksoy, M., 1980, Different types of malignancies due to occupational exposure to benzene: a review of recent observations in Turkey, *Environ. Res.*, 23:181-190.

Aksoy, M., Erdem, S., and Doncol, S., 1976, Types of leukemia in chronic benzene poisoning. A study in thirty-four patients, *Acta Haematol.*, 55:65-72.

Aksoy, M., Erdem, S., and Dincol, G., 1974, Leukemia in shoe-workers exposed chronically to benzene, *Blood*, 44:837-841.

Albert, R.E., 1979, Carcinogen Assessment Group's final report on population risk to ambient benzene exposures. EPA-450/5-80-004. U.S. Environmental Protection Agency, Office of Air Quality Planning and Standards, Research Triangle Park, NC.

Anderson, E., and Carcinogen Assessment Group of the U.S. Environmental Protection Agency, 1983, Quantitative approaches in use to assess cancer risk, *Risk Anal.*, 3:277-295.

Armitage, P., and Doll, R., 1961, Stochastic models for carcinogenesis, *in*: "Proceedings of the Fourth Berkeley Symposium on Mathematical Statistics and Probability," 4:19-38.

Arp, E.W., Wolf, P.H., and Checkoway, H, 1983, Lymphocytic leukemia and exposures to benzene and other solvents in the rubber industry, *J. Occup. Med.*, 25:598-602.

Austin, H., Delzell, E., and Cole, P., 1988, Benzene and leukemia. A review of the literature and a risk assessment, *Amer. J. Epidemiol.*, 127:419-439.

Bond, C.G., McLaren, E.A., Baldwin, C.L., 1986, An update of mortality among chemical workers exposed go benzene, *Br. J. Ind. Med.*, 43:685-691.

Brett, S.M., Rodricks, J.V., Chinchilli, V.M., 1989, Review and update of leukemia risk potentially associated with occupational exposure to benzene, *Environ. Health Perspect.*, 82:267-281.

Checkoway, H., Wilcosky, T., Wolf, P, 1984, An evaluation of the associations of leukemia and rubber industry solvent exposures, *Amer. J. Ind. Med.*, 38:225-234.

Crump, K., and Allen, B., 1984, "Quantitative Estimates of Risk of Leukemia from Occupational Exposure to Benzene," prepared for U.S. Occupational Safety and Health Administration, Washington, D.C.

Crump, K., and Howe, R., 1984, The multistage model with a time-dependent dose pattern: applications to carcinogenic risk assessment, *Risk Anal.*, 4:163-176.

Crump, K., Howe, R., Allen, B., and Crockett, P., 1984, "Application of the Armitage-Doll Model of Cancer to Epidemiologic Data," prepared for Battelle Laboratories and U.S. Environmental Protection Agency, Contract No. 68-01-6721.

Crump, K., Hoel, D., Langley, H., and Peto, R., 1976, Fundamental carcinogenic processes and their implications to low dose risk assessment. *Cancer Res.*, 36:2973-2979.

Environmental Protection Agency (EPA), 1985, "Interim Quantitative Cancer Unit Risk Estimates Due to Inhalation of Benzene," EPA-600/X-85-022, United States Environmental Protection Agency, Washington, D.C.

Environmental Protection Agency (EPA), 1979, "Carcinogen Assessment Group's Final Report on Population Risk to Ambient Benzene Exposures," EPA-450/5-80-004, United States Environmental Protection Agency, Washington, D.C.

Girard, P.R., and Revol, L., 1970, La fréquence d'une exposition benzénique au cours des hémopathies graves, *Nouv Rev Fr Hématol.*, 10:477-484.

Ishimaru, T., Okada, H., Tomiyasu T, 1971, Occupational factors in the epidemiology of leukemia in Hiroshima and Hagasaki, *Amer. J. Epidemiol.*, 93:157-165.

Kipen, H.M., Cody, R.P., Goldstein, B.D., 1989, Use of longitudinal analysis of peripheral blood counts to validate historical reconstruction of benzene exposure, *Environ. Health Perspect.*, 82:199-206.

Kipen, H.M., Cody, R.P., Crump, K.S., Allen, B.C., and Goldstein, B.D., 1988, Hematologic effects of benzene: a thirty-five year longitudinal study of rubber workers, *Toxicol. Ind. Health,* 4:411-430.

Kipen, H.M., Cody, R.P., and Goldstein, B.D., 1987, "Report of Pliofilm Cohort Hematology," OSHA Docket H-059C, Exhibit 267.B.

Kipen, H.M., Cody, R.P., and Goldstein, B.D., 1986, "Preliminary Report of Pliofilm Cohort Hematology," OSHA Docket H-059C, Exhibit 204-7.

Linos, A., Kyle, R.A., O'Fallon, W.M., 1980, A case-control study of occupational exposures and leukaemia, *Int. J. Epidemiol.*, 9:131-135.

National Academy of Sciences (NAS), 1990, "Health Effects of Exposure to Low Levels of Ionizing Radiation, BEIR V," National Academy Press, Washington, D.C.

National Academy of Sciences (NAS), 1977, "Drinking Water and Health," National Academy of Sciences, Washington, D.C.

Occupational Safety and Health Administration (OSHA), 1985, Occupational exposure to benzene: proposed rule and notice of public hearing, *Fed. Reg.*, 50(237):50512-50586. December 10, 1985.

Occupational Safety and Health Administration (OSHA), 1987, Occupational exposure to benzene; final rule, *Fed. Reg.*, 52(176):34460-34578. September 11, 1987.

Ott, M.G., Townsend, J.C., Fishbeck, W.A., 1978, Mortality among individuals occupationally exposed to benzene, *Arch. Environ. Health*, 33:3-10.

Paustenbach, D.J., Price, P.S., Ollison, W., Jernigan, J.D., Bass, R.D., and Peterson, H.D., 1991, A re-evaluation of benzene exposure for the pliofilm (rubber worker) cohort (1936-1976), (In review: *J. Toxicol. Environ. Health*, 1992).

Rinsky, R.A., Young, R.J., and Smith, A.B., 1981, Leukemia in benzene workers, *Amer. J. Ind. Med.*, 2:217-245.

Rinsky, R.A., Smith, A.B., Hornung, R., Filloon, R.G., Young, R.J., Okun, H.A., and Landrigan, P.J., 1986, "Benzene and Leukemia: An Epidemiologic Risk Assessment," National Institute for Occupational Safety and Health, Cincinnati, OH.

Rinsky, R.A., Smith, A.B., Hornung R., 1987, Benzene and leukemia: an epidemiologic risk assessment, *N. Engl. J. Med.*, 316:1044-1050.

Rushton, L., and Alderson, M.R., 1981, A case-control study to investigate the association between exposure to benzene and deaths from leukaemia in oil refinery workers, *Br. J. Cancer*, 43:77-84.

Selling, L., 1916, Benzol as a leukotoxin. Studies on the degeneration and regeneration of the blood and hematopoietic organs, *Johns Hopkins Hosp. Rep.*, 17:83-142.

Swaen, G.M., and Meijers, J.M., 1989, Risk assessment of leukaemia and occupational exposure to benzene, *Br. J. Ind. Med.*, 46:826-830.

Thorslund, T.W., Hegner, R.E., Anver, M.R., and Voytek, P.R., 1988, "Quantitative Re-evaluation of the Human Leukemia Risk Associated with Inhalation Exposure to Benzene," prepared by Clement Associates, Inc., Fairfax, VA.

White, M.C., Infante, P.F., and Chu, K.C., 1982. A quantitative estimate of leukemia mortality associated with occupational exposure to benzene. *Risk Anal.*, 2:195-204.

Wong, O., 1983, An industry-wide mortality study of chemical workers occupationally exposed to benzene. Environmental Health Associates, Inc., Berkeley, CA. Submitted to Chemical Manufacturers Association, unpublished.

Yin, S.-N., Li, G.-L., Tain, F.-D., Fu, Z.-I., Jin, C., Chen, Y.-J., Juo, S.-J., Ye, P.-Z., Zhang, J.-Z., Wang, G.-C., Zhang, X.-C., Wu, H.-N., and Zhong, Q.-C, 1989, A retrospective cohort study of leukemia and other cancers in benzene workers, *Environ. Health Perspect.*, 82:207-213.

Yin, S.-N., Li, G.-L., Tain, F.-D., Fu, Z.-I., Jin, C., Chen, Y.-J., Luo, S.-J., Ye, P.-Z., Zhang, X.-C., Wu, H.-N., Zhong, Q.-C., Zhang, J.-Z., and Wang, G.-C., 1987. Leukemia in benzene workers: a retrospective cohort study, *Br. J. Ind. Med.*, 44:124-128.

COMPARATIVE METABOLISM AND GENOTOXICITY DATA ON BENZENE: THEIR ROLE IN CANCER RISK ASSESSMENT

Sandro Grilli,[2] Silvio Parodi,[1,3] Maurizio Taningher[1,3] and Annamaria Colacci[2,4]

[1]Centro Interuniversitario per la Ricerca sul Cancro
[2]Istituto di Cancerologia, Bologna
[3]Istituto Nazionale per la Ricerca sul Cancro (IST), Genova
[4]IST-Biotechnology Satellite Unit-Bologna

INTRODUCTION

Benzene is one of the chemicals with the largest production and utilization, with widest environmental spread and availability of animal and human toxicology data (IARC, 1982a). It is, therefore, a model compound which requires metabolic activation in order to exert toxic effects.

Recent reviews by Grilli et al. (1987) and Parodi et al. (1989) have evaluated the genotoxicity and carcinogenicity of benzene with special reference to dose-response relationships. They showed that inhalation exposure above 30-100 ppm leads to saturation of the ability to metabolically activate benzene in various species. This finding, along with a linear relationship between benzene dosage and DNA adducts formation in rat organs *in vivo* within a wide range of low and very low doses (Mazzullo et al., 1989), must be carefully considered in benzene cancer risk assessment.

The CRA process began in the middle '70s for human exposure and in the early '80s for results of carcinogenicity on rodents. Many data have been reported in the literature as to the initiating ability of benzene whereas very few data have been produced on its possible promoting effect.

In this paper, the main information about benzene metabolism and mutagenicity is updated and reviewed in the perspective of evaluating its ability to initiate the process of chemical carcinogenesis. Finally, animal cancer risk estimates are compared to those derived from epidemiological studies of occupationally exposed persons shown in the most recent review by Parodi et al. (1992). The animal estimates derive from long-term assays in rodents. The process of extrapolation from animal to humans, takes into ac-

Table 1. Short-term assays of potential carcinogenicity of benzene: evaluation of most genotoxicity data from literature

Endpoint	DNA damage (+)	(-)	Mutation (+)		(-)	Chromosome (+)		(-)	other (+)		(-)
Prokaryotes	0	pol A	2	1[a]	ST	1[b]						
				0	ST, BS	7						
				0	ST	5[c]						
				1[d]	EC	0						
				0	ST	1*						
				0	EC	1*						
Fungi/green plants			1[e]	SC	1	3[c]	A	1[c]			
			5[c]	GM, GC,CO	16[c]	0	CA	2†			
Insects			2	TRD	0	0	HT	1			
			1	RM	2						
			2[c]	RM	3[c]						
Amphibians											
Mammalian cells (in vitro)	1[f]	MtMBI	0	0	L5178Y	1	1[b]	CA	0	0	MC	3[c]
	2[g]	MMBI	0	6[c]	L5178Y, V79	9[c]	1[b]	CA	0	3[c]	TR	5[c]
	1[c]	AE	2[c]	0	CHO,HL BALB/c 3T3	1[h]	7	CA	1	2[h]	TR	0
	2[c]	UDS	4[c]				3[c]	CA	4[c]			
	1[i]		0	0	CHO, L5178Y	3*	2[c]	A/P	3[c]			
	0	UDS	1*				2	SCE	1			
							0	SCE	6[c]			
Mammals (in vivo)	1[j]		1[k]	0		1[l]	0	MN	1[c]	0	DL	1
	6[g,m]		0	1‡		0	13	CA	1			
	2[o]		0				1[n]	CA	0			
	1[q]		0				4[p,‡]	CA	0			
	0	UDS	1[s]				6[r,s]	MN	0			
	0	AE	3[#]				11	MN	0			
							28[t]	MN	2[u]			
							0	SCE	1*			

(Cont'd)

Table 1. (Continued)

Humans (in vivo)	19	CA	2		
	4[b]	CA	1[v]		
	1[w]	CA	0		
	1[y]	CA	1[x]		
	0	SCE	2		
	0	SCE	2[x]		
Class positivity:	17/31 = 54.8%		20/71 = 28.2%	108/140 = 77.1%	5/14 = 35.7%
Overall activity:				150/256 = 58.6%	

A=aneuploidy; AE = alkaline elution; BS = Bacillus subtilis; CA = chromosome aberrations; CHO = chinese hamster ovary; CO = crossing over; DL = dominant lethal test (rat); GC = gene conversion; GM = gene mutation; HL = human lymphocytes; HT = D. melanogaster, heritable translocations; MC = metabolic cooperation; MMBI = microsome-mediated binding to calf thymus DNA; MN = micronucleus test; MtMBI = mitochondria-mediated binding to mitochondrial DNA; P = polyploidy; pol A = DNA polymerase-deficient and proficient Escherichia coli; RM = recombination and mutation; SC = Saccharomyces cerevisiae; SCE = sister chromatid exchange; ST = salmonella typhimurium; TR = cell transformation; TRD = Trasdecantia assay; UDS = unscheduled DNA synthesis. Number of short-term tests either positive (+) or negative (-) are reported. Unless otherwise indicated, data are taken from IARC (1982a,b), ECETOC (1984), Dean (1985) and IARC (1987) reviews. Data lacking in experimental details, equivocal for the authors or for IARC working group or subsequently changed by their authors after having modified experimental design, have not been considered.

[a]From Glatt et al. (1989). [b]From Lee et al. (1983). [c]From Table 1 of ICPS collaborative study of Ashby et al. (1985) .
[d]From. Rossman et al. (1989). [e]From Morita et al. (1989). [f]From Kalf et al. (1985). [g]From Arfellini et al. (1985); Krewet et al. (1990), DNA adducts
[h]From Fitzgerald et al. (1989); Amacher and Zalljaadt (1983). [i]From Reddy et al. (1989a), DNA adducts in Zymbal gland cells in vitro
[j]From Parodi et al. (1983a) alkaline elution. [k]From Parodi et al. (1983b) viscosimetric unwinding.
[l]Point mutation in rat chromosome. [m]From Arfellini et al. (1985); Mazzullo et al., (1989) DNA adducts in various organs; Lutz and Schlatter (1977) adducts in rat liver; Reddy et al. (199oa) DNA adducts in various organs; Krewet et al. (1990). [n]From Gad-el-Karim et al. (1984)
[o]From Snyder et al. (1989), DNA adducts in bone marrow; Reddy et al. (1989b) DNA adducts in Zymball glands cells
[p]From Legator, (personal communication) chromosome aberrations in mouse lymphocytes after inhalation; Rithidech et (1987); Au et al., (1988); Au et al. (1990b), mouse lung macrophages. [q] From Bauer et al. (1989) adducts to DNA. [r]From Sabourin et al. (1990); Angelosanto et al. (1990); Au et al. (1990a); Au et al. (1991). [s]From Steimetz et al. (1990). [t]From Choy et al. (1985) and MacGregor et al. (1990) (1 positive assay); Mohtashamipur et al. (1987) (1 positive assay); Harper et al., 1984 (1 positive assay); Harper and Legator (1987) (1 positive assay); Ciranni et al. (1988) (2 positive assay); Harper (1989a) (2 positive assays in two different strains); Pirozzi et al. (1989) (2 positive assays in two different strains); Anwar et al. (1989a, 1989b) (2 positive assay); Harper (1989a) (2 positive assays in two different strains); Harper et al. (1989b) (2 positive assays); Suzuki et al. (1989) (4 positive assays by oral and ip administration to two different strains); Tice et al (1988, 1989) (3 positive assays); Sutou (1990) (3 positive assay). [u]From Harper et al. (1989b) (pregnant mice and feti). [v]From Jablonicka et al. (1987). [w] From Yardley-Jones et al. (1990). [y] Sasadiek et al. (1989)
[x]From Yardley-Jones et al. (1988); Seiji et al. (1990). *Oberly et al. (1990), s. cerevisiae chromosome loss. #Lee and Garner (1991, bone marrow cells of male and female mice, treated p.o. or i.p. *Oberly et al. (1990), UDS in rat hepatocytes; SCE in Chinese hamster bone marrow cells. [t]Au et al. (1990b), 6-thioguanine resistence, mouse spleen lymphocytes, chromosome aberrations in mouse lung macrophages.

Table 2. Dose-response relationship for the clastogenic activity of benzene in rodents

Species and strain	Route of administration	End point and dose unit	Effect*/dosage ratio at different dosage		
			Low	Medium	High
MOUSE					
Male Swiss	Oral[b]	Micronuclei/1000 PCE, mg/kg	4-2.4/8.8[c] = 0.18	6-2.4/88 = 0.04	28.7-2.4/880 = 0.03
Male Swiss	Oral[b]	Chromosome aberrations per metaphase, mg/kg	0.06-0.03/8.8[c]=0.0034	0.16-0.03/88=0.0015	1.1-0.03/880=0.0012
Male B6C3F1	Oral[d,e]	Micronuclei/1000 NCE, mg/kg	2.5-1.2/25[f]=0.052 6.5-1.2/200=0.026	4.1-1.2/50=0.058 8.4-1.2/400=0.018	5.8-1.2/100=0.046 9.3-1.2/600=0.013
Male Swiss ICR	Oral[g]	Chromosome aberrations per metaphase, mg/kg	2-1/36.6[h]=0.057	6-1/73.2=0.083	14.5-1/146.4=0.092
Male C57Bl/6	Oral[i]	Micronuclei/1000 PCE, mg/kg	20/220[j]=0.091	33.3/440=0.076	43.3/880=0.049
Male DBA/2	Oral[i]	Micronuclei/1000 PCE, mg/kg	26.7/220[j]=0.121	46.7/440=0.106	66.7/880=0.076
Male Swiss ICR	Oral[k]	Micronuclei/1000 PCE, mg/kg	23.9-3.4/220[j]=0.093	47.9-3.4/440=0.10	56.2-3.4/880=0.060
Male Swiss ICR	Oral[i]	Micronuclei/1000 PCE, mg/kg		19.9-3.4/440=0.038	22.73-3.4/880=0.022
Male Swiss ICR	Oral[l]	Micronuclei/1000 NCE, mg/kg	11.2-4.6/36.6[b]=0.180	22.6-4.6/73.2=0.246	40-4.6/146.4=0.242
Male ICR	Oral[m]	Micronuclei/1000 PCE, mg/kg	27-2.5/880[n]=0.028	64-2.5/1320=0.047	35-2.5/1760=0.018
Female ICR	Oral[m]	Micronuclei/1000 PCE, mg/kg	12-0.9/880[n]=0.013	21.5-0.9/1320=0.015	10.5-0.9/1760=0.0055
Male Ms/Ae	Oral[o,p]	Micronuclei/1000 PCE, mg/kg	24-4.5/500[a]=0.039 33-4.5/4000=0.0069	50.3-4.5/1000=0.046	70-4.5/2000=0.033
Male CD-1	Oral[o,p]	Micronuclei/1000 PCE, mg/kg	16-0.8/500[a]=0.030 18-0.8/4000=0.0043	21.3-0.8/1000=0.020	27.5-0.8/2000=0.013
Male NMRI	Intraperitoneal[r]	Micronuclei/1000 PCE, mg/kg	4.4-1.8/132=0.020	8.1-1.8/264=0.024	12.4-1.8/528=0.020

266

Male Swiss-Webster	Intraperitoneal[a]	Micronuclei/1000 PCE, mg/kg		14.2-2.8/400=0.028	12.2-4.2/1000=0.008
Male C57Bl/6	Intraperitoneal[a]	Micronuclei/1000 PCE, mg/kg		17.7-4.7/400=0.032	25.5-4.5/600=0.035
Male Ms/Ae	Intraperitoneal[p,t]	Micronuclei/1000 PCE, mg/kg	45-2.5/250=0.17	12-2.5/500=0.019	20.5-2.5/1000=0.018
Male CD-1	Intraperitoneal[p,t]	Micronuclei/1000 PCE, mg/kg	12-2.5/2000=0.005 1.8-0.8/250=0.004	3.5-0.8/500=0.005	4.8-0.8/1000=0.004
Male CD-1	Intraperitoneal[‡]	Micronuclei/1000 PCE, mg/kg	2-0.8/2000=0.0006 5.3-1/250=0.017	6.7-1/500=0.011	10.8-1/1000=0.0098
Male CD-1	Intraperitoneal[‡]	Micronuclei/1000 PCE, mg/kg	5.7-1/250=0.019	8-1/500=0.014	14-1/1000=0.013
Male CD-1	Intraperitoneal[‡]	Micronuclei/1000 PCE, mg/kg	3.3-1/125=0.018	15.2-1/250=0.057	6.3-1/500=0.011
Male DBA/2	Inhalation[u]	SCE/metaphase, ppm	7.6-5.9/10=0.17	9.5-5.9/100=0.036	13.8-5.9/1000=0.0079
Male DBA/2	Inhalation[u]	Micronuclei/1000 PCE, ppm	9-2.1/10=0.67	20.3-2.1/100=0.18	28.1-2.1/1000=0.026
MaleCD-1	Inhalation[v]	% Lung	5.7 - 1.2/O.1 = 45 macrophages with chromosome anomalies, ppm	6.8 - 1.2/1 = 5.6	
RAT					
Male Wistar	Inhalation	% Cells with chromosome anomalies, ppm 6 hr	3.03-1.54/10=0.15	4.45-1.54/100=0.029	7.81-1.54/1000=0.0063
Male S.-D.	Inhalation	SCE/metaphase, ppm 6hr	10.4-8.6/10=0.18	11.1-8.6/30=0.083	
Male S.-D.	Inhalation[u]	Micronuclei/1000 PCE, ppm 6h	6.2-2.0/10=0.42	7.6-2.0/30=0.19	
S.-D.	Oral*	% Cells with chromosome aberrations, mg/kg	1.8-1/250=0.0032	2.33-1/500=0.0027	

(Cont'd)

Table 2. (Continued)

PCE=polychromatic erythrocytes, NCE = normochromatic erythrocytes; SCE = sister chromatid exchanges
[a]Treated values minus control values [b]Two doses were administered 24 hr apart and animals were sacrificed 30 hr after the first dose (Gad-el-Karim et al., 1984). [c]Corresponding to 3.76 ppm according to EPA (Lee et al., 1983) assuming a respiratory absorption coefficient of 0.5 and a ventilation capacity of 0.03 L/min for the mouse (Gold et al., 1984). [d]Subchronic treatment within NTP study. Micronuclei were measured after exposure for 5 days/week for 4 months (Choy et al, 1985; MacGregor et al., 1990). [e]Six dosage points. [f]Corresponding to 10.7 ppm according to EPA (Lee et al., 1983) assuming a respiratory absorption coefficient of 0.5 and a ventilation capacity of 0.03 L/min for the mouse (Gold et al., 1984). [g]Animals were treated daily for two weeks with no treatment on days 5 and 10 (Rithidech et al., 1987). [h]Corresponding to 15.6 ppm according to EPA (Lee et al., 1983) assuming a respiratory absorption coefficient of 0.5 and a ventilation capacity of 0.03 L/min for the mouse (Gold et al., 1984). [i]Animals were sacrificed 24 hr after the single treatment (Harper and Legator, 1987). [j]Corresponding to 93.9 ppm according to EPA (Lee et al., 1983) assuming a respiratory absorption coefficient of 0.5 and a ventilation capacity of 0.03 L/min for the mouse (Gold et al., 1984). [k]Animals were sacrificed 24 hr after the single treatment (Harper et al., 1984). [l]Animals were treated daily for two weeks with no treatment on days 5 and 10 (Rithidek et al., 1988). [m]Animals were orally treated with a single dose. Micronuclei were determined 24 hr after the treatment. For maternal and fetal challenge, pregnant females were treated on days 16-17 of gestation and sacrificed on days 17-18 (Harper et al. 1989b). [n]Corresponding to 375.9 ppm according to EPA (Lee et al, 1983) assuming a respiratory absorption coefficient of 0.5 and a ventilation capacity of 0.03 L/min for the mouse (Gold et al., 1984). [o]Animals were orally treated with a single dose. Micronuclei were determined 24 hr after the treatment (Suzuki et al., 1989). [p]Four dosage points. [q]Corresponding to 213.6 ppm according to EPA (Lee et al., 1983) assuming a respiratory absorption coefficient of treatment (Suzuki et al., 1989). [p]Four dosage points. [q]Corresponding to 213.6 ppm according to EPA (Lee et al., 1984). [r]Animals were treated with two ip administrations 24 hr apart and sacrificed 30 hr after the first injection (Mohtashamipur et al., 1987). [s]Animals were i.p. treated daily for two days and were sacrificed 17 hr after the last treatment (Pirrozzi et al., 1989). [t]Animals were i.p. treated with a single dose. Micronuclei were determined 24 hr after the treatment (Suzuki et al. 1989). [u]Exposure of 6 hr (Erexson et al., 1986). [v]Exposure of 24 hr/day, 7 days/week for 6 weeks (Au et al., 1988). [w]Exposure of 6 hr (Styles and Richardson, 1984). [x]Animals were treated i.p. and micronuclei were determined 24 hr after the treatment (Japanese collaborative study, Sutou et al. 1990). *Animals were orally exposed 5 days a week for 5 weeks and chromosome aberrations were counted in spleen lymphocytes (Au et al., 1990b).

count interspecies differences in benzene pharmacokinetics (PK).

GENOTOXICITY

Many reviews are available on benzene genotoxicity, the IARC evaluation being one of the most recent and complete (IARC, 1987). Here reported in Table 1 is the overall pattern of results obtained in mutagenicity testing (256 different experiments measuring various endpoints). Benzene is positive in about 80% of clastogenicity tests and in about 55% of tests measuring DNA damaging activity *in vivo* and *in vitro*. Thus, benzene is clearly a clastogenic and a genotoxic agent, whereas it does not induce gene mutations according to Ashby's (1985) criterion that a positive response rate of about 20%, when many tests of gene mutations have been performed, is essentially a negative response. The updated evaluation of benzene mutagenicity rather agrees with previous analyses by Grilli et al. (1987) and by Parodi et al. (1989). They were based on a lower number of experiments.

Dose-response relationship

Table 2 shows the available data on induction of chromosome abnormalities as a function of administered dose in both sexes of rats and mice treated with benzene by different routes. Trend to saturation in the efficiency of clastogenic response, i.e. SCE, micronuclei in PCE and chromosome aberrations, was observed in rats exposed to 30-100 ppm by inhalation. In the mouse, a similar saturation of SCE or micronucleated PCE induction was reached between 30 and 100 ppm except for induction of chromosome abnormalities in lung macrophages where a trend to saturation was already observable in animals exposed to a subchronic inhalation of dose of 1 ppm benzene, 7 hr/day, 7 day/week for 6 weeks (Au et al., 1988).

Looking at Table 2, for PCE micronuclei induced in mice, a sufficient number of results, obtained at different dosages, was available both for oral and for intraperitoneal administration (a total number of 61 different dosage points in 19 different experiments). For these 61 dosage points we have correlated the $\log_{10}$ of potency of the response, defined as [(no. of micronuclei per 1,000 PCE)/(dosage in mg/kg)] with $\log_{10}$ (dosage in mg/kg). The equation of the regression line was:

$$y = -0.086 - 0.566 \times x$$

with a correlation coefficient $r = -0.551$ which is statistically significant with $p < 0.001$. The correlation was even better ($r = -0.70$, $p < 0.001$) considering only 37 couples of data related to oral administration. Thus, a general overview of all results obtained strongly suggests a saturation of metabolism, approximately above 50 mg/kg (gavage) in the most susceptible strains of mice. The average result indicates that for a 10 times increase in dosage (for instance 50 to 500 mg/kg) response increases by only about 2-3 times. The dosage of 50 mg/kg administered to the mouse by the oral route corresponds, approximately, to 6-hr inhalation exposure to either 31 ppm, according to Henderson

Table 3. **Toxic effects of a few benzene metabolites**

Trans,Trans-muconaldehyde:
- acts as direct alkylating agent and produces three *in vitro* adducts with d-guanosine
- lowers fluidity of macrophage and neutrophilic granulocyte membrane
- induces loss of NADPH-dependent oxidase activity in macrophage and neutrophilic granulocyte membrane
- inhibits both *in vitro* and *in vivo* production of erythroid stem cells (CFU-E and CFU-GM) in bone marrow of CD-1 mice
- lowers RBC number, hemoglobin concentration, hematocrite, liver content of SH- groups
- reduces bone marrow cells in the mouse by inhibiting ^{59}Fe uptake into hemoglobin of RBC
- is further metabolized to non-toxic t,t-muconic acid

From: Witz et al.,(1985); Sabourin et al.,(1988); Latriano et al.,(1989), Snyder et al., (1989); Witz et al ., (1989) .

1,4-Benzoquinone:
- interferes with microtubule assembly
- inhibits DNA and RNA synthesis
- binds to proteins and DNA where it induces single strand breaks
- induces chromosome fragmentation during cell division
- inhibits cell division by inducing pycnosis of metaphase chromosomes
- interferes with growth of bone marrow stromal cells
- interferes with blastogenesis and lymphocyte agglutination

From: Smith et al ., (1989)

---

et al. (1989), or 15 ppm, according to U.S. EPA computations based on Lee et al. (1983), assuming a respiratory absorption coefficient of 0.5 and a ventilation capacity of 0.03 l/min (Gold et al., 1984).

Briefly, the findings obtained after inhalation confirm previous analyses in mice and rats by Grilli et al. (1987) and Parodi et al. (1989). Interestingly, Legator (1990), who studied chromosome aberrations induced in spleen lymphocytes by inhalation exposure of mice to benzene for 3-6 weeks, found a linear dose-response relationship in the range 0.04-0.15 ppm. The same Authors (Au et al., 1990b) observed chromosome aberrations in both spleen lymphocytes and lung macrophages in mice and 6-thioguanine mutants in spleen lymphocytes of the same animals also at 0.04 ppm inhalation.

Many considerations would be possible regarding the administration route (inhalation is probably more efficient than other routes because peak concentrations are avoided) (Ciranni et al., 1988a; Harper et al., 1989b; Pirrozzi et al., 1989; Suzuki et al., 1989; Tice et al., 1989), differences in strain (Harper and Legator, 1987; Harper et al., 1989a; Suzuki et al., 1989; Tice et al., 1989), and sex (males are probably more sensi-

Table 4. Genotoxicity of benzene metabolites[a]

	Phenol	Hydroquinone	Catechol	1,4-benzoquinone	1,2,4-benzenetriol	Antidiol epoxide	t,t-muconaldehyde	t,t-muconic acid
S. typh.	+ (TA1538 +S9)	+ (TA1535) −S9	−			+ (various strains) +S9	+ TA 104, ±S9; TA 97, TA 100, TA 102, +S9; − (TA 98, TA 1535)	− (various strains, ±S9)
E.coli WP2	−	−	−					±
6-TG mutants in V79 cells	+	+	+	+		+	+	
HGPRT mutants in CHO cells		+	+		+			
D. melanogaster (SLRL)		−	+	+			−	
in vitro adducts with Zymbal gland DNA	−	−	−	−	+			
DNA damage (c-Ha-ras -1, in vitro)				+	+			
DNA single-strand breaks in L5178Y cells	−			+	+			
DNA single-strand breaks in mouse bone marrow cells in vitro				+	+			
UDS in HPC (rat liver)							−	
DNA adducts in rat liver, spleen, bone marrow and Zymbal glands	−		+	+	+	+		−
SCE in human lymphocyte in vitro	±	+	+	+	+			
SCE in V79 cells		+	+	−	+			

(Cont'd)

Table 4. (Continued)

SCE in mouse bone marrow (in vivo)				+
Micronuclei in V79 cells (in vitro)	+	+	+	+
Micronuclei in human lymphocytes in vitro	+	+		
Chromosome aberration in human lymphocytes (in vitro)		−		
Micronuclei (mouse, in vivo)	+			−
Chromosome aberrations (mouse spermatogones, in vivo)	+			

For benzene genotoxicity see Table 1.
SCE = sister chromatid exchange; SLRL = sex linked recessive lethal; S9 = hepatic metabolizing system.
+ = positive result.
− = negative result.
± = equivocal result.
ᵃ Data here reported are taken from: Barale et al. (1990); Fishbein (1988); Glatt et al. (1989); Glatt and Witz (1990); Lee and Garner (1991); Ludewig et al. (1989; 1990); Reddy et al. (1990b); Robertson et al. (1990); Rossman et al. (1989); Smith et al. (1989); Witz et al. (1990a); Yager et al. (1990).

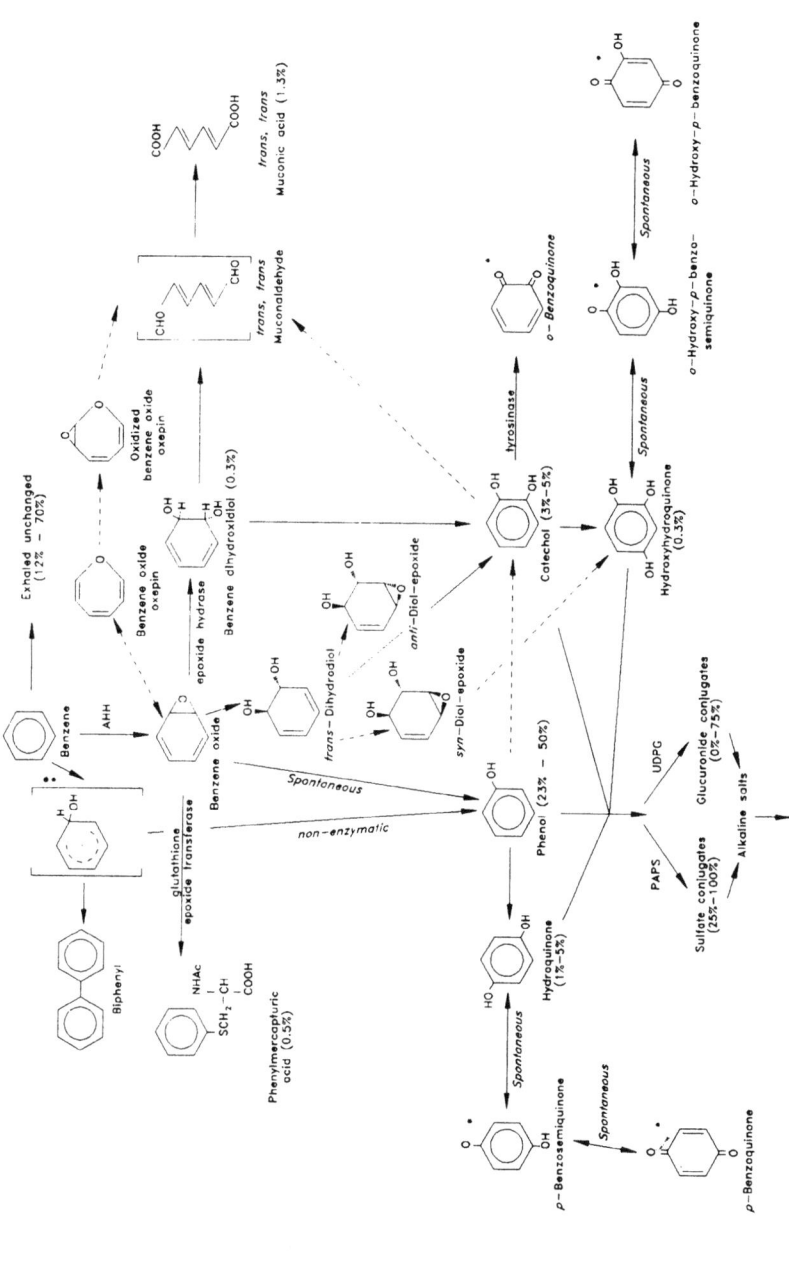

Fig. 1. Pathways of benzene metabolism and excretion. Values in parentheses are percentages of metabolic products detected in urine of animals (rabbits, rats, mice, dog) or humans. Asterisks denote putative or demonstrated alkylating activity toward intracellular nucleophiles. AHH, aryl hydrocarbon hydroxylase; UDPG, uridine diphosphate glucoronyl transferase; PAPS, 3-phosphoadenosine-5-phosphosulfate. Dashed lines indicate putative pathways. From: Kalf (1987), Huff (1989) and Glatt et al. (1989), with modifications.

tive than females to micronuclei induction)(Ciranni et al., 1988b; Harper et al., 1989b; Tice et al., 1989). Pregnant females and fetuses are either much less susceptible (Ciranni et al., 1988b) or insensitive to benzene (Harper et al. 1989b).

An important question is the following: subchronic and chronic exposures are increasing sensitivity for some test of genotoxicity? A real accumulation of effects seems absent for micronuclei (Tice et al., 1988, 1989; Rithidech et al., 1988, MacGregor et al., 1990). For chromosomal aberrations a very high sensitivity (detection level at about 1 ppm) was described in mice by Au et al. (1988) after inhalation exposure for about six weeks. Also Picciano (1980) found a very high sensitivity in humans (detection level at 1 < ppm) after chronic exposure. However, for chromosomal aberrations we were not

Table 5. Summary of data concerning benzene metabolism in rodents

Level of exposure	Dose-response relationship	Reference
Inhalation		
Above 125 ppm (rat)	supralinear	Gut and Frantik (1980)
Above 130 ppm (rat, mouse)	supralinear	Sabourin et al. (1987)
30 ppm and above (rat)	supralinear	Ulanova and Avilova (1987)
30-50 ppm and above (rat)	supralinear	Grilli et al. (1987)
Up to 30 ppm (rat)	linear	Jongeneelen et al. (1987)
Above 100 ppm (rat)	supralinear	Stommel et al. (1989)
Above 50 ppm (rat, mouse)	supralinear	Sabourin et al. (1989); Henderson et al.(1989)
Above 199 ppm (mouse, phenylsulfate)	supralinear	Wells and Nerland (1991)
Above 527 ppm (mouse, phenylglucuronide)	supralinear	Wells and Nerland (1991)
Oral administration		
Above 50 mg/kg (rat, mouse)	supralinear	Sabourin et al. (1987)
Above 10 mg/kg (rat, mouse)	supralinear	Sabourin et al. (1989) Henderson et al.(1989)
Above 40 mg/kg (rat and mouse, hemoglobin adducts)	supralinear	Sun et al. (1990)
Sc administration		
Above 440 mg/kg (adducts with mouse liver and bone marrow proteins)	supralinear	Snyder et al. (1978)
Ip administration		
Above 48.6 mg/kg (DNA adducts in rat liver)	supralinear	Mazzullo et al. (1989)
Above 4.86 mg/kg (adducts with rat liver proteins)	supralinear	Mazzullo et al. (1989)
Above 220 mg/kg (DBA/2N and C57BL/6 mice, excretion of urinary muconic acid)	supralinear	Witz et al. (1990b)

able to find an experiment directly suggesting increased sensitivity for exposures longer than acute.

The relationship between employed dosages and DNA adducts in various organs of rats *in vivo* and genotoxicity of the main benzene metabolites are discussed in the following metabolism section.

METABOLISM

Benzene requires metabolic activation in order to exert toxic effects on various targets, including bone marrow, in animals and humans. Its metabolism has been extensively studied in various species and also in man, so that reviews by many authors are available (IARC, 1982; Kalf, 1987; Marcus, 1987; Glatt et al., 1989; Henderson et al., 1989; Huff et al., 1989). Benzene metabolism does occur mainly in the liver. Bone marrow appears to be capable of metabolizing benzene to a lesser extent. Indeed, V_{max} (the Michaelis-Menten metabolism rate of benzene) in bone marrow is 4-8% of V_{max} in rat or mouse liver (Travis et al., 1990). In the rabbit, benzene oxide is formed by microsomal cyt. P450 IIE1, i.e. the same cytochrome able to metabolize ethanol, acetone and dimethylnitrosamine. Moreover, benzene pretreatment can induce its own metabolism (Schnier et al., 1989).

From a chemical point of view, benzene is a relatively simple molecule, however its metabolic pathways are rather complex. The most updated scheme on overall pathways of benzene biotransformation is given in Fig. 1. Briefly, benzene is first transformed into benzene oxide which rearranges to phenol, conjugates with GSH to form premercapturic acid or sums up water to give a dihydrodiol derivative which can undergo either oxidation to catechol or hydroxylation to 1,2,4-benzenetriol.

Phenol is further metabolized to hydroquinone which undergoes autooxidation leading to 1,4-benzoquinone formation (Fishbein, 1988). Biotransformation of phenol into 1,4-benzo-quinone is carried out also by peroxidases present in the bone marrow (Smith, 1989) and in the macrophages (Kalf et al., 1989). Conversion of hydroquinone into 1,4-benzoquinone is enhanced by phenol, which is also transformed into 4,4'-diphenoquinone, and by catechol, resorcinol, o-, m- and p-cresols, guaiacol, histidine and imidazole (Subrahmanyam et al., 1991). Dihydrodiol derivative can be also transformed into diol epoxides which give rise to 1,2,4-benzenetriol. All these metabolites do have structural ring of benzene. Finally, benzene ring opening leads to formation of trans,trans-muconaldehyde (t,t-muconaldehyde) which is then transformed into trans,trans-muconic acid (t,t,-muconic acid).

Some metabolites, like phenol or catechol, derive from detoxification pathways and are excreted after conjugation with glucuronic acid or sulphate or as mercapturate (Nerland and Pierce, 1990). Other metabolites, e.g. t,t-muconaldehyde, 1,4-benzoquinone, even if produced to a lesser extent, can play a remarkable role in induction of toxic, genotoxic and, possibly, carcinogenic effects. The toxic effects of t,t-muconaldehyde and 1,4-benzoquinone, which are two of the major toxic metabolites of benzene, are reported in Table 3. These metabolites are almost equitoxic with benzene in the ability to inhibit erythropoiesis in the mouse (Witz et al.1985).

Table 6. Human cancer risk estimates derived from long-term assays of carcinogenicity in rodents by means of a linearized multistage model which takes into account the internal dose of benzene calculated according to a pharmacokinetic approach.

1- Bailer and Hoel (1989)
- Data source: NTP study by gavage (Huff, 1986)
- Target: all squamous cell carcinomas
- Benzene metabolites: total metabolites from Sabourin et al. (1987) and also hydroquinone alone or t,t-muconic acid alone (Sabourin et al., 1989)
- Human equivalence: 1 ppm (8 hr/day) = 0.456 mg/kg/day assuming lung absorption of 100%
- Excess cancer risk for 1,000 persons occupationally exposed at 1 ppm for a working lifetime (45 years) (45 ppm.years) after extrapolation from total amount of all squamous cell carcinomas induced by benzene in:

(UCL)	Best fitting	Upper 95% Conf. Limit
Male mouse		
Administered dose	0.69	3.52[a]
Internal dose (total metabolites)	4.21	17.62
Internal dose (hydroquinone)	4.4	
Internal dose (t,t-muconic acid)	4.1	
Male rat		
Administered dose	1.0	1.3
Internal dose (total metabolites)	2.9	3.9
Internal dose (hydroquinone)	5.4	
Internal dose (t,t-muconic acid)	3.4	

[a]Corresponding to excess risk of 7.7×10^{-6} for a person exposed lifetime to a benzene air concentration of 1 $\mu g/m^3$.

2- Beliles and Totman (1989):
- Data source: long-term assays by inhalation of Maltoni et al. (1982) and Snyder et al. (1980) and by gavage (NTP study: Huff, 1986)
- Target: Zymbal gland and hemopoietic tumors
- Benzene metabolites: total metabolites from Sabourin et al. (1987)
- Human equivalence: on the basis of body weight ratio$^{0.74}$ (value measured by inhalation experiments):
- Excess cancer risk for 1,000 persons occupationally exposed at 1 ppm for a working lifetime (45 ppm.years) after extrapolation from Zymbal gland and hemopoietic tumors induced by benzene in:

(Cont'd next page)

(Table 6, Cont'd from previous page)

Internal Dose

Total Metabolites	Target	Best Fitting	UCL (95%)
Mouse (inhalation, Snyder et al., 1980)	Leukemia	1.4	4.8[a]
Rat (inhalation, Maltoni et al., 1982)	Zymbal glands	1.3	2.5
Mouse/Rat (oral, Huff, 1986)	Zymbal glands	0.6	1.1

[b]corresponding to excess risk of 10.47×10^{-6} for a person exposed lifetime to a benzene air concentration of 1 $\mu g/m^3$.

3-A. Excess leukemia risk for 1,000 persons occupationally exposed to 1 ppm for a working lifetime (45 ppm.years) derived from epidemiological studies (Infante, 1987, who presented a 1984 updated review risk estimates):

0.95 - 17.4[c]

[c]Corresponding to excess risk of $2.07 - 37.97 \times 10^{-6}$ for a person exposed lifetime to a benzene air concentration of 1 $\mu g/m^3$.

3-B. Excess leukemia risk for 1,000 persons occupationally exposed to 1 ppm for a working lifetime (45 ppm. years) derived from more recent elaborations of data from epidemiological studies:

Rinsky et al. (1987)	5.3
Austin et al. (1988)	7.95
Brett et al. (1989)	0.5 - 6.4[d]

[d]corresponding to excess risk of $1.1-13.97 \times 10^{-6}$ for a person exposed lifetime to a benzene air concentration of 10 $\mu g/m^3$. Cancer risk estimated by WHO (1987 guideline) for environmental exposure is: 4×10^{-6} for a person lifetime exposed by inhalation to a benzene air concentration of 1 $\mu g/m^3$.

(Cont'd on the next page)

Table 6 (Continued from previous page)

4- Other estimates of excess cancer risk per 1,000 persons occupationally exposed to 1 ppm for a working lifetime (45 ppm.years) derived from mathematical models, which does not account for pharmacokinetics, applied to data from long term assays of carcinogenicity in rodents:

Animals	Tumors	Route	Risk estimate
Male rat (Huff, 1986)	all squamous cell carcinomas	gavage	0.87
Male mouse (Huff, 1986)	all squamous cell carcinomas	gavage	2.25
Male and female rats (Maltoni et al., 1983)	leukemias	inhalation	0.17
Male mouse (Goldstein et al., 1982)	leukemias	inhalation	0.045

From: McMichael (1988), who presents the estimates given by Crump and Allen (1984), modified to the risk associated with a cumulative exposure of 45 ppm.years.

Moreover, hydroquinone, whose conjugates concentrate in bone marrow, and t,t-muconaldehyde reciprocally enhance their inhibitory effects on ^{59}Fe uptake into erythrocyte hemoglobin (Snyder et al., 1989). Hydroquinone and catechol enhance the reduction of erythrocyte production induced by phenol in female mice (Guy et al., 1990).

Table 4 shows genotoxicity of the main metabolites of benzene. In the mutagenesis testing, positive results have been found with 1,2,4-benzenetriol, benzene diolepoxide and, mainly, hydroquinone, 1,4-benzoquinone. t,t-Muconaldehyde induces cytogenetic damage in mammalian cells and is non-mutagenic or very weakly mutagenic in bacteria whereas its mutagenic potency is higher in mammalian cells (Glatt and Witz, 1990; Witz et al., 1990a). The major final carcinogenic metabolite for benzene has not yet been identified, nevertheless Busby et al. (1990) have shown that a racemic mixture of anti--diolepoxides, derived from cis- and trans-1,2-dihydroxy-1,2-dihydrobenzene, is very active in inducing dose-dependent increases of lung tumor incidence and multiplicity in newborn BLU:Ha mice.

Synergistic effects have been reported between hydroquinone and phenol in inducing bone marrow micronucleated cells in the mouse (Barale et al., 1990) as well as between hydroquinone and catechol for the induction of micronucleated cells when using human blood peripheral lymphocytes cultured *in vitro* (Robertson et al., 1990).

The ultimate active metabolite of benzene binds to DNA both *in vitro* and *in vivo* in rat liver (Lutz and Schlatter, 1977; Arfellini et al., 1985; Mazzullo et al., 1989), in mouse liver and in various organs of either species. No major differences in the extent of covalent interaction with DNA, i.e. covalent binding index (CBI), were however observed for different organs (Arfellini et al., 1985).

CBI values were low and typical of carcinogens exhibiting weak initiating activity. Interaction between benzene and mouse liver or bone marrow DNA was found by Rushmore et al. (1984). Reddy et al. (1990a) have measured DNA adducts in liver DNA as well as in oral cavity and mammary gland DNA. However, covalent binding to Zymbal gland DNA, as well as to liver, spleen and bone marrow DNA, was not observed (Reddy et al., 1990b). Thus, binding to DNA is insufficient to explain target specificity. Specific benzene-DNA adducts have not yet been identified. A possible adduct, which has been identified after *in vitro* reaction of benzene metabolites, hydroquinone and 1,4-benzoquinone, with d-guanosine, derives from binding of the quinone derivative with the cyclic N^1 and exocyclic N^2 belonging to the amino group of guanine, i.e. (3'OH)benzentheno(1,N^2) deoxyguanosine (Jowa et al., 1986; Bauer et al., 1989; Jowa et al., 1990). Such an adduct is probably formed also in rabbit liver DNA *in vivo*, among others (Bauer et al., 1989). N^7-phenylguanine is formed in a microsomal system *in vitro* and in the urine of rats which underwent ip administration of benzene (Krewet et al., 1990). ^{32}P-postlabeling analysis by Reddy et al. (1989a,b) showed very low amount of DNA adducts (0.5-4 per 10^9 nucleotides) in rat Zymbal gland DNA *in vivo* and *in vitro*. Probably, this technique is very useful in quantitating polynuclear aromatic hydrocarbon-DNA adducts whereas it is not the best approach to measure benzene-DNA adducts, especially if they are produced by the polar metabolite t,t-muconaldehyde, the only one lacking the aromatic ring structure. Moreover, under comparable experimental conditions, 1,4-benzoquinone induced about 2 adducts per 10^6 nucleotides in Zymbal gland DNA *in vitro*, hydroquinone was 2-fold less effective and 1,2,4-benzenetriol, phenol or catechol were much less active (1 order of magnitude). Similar results were reported later by Reddy et al. (1990b).

Dose-response relationships

Many studies performed in various experimental models and species are available for evaluation. The first analysis of literature data by Grilli et al. (1987) indicated that benzene metabolism approaches saturation above 30-50 ppm in rats exposed by inhalation to either an initial concentration of chemical in a closed chamber or a constant concentration in air. This result has been confirmed in the most recent analysis by the same authors (Parodi et al., 1989) who reviewed inhalation data presented by Gut and Frankin (1980), Grilli et al. (1987), Jongeneelen et al. (1987), Sabourin et al. (1987), Ulanova and Avilova (1987). Sabourin et al. (1987) also showed saturation of metabolizing activity, more evident in the mouse than in the rat, above 50 mg/kg of benzene administered by the oral route.

More recently, Stommel et al. (1989) found a linear relationship between benzene exposure in the range 1-100 ppm and excretion of S-phenylmercapturic acid, a non-toxic metabolite of benzene, in Wistar rats at 32 hr after exposure. Tendency to saturation was evident at 500 ppm. Wells and Nerland (1991) found a trend to saturation in the formation of phenyl sulfate and phenyl glucuronide above 199 and 527 ppm, respectively, in Swiss mice exposed via inhalation to 55-2290 ppm benzene.

Sabourin et al. (1989) and Henderson et al. (1989) investigated metabolite formation in F344/N rats and in B6C3F1 mice receiving either 1, 10, or 200 mg/kg of benzene by gavage or exposed to 5, 50 or 600 ppm of benzene by inhalation. Saturation of blood concentration of phenyl sulfate and t,t-muconic acid begins above 50 ppm (inhalation) and above 10 mg/kg (gavage). Above these levels, even hydroquinone or t,t-muconaldehyde formation in the liver of both species is no longer linear. Thus, the proportion of metabolites from toxic pathways of benzene, i.e. hydroquinone and t,t-muconic acid, decreases as benzene dose increases. This fact could lead to underestimation in cancer risk estimates which are usually calculated on the basis of tumor incidences observed in rodents exposed to high dosages of benzene, where the proportion of detoxifying metabolism of benzene is higher.

In vivo interaction of benzene with DNA and proteins has also been measured as a function of the administered dosage. After ip administration to Wistar rats, Mazzullo et al. (1989) found a linear relationship between liver DNA adducts and benzene dosage within a wide range of benzene dosages, including very low ones. Very similar results were observed in the lung, spleen, kidney and stomach. Trend to saturation of the binding ability was observed in liver above 48.6 mg/kg in the case of DNA and above 4.86 mg/kg for RNA and proteins. Dose-response relationships for benzene-DNA interaction can be used as a surrogate of dose-tumor incidence relationship for low and very low doses only if the mechanisms of DNA repair respond linearly to different levels of damage and low doses of benzene do not play a significant promoting role in chemical carcinogenesis. These doses for which adducts have been detected are much lower than those capable of inducing a statistically significant excess of tumor incidence in long-term assays of carcinogenicity.

As to hemoglobin adducts formation, a similar picture was reported by Sun et al. (1990) who found supralinearity above 40 mg/kg benzene after oral administration of 0.008-800 mg/kg to F344/N rats and B6C3F1 mice. On the other hand, Witz et al. (1990b) found the same trend to saturation above 220 mg/kg benzene after i.p. administration of 0.5- 880 mg/kg to two strains of mice.

An updated evaluation of dose-response relationship is shown in Table 5 and confirms previous analyses indicating that benzene metabolic activation approaches saturation at air concentrations above 30-50 ppm and at oral dosages higher than 10-50 mg/kg. As previously stated, long-term assays of carcinogenicity have been carried out in rodents using higher dosages of benzene.

In order to compare absorbed doses of benzene after administration by two different routes, gavage and inhalation, general mathematical equations, as described by Beliles and Totman (1989), can be employed. These formulas take into account body surface and metabolic rates in different species and are routinely used in quantitating human cancer risk through interspecies extrapolation. However, Henderson et al. (1989) have proposed safer equivalents of absorbed dose based on total metabolites measured for each administration route. It should be emphasized, nevertheless, that the relative proportion of benzene metabolites varies according to route of administration, species and organs, as analyzed by Sabourin et al. (1989) (liver, lung, blood, bone marrow and urine) and by Low et al. (1989) (liver, blood, bone marrow, Zymbal and mammary glands, nasal and oral cavities and kidney). Therefore, some degree of uncertainty remains in such an approach until the final carcinogen for benzene is identified.

Species-specific differences

As for benzene metabolic pathways, the formation of phenol, catechol and hydroquinone are rather similar in animals and in man from both the qualitative and quantitative points of views. For instance, benzene metabolism in terms of relative and absolute production of phenol, catechol, quinol, 1,2,4-benzenetriol and t,t,-muconic acid in the rabbit treated by oral administration is similar to that from human exposure by inhalation (Inoe et al., 1989). Nevertheless, some quantitative differences among species, strains or sexes exist (Bradfuehrer et al., 1990). Species-specific variations in the capability to metabolize benzene have been well demonstrated even in studies performed on animal species identical with those that underwent long-term assays of carcinogenicity.

Sabourin et al. (1987) showed that saturation of formation of urinary plus fecal metabolites is more evident in B6C3F1 mice than in F344/N rats exposed by inhalation because benzene absorption in mice is higher than in rats which present a lower ventilation rate.

Sabourin et al. (1988) found differences among F344/N rats and B6C3F1 mice in ability to produce t,t-muconic acid, hydroquinone glucuronide and hydroquinone monosulfate in liver, lung and blood. By contrast, they found differences in the formation of neither phenyl sulfate nor other conjugates of phenol and catechol. After single inhalation exposure to 50 ppm for 6 hr, higher amounts of toxic benzene metabolites have been measured in the mouse with respect to the rat, which is less susceptible to carcinogenesis by benzene. This investigation has been successively enlarged and completed by Sabourin et al. (1989) and by Medinsky et al. (1989a,b) who, aiming at developing a physiologically-based pharmacokinetic (PBPK) model, compared benzene metabolism in the same two species after oral administration or inhalation. Up to 1000 ppm, mice can bioactivate higher amounts (2-3-fold) of inhaled benzene than rats do. Hydroquinone and t,t-muconic acid quantities are 4 - 10-fold higher, on the basis of mg/kg b.wt. dosage, in the mouse than in the rat. These differences were more evident after inhalation exposure.

Sabourin et al. (1990) showed that the extent of benzene-hemoglobin adducts is similar in F344/N rats and in B6C3F1 mice after repeated inhalation exposures to 600 ppm, 6 hr/day, 5 day/week for 3 weeks, even though the rate of benzene metabolism is much higher in mice than in rats. From a qualitative point of view, the only effects of benzene pretreatment on the metabolic profile in rat or mouse urine were slight shifts from glucuronation to sulfation in mice and from sulfation to glucuronation in rats.

In mice, strain-specific quantitative, but not qualitative variations, in benzene metabolism are observed (Harper et al., 1989a; Longacre, 1980; Twerdok e Trush, 1990; Witz et al., 1990b).

Differences in benzene metabolite formation due to dose rate variation have been reported by Snyder et al. (1978), by Sabourin et al. (1989) and Henderson et al. (1989). They found that the proportion of hydroquinone and t,t-muconic acid in blood, bone marrow and liver of rats and, especially, of mice decreases as the dose rate of benzene administration increases. The relative proportion of t,t-muconic acid is higher in mice than in rats.

Pharmacokinetic (PK) models

Benzene biotransformation follows a first order kinetics up to 10-50 mg/kg b. wt. (gavage) or 30-50 ppm (inhalation) and zero order kinetics (saturation) above these doses or concentration levels. Thus, incorporation of PK into risk assessment techniques is needed in order to substitute internal dose, i.e. the absorbed dose, or rather the really metabolized dose on the target site, to the external dosage (administered dose or exposure level).

The first PK models, based on Michaelis-Menten equation, have been integrated with mathematical, mainly linearized multistage, models employed in assessing cancer risks by vinyl chloride (Gehring et al., 1978; 1979; Bois et al., 1985) and by dichloromethane (Reitz et al., 1987; 1988). PK models assume that K_m (the Michaelis constant) is a constant parameter in interspecies extrapolation. However, in the case of benzene, K_m values (μg/l) have been found to slightly differ in the two species after oral administration (137 and 103 for rat and mouse, respectively) or inhalation (339 and 495 for rat and mouse, respectively) (Beliles and Totman, 1989). Nevertheless, the amounts of benzene metabolites measured in the rat or in the mouse agree with those predicted by the general available equations.

Benzene metabolism in the mouse is more easily saturable than in the rat and better resembles human metabolism according to Bailer and Hoel (1989). Therefore, human cancer risk estimates obtained from mice treated by gavage are more acceptable than those derived from rats. Cancer risk estimates given by the authors, based on a PK model considering total metabolites as internal dose, are about 5-fold higher than those simply based on the dosages administered to animals. These last estimates, on the other hand, are 1 order of magnitude lower than cancer estimates directly derived from approaches of analytical epidemiology (Table 6). Despite few different assumptions concerning the type of experimental tumor utilized for risk assessment, the estimates of Bailer and Hoel (1989) agree well with those of Beliles and Totman (1989) who also applied PK models (Table 6).

Bailer and Hoel (1989) have also produced risk estimates based on PK applied to a linearized multistage model for two specific markers of toxic pathways of benzene metabolism, i.e. hydroquinone and t,t-muconic acid. These estimates, reported in Table 6, are rather similar to those calculated by the same authors on internal doses of total metabolites. Thus, most recent risk estimates from long-term assays corrected by PK models in extrapolating from high to low doses and to human exposures agree rather well with estimates from human exposures.

PB-PK models are necessary in order to achieve better extrapolations among different routes of administration or species. These models, which have already been used for dichloromethane (Reitz et al., 1988), perchloroethylene (Ward et al., 1988) and carbon tetrachloride (Paustenbach et al., 1988), by performing computer-assisted calculations, can simulate material partition among various compartments, organs and tissues as well as metabolic transformation routes.

PB-PK models utilized in cancer risk assessment of dichloromethane led to marked reduction (at least 2 orders of magnitude) of risk estimates simply obtained from the administered dose (Reitz et al., 1988).

An analogous model for benzene is still under validation (Medinsky et al., 1989a,b; Travis et al., 1990). The PB-PK model of Travis et al. (1990) considers that benzene metabolism does occur in liver and also in bone marrow. The PB-PK model by Medinsky et al. (1989a,b) does not take into account the role played by bone marrow metabolism. Anyway, computer assisted simulations are consistent with data from human and animal exposure by oral, ip, and inhalation routes.

Application of PB-PK models to overall data from long term carcinogenicity testing of benzene in rodents would give even better estimates of cancer risk extrapolated to humans. Thus, their internal consistency and comparability of their results with excess risk estimates from epidemiological data would be significantly improved. The goal of validation of PB-PK models will be likely to be reached in the near future since the chemical industry will present detailed data on mechanism of action and on comparative PK for new chemicals, of high commercial or industrial interest, showing some species-, organ- or sex-specific carcinogenic properties after administration of very high doses.

CONCLUSIONS

The bulk of information on benzene metabolism and mutagenicity reviewed here gives rise to biological, mathematical and toxicological considerations leading to human cancer risk estimates from experimental models which are rather consistent with those directly measured in humans by means of analytical epidemiology. The latter are reviewed in this volume by Parodi et al. (1992). Indeed, PK models indicate an excess cancer risk of few cases per 1,000 persons occupationally exposed to 1 ppm for a working life-time (45 years) just as epidemiological studies do.

ACKNOWLEDGMENTS

The authors are sincerely grateful to Dr. Pier-Luigi Lollini for his invaluable help in the preparation of this manuscript and for drawing up the metabolic pathway of benzene.

REFERENCES

Amacher, D.E., and Zallijaadt, I., 1983, The morphological transformation of Syrian hamster embryo cells by chemicals reportedly non mutagenic to Salmonella typhimurium, Carcinogenesis, 4: 291.

Angelosanto, F.A., Kroth, M.D., Blackburn, G.R., and Mackerer, C.R., 1990, Benzene induces a dose-dependent increase in micronuclei in Zymbal gland cells of rats, Proc. Am. Assoc. Cancer Res., 31:101.

Anwar, W.A., Au, W.W., Legator, M.S., and Sadagopa Ramanujam, V.M., 1989a, Effect of dimethyl sulfoxide on the genotoxicity and metabolism of benzene in vitro, Carcinogenesis, 10:441.

Anwar, W.A., Au, W.W., Sadagopa Ramanujam, V.M., and Legator, M.S., 1989b, Enhancement of benzene clastogenicity by prazinquantel in mice, Mutat. Res., 222:-283.

Arfellini, G., Grilli, S., Colacci, A., Mazzullo, M., and Prodi, G., 1985, *in vivo* and *in vitro* binding of benzene to nucleic acids and proteins of various rat and mouse organs, Cancer Lett., 28:159.

Ashby, J., de Serres, F.J., Draper, M., Ishidate, M., Jr, Margolin, B.M., Matter, B., and Shelby, M.D., 1985, Overview and conclusions of the IPCS collaborative study on *in vitro* assay systems, in: "Progress in Mutation Research", 5:117, Ashby J., de Serres F.J. et al., eds., Elsevier Science Publishers, Amsterdam (WHO Geneva).

Au, W.W., Bibbins, P., Ward, J.B., Jr., and Legator, M.S., 1988, Development of a rodent lung macrophage chromosome aberration assay, Mutat. Res., 208:1.

Au, W.W., Anwar, W.A., Hanania, E., and Sadagopa Ramanujam, V.M., 1990a, Antimutagenic effects of dimethyl sulfoxide on metabolism and genotoxicity of benzene *in vivo*, in: "Antimutagenesis and Anticarcinogenesis Mechanisms II". Kuroda Y., Shankel D.M. and Waters M.D. (eds). Basic Life Sciences, 52: 389.

Au, W.W., Cantelli Forti, G., Hrelia, P., and Legator, M.S., 1990b, Cytogenetic assays in genotoxic studies: somatic cell effects of benzene and germinal cell effects of dibromochloropropane, Teratog. Carcinog. Mutag., 10:125.

Au, W.W., Anwar, W., Paolini, M., Ramanujam, S., and Cantelli Forti, G., 1991, Mechanism of clastogenic and co-clastogenic activity of cremophore with benzene in mice, Carcinogenesis, 12:53.

Austin, H., Delzell, E., and Cole, P., 1988, Benzene and leukemia. A review of the literature and a risk assessment, Am. J. Epidem., 127:419.

Bailer, J., and Hoel, D.G., 1989, Metabolite-based internal doses used in a risk assessment of benzene, Environ. Health Perspect., 82:177.

Barale, R., Marrazzini, A., Betti, C., Vangelisti, V., Loprieno, N., and Barrai, I., 1990, Genotoxicity of two metabolites of benzene: phenol and hydroquinone show strong synergistic effects *in vivo*, Mutat. Res., 244:15.

Bauer, H., Dimitriadis, E.A., and Snyder, R., 1989, An *in vivo* study of benzene metabolite DNA adduct formation in liver of male New Zealand rabbits, Arch. Toxicol., 63:209.

Beliles, R.P., and Totman, L.C., 1989, Pharmacokinetically based risk assessment of workplace exposure to benzene, Regul. Toxicol. Pharmacol., 9:186.

Bois, F., Crouch, E.A.C., Wilson, R., and Vasseur, P., 1985, Integration of pharmacokinetics with multistage carcinogenesis modeling in cancer risk assessment. Presented at Conference of Current Perspectives on Animal Selection and Extrapolation in Toxicity Testing and Human Risk Assessment. Environmental Health Laboratory, Monsanto Company, St. Louis, MO, USA, October 1985.

Brett, S.M., Rodricks, J.V., and Chinchilli, V.M., 1989, Review and update of leukemia risk potentially associated with occupational exposure to benzene, Environ. Health Perspect., 82:267.

Brodfuehrer, J.I., Chapman, D.E., Wilke, T.J., and Powois, G., 1990, Comparative studies of the *in vitro* metabolism and covalent binding of 14C-benzene by liver slices and microsomal fraction of mouse, rat and human, Drug Metab. Disp., 18:20.

Busby, W.F.Jr., Wang, J.S., Stevens, E.K., Padykula, R.E., Aleksejczyk, R.A., and Berchtold, G.A., 1990, Lung tumorigenicity of benzene oxide, benzene dihydrodiols and benzene diolepoxides in the BLU:Ha newborn mouse assay, Carcinogenesis, 11:1473.

Choy, W.N., MacGregor, J.T., Shelby, M.D., and Marnopot, R.R., 1985, Induction of micronuclei by benzene in B6C3F1: retrospective analysis of peripheral blood smears from the NTP carcinogenesis bioassay, Mutat. Res., 143:55.

Ciranni, R., Barale, R., Ghelardini, G., and Loprieno, N., 1988a, Benzene and the genotoxicity of its metabolites. II. The effect of the route of administration on the micronuclei and bone marrow depression in mouse bone marrow cells, Mutat. Res., 209:23.

Ciranni, R., Barale, R., Marrazzini, A., and Loprieno, N., 1988b, Benzene and the genotoxicity of its metabolites. I. Transplacental activity in mouse fetuses and their dams, Mutat. Res., 208:61.

Crump, K.S., and Allen, B.C., 1984, Quantitative estimates of risk of leukemia from occupational exposure to benzene. Occupational Safety and Health Administration, docket H-059, exhibit 152, May 1984.

Dean, B.J., 1985, Recent findings on the genetic toxicology of benzene, toluene, xylenes and phenols, Mutat. Res., 154:153.

ECETOC, 1984, A review of recent literature on the toxicology of benzene, European Chemical Industry Ecology and Toxicology Center, Bruxelles, Technical Report No. 16.

Erexson, G.L., Wilmer, J.L., Steinhagen, W.H., and Kligerman, A.D., 1986, Induction of cytogenetic damage in rodents after short-term inhalation of benzene, Environ. Mutag., 8:29.

Fishbein, L., 1988, Genetic Effects of Benzene, Toluene and Xylene, IARC Sci. Publ. 85:19. International Agency for Research on Cancer, Lyon.

Fitzgerald, D.J., Piccoli, C., and Yamasaki, H., 1989, Detection of non-genotoxic carcinogens in the BALB/c 3T3 cell transformation/mutation assay system, Mutagenesis, 4:286.

Gad-el-Karim, M.M., Harper, B.L., and Legator, M.S., 1984, Modifications in the myeloclastogenic effect of benzene in mice with toluene, phenobarbital, 3-methylcholanthrene, Aroclor 1254 and SKF 525-A, Mutat. Res.135:225.

Gehring, P.J., Watanabe, P.G., and Park, C.N., 1978, Resolution of dose-response toxicity data for chemicals requiring metabolic activation: example-vinyl chloride, Toxicol. Appl. Pharmacol., 44:581.

Gehring P.J., Watanabe P.G., and Park C.N., 1979, Risk of angiosarcoma in workers exposed to vinyl chloride as predicted from studies in rats, Toxicol. Appl. Pharmacol., 49:15.

Glatt, H., Padykula, R., Berchtold, G.A., Ludewig, G., Platt, K.L., Klein, J., and Oesch, F., 1989, Multiple activation pathways of benzene leading to products with varying genotoxic characteristics, Environ. Health Perspect., 82:81.

Glatt, H., and Witz, G., 1990, Studies on the induction of gene mutations in bacterial and mammalian cells by the ring-opened benzene metabolites trans,trans-muconaldehyde and trans,trans- muconic acid, Mutagenesis, 5:263.

Gold, L.S., Sawyer, C.B., Magaw, R., Backman, G.M., de Viciana, M., Levinson, R., Hooper, N.K., Havender, W.R., Bernstein, L., Peto, R., Pike, M.C., and Ames, B.N., 1984, A carcinogenic potency database of the standardized results of animal bioassays, Environ. Health Perspect., 58:9.

Goldstein, B.D., Snyder, C.A., Laskin, S., Bromberg, I., Albert, R.E., and Nelson, N., 1982, Myelogenous leukemia in rodents inhaling benzene, Toxicol. Lett., 13:169.

Grilli, S., Lutz, W.K., and Parodi, S., 1987, Possible implications fron results of animal studies in human risk estimations for benzene: nonlinear dose-response relationship due to saturation of metabolism, J. Cancer Res. Clin. Oncol., 113:349.

Gut, I., and Frantik, E., 1980, Kinetics of benzene metabolism in rats in inhalation exposure, Arch. Toxicol. (suppl.), 4:315.

Guy, R.L., Dimitriadis, E.A., Hu, P., Cooper, K.R., and Snyder, R., 1990, Interactive inhibition of erythroid ^{59}Fe utilization by benzene metabolites in female mice, Chem.-Biol. Interactions, 74:55.

Harper, B.L., Sadagopa Ramanujam, V.M., Gad-el-Karim, M.M., and Legator, M.S., 1984, The influence of simple aromatics on benzene clastogenicity, Mutat. Res., 128:105.

Harper, B.L., and Legator, M.S., 1987, Pyridine prevents the clastogenicity of benzene but not of benzo[a]pyrene or cyclophosphamide, Mutat. Res., 179:23.

Harper, B.L., Ramanujam, V.M.S., Kurosky, L., and Legator, M.S., 1989a, Genetic effects of benzene and radiation in ICR and X/Gf mice, Cancer Res., 48:59.

Harper, B.L., Sadagopa Ramanujam, V.M., and Legator, M.S., 1989b, Micronucleus formation by benzene, cyclophosphamide, benzo(a)pyrene, and benzidine in male, female, pregnant female, and fetal mice, Teratog. Carcinog. Mutag., 9:239.

Henderson, R.F., Sabourin, P.J., Bechtold, W.E., Griffith, W.C., Medinsky, M.A., Birnbaum, L.S., and Lucier G.W., 1989, The effect of dose, dose rate, route of administration and species on tissue and blood levels of benzene metabolites, Environ. Health Perspect., 82:9.

Huff, J.E., 1986, Toxicology and carcinogenesis studies of benzene (CAS No. 71-43-2) in F344/N rats and B6C3F1 mice (Gavage studies), DHHS, National Toxicology Program/National Institute of Environmental Health Sciences, Research Triangle Park, NC, USA, Technical report No. 289.

Huff, J.E., Haseman, J.K., DeMarlni, D.M., Eustis, S., Maronpot, R.R., Peters, A.C., Persing, R.L., Chrisp, C.E., and Jacobs, A.C., lg89, Multiple-site carcinogenicity of benzene in Fischer 344 rats and B6C3F1 mice, Environ. Health Perspect., 82:125.

IARC Monographs on the Evaluation of the Carcinogenic Risk of Chemicals to Humans, 1982a, "Benzene", International Agency for Research on Cancer, Lyon, 29:93.

IARC Monographs on the Evaluation of the Carcinogenic Risk of Chemicals to Humans, 1982b, International Agency for Research on Cancer, Lyon, Suppl., 4:56.

IARC Monographs on the Evaluation of the Carcinogenic Risk of Chemicals to Humans, 1987, International Agency for Research on Cancer, Lyon, Suppl., 6:91,

Infante, P.F., 1987, Benzene toxicity: studying a subject to a death, Am. J. Ind. Med., 11:599.

Inoe, O., Seiji, K., Nakatsuka, H., Watanabe, T., Yin, S.N., Li, G.-L., Cai, S.-X., Jin, C., and Ikeda, M., 1989, Excretion of 1,2,4-benzenetriol in the urine of workers exposed to benzene, Br. J. Ind. Med., 46:559.

Jablonicka, A., Vargova, M., and Karelova, J., 1987, Cytogenetic analysis of peripheral blood lymphocytes in workers exposed to benzene, J. Hyg. Epidem. Microbiol. Immunol., 31:127.

Jongeneelen, F.J., Dirven, H.A.A.M., Leijdekkers, C.M., Henderson, P.T., Brouns, R.M.E., and Halm, K., 1987, S-Phenyl-N-acetylcysteine in urine of rats and workers after exposure to benzene, J. Anal. Toxicol., 11:100.

Jowa, L., Winkle, S., Kalf, G., Witz, G., and Snyder, R., 1986, Deoxyguanosine adducts formed from benzoquinone and hydroquinone, in: "Biological Reactive Intermediates III: Mechanisms of Action in Animal Models and Human Disease" Kocsis

J.J., Jollow D.J., Witmer C.M., Nelson J.O. and Snyder R. (eds), pp. 825-832, Plenum Press, New York.

Jowa, L., Witz, G., Snyder, R., Winkle, S., and Kalf, G.F., 1990, Synthesis and characterization of deoxyguanosine-benzoquinone adducts, J. Appl. Toxicol., 10:47.

Kalf, G.F., Snyder, R., and Rushmore, T.H., 1985, Inhibition of RNA synthesis by benzene metabolites and their covalent binding to DNA in rabbit bone marrow mitochondria in vitro, Am. J. Ind. Med., 7:485.

Kalf, G.F., 1987, Recent advances in the metabolism and toxicity of benzene, CRC Critical Rev. Toxicol., 18:141.

Kalf, G.F., Schlosser, M.J., Renz, J.F., and Pirrozzi, S.J., 1989, Prevention of benzene-induced myelotoxicity by nonsteroidal anti-inflammatory drugs, Environ. Health Perspect., 82:57.

Krewet, E., Verkoyen, C., Muller, C., and Norpoth, K., 1990, Studies on the formation of different phenylguanines during benzene metabolism, Proc. Am. Assoc. Cancer Res., 31:96.

Latriano, L., Witz, G., Goldstein, B.D., and Jeffrey, A.M., 1989, Chromatographic and spectrophotometric characterization of adducts formed during the reaction of trans,trans-muconaldehyde with ^{14}C-deoxyguanosine-5'-phosphate, Environ. Health Perspect., 82:249.

Lee, S.D., Dourson, M., Mukerjee, D., Stara, J.F., and Kawecki, J., 1983, Assessment of benzene health effects in ambient water, in: "Advances in Modern Environmental Toxicology" Mehlman M.A. (ed.), 4:91, Princeton Scientific Publishers, Pricenton.

Lee, E.W., and Garner, C.D., 1991, Effects of benzene on DNA strand breaks in vivo versus benzene metabolite-induced DNA strand breaks in vitro in mouse bone marrow cells, Toxicol. Appl. Pharmacol., 108:497.

Legator, M.S. (personal communication, Bologna, October 19, 1990) .

Longacre, S.L., Kocsis, J., and Snyder, R., 1980, Benzene metabolism and toxicity in CD-1, C57/B6 and DBA/2N mice, in: "Microsomes, Drug Oxidation and Chemical Carcinogenesis". M.J. Coon, A.H. Conney, R.W. Estabrook, H.V. Gelboin, J.R. Gillette and P.J. O'Brien (eds), pp. 897-902, Academic Press, New York.

Low, L.K., Meeks, J.R., Norris, K.J., Mehlman, M.A., and Mackerer, C.R., 1989, Pharmacokinetics and metabolism of benzene in Zymbal gland and other key target tissues after oral administration in rats, Environ. Health Perspect., 82:215.

Ludewig, G., Dogra, S., and Glatt, H., 1989, Genotoxicity of 1,4-benzoquinone and 1,4-naphthoquinone in relation to effects on glutathione and NAD(P)H levels in V79 cell, Environ. Health Perspect., 82:223.

Ludewig, G., Platt, K.L., Oesch, F., and Glatt, H.R., 1990, Quinones derived from benzene and polycyclic aromatic hydrocarbons: genotoxicity profile and cytotoxicity, Mutagenesis, 5:626.

Lutz, W.K. , and Schlatter, C., 1977, Mechanism of the carcinogenic action of benzene: irreversible binding to rat liver DNA, Chem.-Biol. Interact., 9:253.

MacGregor, J.T., Wehr, C.M., Henika, P.R., and Shelby, M.D., 1990, The in vivo erythrocyte micronucleus test: measurement at steady state increases assay efficiency and permits integration with toxicity studies, Fund. Appl. Toxicol., 14:513.

Mahtashamipur, E., Strater, H., Triebel, R. and Norpoth, K., 1987, Effect of pretreatment of male NMRI mice with enzyme inducers or inhibitors on clastogenicity of toluene, Arch. Toxicol., 60:460.

Maltoni, C., Conti, B., Cotti, G., and Belpoggi, F., 1983, Benzene: a multipotential carcinogen. Results of long term bioassays performed at the Bologna Institute of Oncology: current results and ongoing research, Am. J. Ind. Med., 4:589.

Maltoni, C., Cotti, G., Valgimigli, L., and Mandrioli, A., 1982, Zymbal gland carcinomas in rats following exposure to benzene inhalation, Am. J. Ind. Med., 3:11.

Marcus, W.L., 1987, Chemical of current interest-benzene, Toxicol. Ind. Health, 3:205.

Mazzullo, M., Bartoli, S., Bonora, B., Colacci, A., Grilli, S., Lattanzi, G., Niero, A., Turina, M.P., and Parodi, S., 1989, Benzene adducts with rat nucleic acids and proteins: dose-response relationship after treatment *in vivo*, Environ. Health Perspect., 82:259.

McMichael, A.J., 1988, Carcinogenicity of benzene, toluene and xylene: epidemiological and experimental evidence, IARC Sci. Publ., 85:3, International Agency for Research on Cancer, Lyon.

Medinsky, M.A., Sabourin, P.J., Lucier, G., Birnbaum, L.S., and Henderson, R.F., 1989a, A physiological model for simulation of benzene metabolism by rats and mice, Toxicol. Appl. Pharmacol., 99:193.

Medinsky, M.A., Sabourin, P.J., Henderson, R.F., Lucier, G., and Birnbaum, L.S., 1989b, Differences in the pathways for metabolism of benzene in rats and mice simulated by a physiological model. Environ. Health Perspect., 82:43

Morita, T., Iwamoto, Y., Shimizu, T., Masuzawa, T., and Yanagihara, Y., 1989, Mutagenicity tests with a permeable mutant of yeast on carcinogens showing false negative Salmonella assay, Chem. Pharm. Bull., 37:407

Nerland, D.E., and Pierce, W.M., 1990, Identification of N-acetyl-S-(2,5-dihydroxyphenyl)-L-cysteine as a urinary metabolite of benzene, phenol and hydroxyquinone, Drug Metab. Dispos., 18:958.

Oberly, T.J., Rexroat, M.A., Bewsey, B.J., Richardson, K.K., and Michaelis, K.C., 1990, An evaluation of the CHO/HGPRT mutation assay involving suspension cultures and soft agar cloning: results for 33 chemicals, Environ. Mol. Mutag., 16:260.

Parodi, S., Taningher, M., and Santi, L., 1983a, Alkaline elution *in vivo*: fluorometric analysis in rats. Quantitative predictivity of carcinogenicity, as compared with other short-term tests, in: "Indicators of Genotoxic Exposure", Bridges B.A., Butterworth B.E. and Weinstein I.B. (eds), Cold Spring Harbor Laboratory, Banbury Rep., 13:137.

Parodi, S., Balbi, C., Abelmoschi, M.L., Pala, M., Russo, P., and Santi, L., 1983b, Studies on DNA damage: discordant responses of rate of DNA disentanglement (viscosimetrically evaluated) and alkaline elution rate, obtained for several compounds, Cell Biophys., 5:285.

Parodi, S., Lutz, W.K., Colacci, A., Mazzullo, M., Taningher, M., and Grilli, S., 1989, Results of animal studies suggest a nonlinear dose- response relationship for benzene effects, Environ. Health Perspect., 82:171.

Parodi, S., Colacci, A., Grilli, S., and Taningher, M., 1992, chapter in this volume.

Paustenbach, D.J., Andersen, M.E., Clewell, H.J., III, and Gargas, M.L., 1988, A physiologically based pharmacokinetic model for inhaled carbon tetrachloride, Toxicol. Appl. Pharmacol., 96:191.

Picciano, D., 1980, Monitoring industrial populations by cytogenetic procedures, in: "Proceedings of a workshop on methodology for assessing reproductive hazards in the workplace", Infante P. and Legator M.S. (eds), U.S. Dept. of Health and Human Services, PHS, CDC, NIOSH, Div. Surveillance, Hazard Evaluation and Field Studies, pp. 293-306.

Pirozzi, S.J., Renz, J.F., and Kalf, G.F., 1989, The prevention of benzene-induced genotoxicity in mice by indomethacin, Mutat. Res., 222:291.

Reddy, M.V., Blackburn, G.R., Irwin, S.E, Kommineni, C.R., Mackerer, C.R., and Mehlman, M.A., 1989a, A method for *in vitro* culture of rat Zymbal gland: use in mechanistic studies of benzene carcinogenesis in combination with ^{32}P-postlabeling, Environ. Health Perspect., 82:239.

Reddy, M.V., Blackburn, G.R., Schreiner, C.A., Mehlman, M.A., and Mackerer, C.R., 1989b, ^{32}P analysis od DNA adducts in tissues of benzene-treated rats, Environ. Health Perspect., 82:253.

Reddy, M.V., Bleicher, W.T., Blackburn, G.R., and Mackerer, C.R., 1990a, Nucleic acid and protein adducts in tissues of benzene-treated rats, Proc. Am. Assoc. Cancer Res., 31:91.

Reddy M.V., Bleicher W.T., Blackburn G.R. and Mackerer C.R., 1990b, DNA adduction by phenol, hydroquinone or benzoquinone *in vitro* but not *in vivo*: nuclease P1-enhanced ^{32}P-postlabeling of adducts as labeled nucleoside bisphosphates, dinucleotides and nucleoside monophosphates, Carcinogenesis, 11:1349.

Reitz, R.H., Nolan, R.J., and Schumann, A.M., 1987, Organohalides, in: "Proceedings of the National Academy of Science Workshop on Pharmacokinetics", Safe Drinking Water Committee, Subcommittee on Pharmacokinetics, Board on Environmental Studies and Toxicology, National Research Council.

Reitz, R.H., Mendrala, A.L., Park, C.N., Andersen, M.E., and Guengerich, F.P., 1988, Incorporation of *in vitro* enzyme data into the physiologically- based pharmacokinetic (PB-PK) model for methylene chloride: implications for risk assessment, Toxicol. Lett., 43:97.

Rinsky, R.A., Smith, A.B., Hormung, R., Fillon, T.G., Young, R.J., Okun, A.H., and Landrigan, P.J., 1987, Benzene and leukemia: an epidemiological risk assessment, New Engl. J. Med., 316:1044.

Rithidech, K., Au, W.W., Sadagopa Ramanujam, V.M., Whorton, E.G., Jr, and Legator, M.S., 1987, Induction of chromosome aberrations in lymphocytes of mice after subchronic exposure to benzene, Mutat. Res., 188:135.

Rithidech, K., Au, W.W., Sadagopa Ramanujam, V.M., Whorton, E.B., and Legator, M.S., 1988, Persistence of micronuclei in peripheral blood normochromatic erythrocytes of subchronically benzene-treated male mice, Environ. Mol. Mutag., 12:319.

Robertson, M.L., Eastmond, D.A., and Smith, M.T., 1990, A mixture of benzene metabolites produces a synergistic genotoxic response in cultured human lymphocytes, Environ. Mol. Mutag., Suppl 15:58.

Rossman, T.G., Klein, C.B., and Snyder, C.A., 1989, Mutagenic metabolites of benzene detected in the microscreen assay, Environ. Health Perspect., 81:77.

Rushmore, T.R., Snyder, R., and Kalf, G.F., 1984, Covalent binding of benzene and its metabolites to DNA in rabbit bone marrow mitochondria *in vitro*, Chem.-Biol. Interact., 49:133.

Sabourin, P.J., Chen, B.T., Lucier, G., Birnbaum, L.S., Fisher, E., and Henderson, R.F., 1987, Effect of dose on the absorption and excretion of ^{14}C-benzene administered orally or by inhalation in rats and mice, Toxicol. Appl. Pharmacol., 87:325.

Sabourin, P.J., Bechtold, W.E., Birnbaum, L.S., Lucier, G., and Henderson, R.F., 1988, Difference in the metabolism of inhaled ^{3}H-benzene by F344/N rats and B6C3F mice, Toxicol. Appl. Pharmacol., 94:128.

Sabourin, P.J., Bechtold, W.E., Griffith, W.C., Birnbaum, L.S., Lucier, G., and

Henderson, R.F, 1989, Effect of exposure concentration, exposure rate and route of administration on metabolism of benzene by F344 rats and B6C3F1 mice, Toxicol. Appl. Pharmacol., 99:421.

Sabourin, P.J., Sun, J.D., MacGregor, J.T., Wehr, C.M., Birnbaum, L.S., Lucier, G., and Henderson, R.F., 1990, Effect of repeated benzene inhalation exposures on benzene metabolism, binding to hemoglobin, and induction of micronuclei, Toxicol. Appl. Pharmacol., 103:452.

Sasiadek, M., Jagielski, J., and Smolik, R., 1989, Localization of breakpoints in the karyotype of workers professionally exposed to benzene, Mutat. Res., 224:235.

Schnier, G.G., Laethem, C.L., and Koop, D.R., 1989, Identification and induction of cytochromes P450, P450 IIE1 and P450 IA1 in rabbit bone marrow, J. Pharm. Exp. Ther., 251:790.

Seiji, K., Jin, C., Watanabe, T., Nakatsuka, H., and Ikeda, M., 1990, Sister chromatid exchanges in peripheral lymphocytes of workers exposed to benzene, trichloroethylene, or tetrachloroethylene, with reference to smoking habits, Intern. Arch. Occup. Environ. Health, 62:171.

Smith, M.T., Yager, J.W., Steinmetz, K.L., and Eastmond, D.A., 1989, Peroxidase-dependent metabolism of benzene's phenolic metabolites and its potential role in benzene toxicity and carcinogenicity, Environ. Health Perspect., 82:23.

Snyder, R., Lee, E.W., and Kocsis, J.J., 1978, Binding of labeled benzene metabolites to mouse liver and bone marrow, Res. Comm. Chem. Path. Pharmacol., 20:191.

Snyder, C., Goldstein, B., Sellakumar, A., Bromberg, I., Laskin, S., and Albert, R., 1980, The inhalation toxicology of benzene: incidence of hematopoietic neoplasms and hematotoxicity in AKR/J and C57BL/6J mice, Toxicol. Appl. Pharmacol., 54:323.

Snyder, R., Dimitriadis, E., Guy, R., Hu, P., Cooper, K., Bauer, H., Witz, G., and Goldstein, B.D., 1989, Studies on the mechanism of benzene toxicity, Environ. Health Perspect., 82:31.

Steimetz, K.L., Garin, K.E., Hamilton, C.M., Bakke, J.P., MacGregor J.T., and Mirsalis, J.C., 1990, Multiple endpoint *in vivo* genetic toxicology assay: evaluation of hepatic unscheduled DNA synthesis, S-phase synthesis and peripheral blood micronuclei following repeated dosing of male B6C3F1 mice, Environ. Mol. Mutag. Suppl. 15:58.

Stommel, P., Muller, G., Stucker, W., Verkoyen, C., Schobel, S., and Norpoth, K., 1989, Determination of S-phenylmercapturic acid in the urine - an improvement in the biological monitoring of benzene exposure, Carcinogenesis, 10:279.

Styles, J.A., and Richardson, C.R., 1984, Cytogenetic effect of benzene: dosimetric studies on rats exposed to benzene vapour, Mutat. Res., 135:203.

Subrahmanyam, V.V., Kolachana, P., and Smith, M.T., 1991, Metabolism of hydroquinone by human peroxidase: mechanisms of stimulation by other phenolic compounds, Arch. Biochem. Biophys., 286:76.

Sun, J.D., Medinsky, M.A., Birnbaum, L.S., Lucier, G., and Henderson, R.F., 1990, Benzene hemoglobin adducts in mice and rats: characterization of formation and physiological modeling, Fund. Appl. Toxicol., 15:468.

Sutou, S., 1990, Single versus multiple dosing in the micronucleus test: the summary of the fourth collaborative study by CSGMT/JEMS.MMS, Mutat. Res., 234: 205.

Suzuki, S., Atai, H., Hatakeyama, Y., Hara, M., and Nakagawa S., 1989, Administration-route-related differences in the micronucleus test with benzene, Mutat. Res., 223: 407.

Tice, R.R., 1988, The cytogenetic evaluation of *in vivo* genotoxic and cytotoxic activity using rodent somatic cells, Cell Biol. Toxicol., 4:475.

Tice, R.R., Luke, A., and Drew, R.T., 1989, Effect of exposure route, regimen, and duration on benzene-induced genotoxic and cytotoxic bone marrow damage in mice, Environ. Health Perspect., 82:65.

Travis, C.C., Quillen, J.L., and Arms, A.D., 1990, Pharmacokinetics of benzene, Toxicol. Appl. Pharmacol., 102:400.

Twerdok, L.E., and Trush, M.A., 1990, Differences in quinone reductase activity in primary bone marrow stromal cells derived from C57BL/6 and DBA/2 mice, Res. Comm. Chem. Path. Pharmacol., 67:375.

Ulanova, I.P., and Avilova, G.G., 1987, Application of toxicokinetics criteria as basis for toxicometric parameters, J. Hyg. Epidemiol. Microbiol. Immunol., 31:113.

Ward, R.C., Travis, C.C., Hetrick, D.M., Andersen, M.E., and Gargas, M.L., 1988, Pharmacokinetics of tetrachloroethylene, Toxicol. Appl. Pharmacol., 93:108.

Wells, M.S., and Nerland, D.E., 1991, Hematotoxicity and concentration-dependent conjugation of phenol in mice following inhalation exposure to benzene, Toxicol. Lett., 56:159.

Whittaker, S.G., Zimmermann, F.K., Dicus, B., Piegorsch, W.W., Resnick, M.A., and Fogal, S., 1990, Detection of induced mitotic chromosome loss in Saccharomyces cerevisiae - An interlaboratory assessment of 12 chemicals, Mutat. Res., 241:225.

WHO Regional Publications, 1987, European series No. 23 "Air quality guidelines for Europe", Copenhagen, pp. 45-58.

Witz, G., Rao, G.S., and Goldstein, B.D., 1985, Short-term toxicity of trans,trans-muconaldehyde, Toxicol. Appl. Pharmacol., 80:511.

Witz, G., Latriano, L., and Goldstein, B.D., 1989, Metabolism and toxicity of trans, trans-muconaldehyde, an open-ring microsomal metabolite of benzene, Environ. Health Perspect., 82:19.

Witz, G., Gad, S.C., Tice, R.R., Oshiro, Y., Piper, C.E., and Goldstein, B.D., 1990a, Genetic toxicity of the benzene trans,trans-muconaldehyde in mammalian and bacterial cells, Mutat. Res., 240:295.

Witz, G., Maniara, W., Mylavarapu, V., and Goldstein, B.D., 1990b, Comparative metabolism of benzene and trans,trans-muconaldehyde to trans,trans-muconic acid in DBA/2N and C57BL/6 mice, Biochem. Pharmacol., 40:1275.

Yager, J.W., Eastmond, D.A., Robertson, M.L., Paradisin, W.M., and Smith, M.T., 1990, Characterization of micronuclei induced in human lymphocytes by benzene metabolites, Cancer Res., 50:393.

Yardley-Jones, A., Anderson, D., Jenkinson, P.C., Lovell, D.P., Blowers, S.D., and Davies, M.J., 1988, Genotoxic effects in peripheral blood and urine of workers exposed to low level benzene, Br. J. Ind. Med., 45:694.

Yardley-Jones, A., Anderson, D., Lovell, D.P., and Jenkinson, P.C., 1990, Analysis of chrom. aber. in workers exposed to low level benzene, Br. J. Ind. Med., 47:48.

DOSE-RESPONSE RELATIONSHIPS FOR BENZENE: HUMAN AND EXPERIMENTAL CARCINOGENICITY DATA

Silvio Parodi[1], Davide Malacarne[2], Sandro Grilli[3],
Annamaria Colacci[2] and Maurizio Taningher[1]

[1]Istituto di Oncologia Clinica e Sperimentale and CIRC
University of Genova - Italy
[2]Istituto Nazionale per la Ricerca sul Cancro, Genova - Italy
[3]Istituto di Cancerologia and CIRC, University of Bologna - Italy

EPIDEMIOLOGICAL DATA

In two previous papers we have discussed problems related to dose-response relationships for benzene in animals and humans.[1,2]

We concluded that the papers of Ott et al.,[3] Rinsky et al.[4] Yin et al.[5] and Vigliani[6] could give some useful information about average levels of exposure in the exposed workers, average length of exposure and relative risk. A summary of these data is reported in Table 1. The data reported for daily levels of exposure around 20-30 ppm are apparently in agreement with a summary of risk assessments based on a linear model, referred to the data of Infante-Rinsky,[3,7,8] as reported by Brett et al.[9]

There is, however, a deficiency in the way we have presented the data. Evaluation of exposure is based on the exposure of cases and not on the average exposure of a well defined cohort. An enrichment of the proportion of high exposures is unavoidable in the cases and they do not represent a correct sampling of the original cohort at risk.

Basing our evaluation of exposure on cases, we were however able to compare the above levels of exposure with the much higher ones reported by Vigliani.[6] When we compared daily levels of exposure around 20-30 ppm with ones around 200-500 ppm, we noticed an apparent flattening of the response (Table 1).

There could be some relationship between this response flattening at high levels of exposure and similar results in mice discussed in an other part of this manuscript. In addition, results obtained in short-term tests of genotoxicity and metabolic studies also seem to suggest saturation of enzymes metabolically activating benzene at high concentrations or dosages[10].

If a careful examination of the most relevant epidemiological studies on benzene is performed,[3,4,6,7,11-14] it appears evident that the most relevant study for analyzing a dose-response relationship for levels of exposure around and below 30 ppm/day is the work of Rinsky et al.[4] We share this opinion with Brett et al.[9] We will devote the next part of our epidemiological analysis to the work of Rinsky et al.[4]

We have utilized the cohort study of 1165 exposed workers, in the interval 1940-1965, reported in Table 2 of Rinsky et al.[4] We have summarized the data of that cohort study in a new table (Table 2), that was the basis of our analysis.

Table 1

Authors	Exposure Averg. Level ppm/day	Averg. Length in Years	Relative Risk	Addit'l *
Ott et al.	16	8.5	3.75	6.4
Rinsky et al.	24	8.7	3.37	3.6
Yin et al.	28	9.0	5.74	5.9
Vigliani	350	15	21	1.1

*Lifetime leukemia cases per 1000 workers exposed to 45 ppm yrs BZ; from linear extrapolation assuming a background lifetime probability of 0.007

As reported, the cohort corresponding to the highest level of exposure (27 times the lowest level of exposure) is about 40 times smaller than the cohort exposed to the lowest level, but it is associated with even more leukemias than the largest cohort. Anyway the overall number of leukemias involved is very small (2,2,2 and 3 for each cohort respectively).

From a statistical point of view the appearance of leukemias in exposed workers is a rare event and is subject to a Poisson distribution, as a first approximation. Assuming that the number of observed leukemias in a given cohort is the closest value to the mean of a Poisson distribution, for the ideal population of which the real cohort is a small sample, then, if, for instance, $\mu=2$, then $P(0)=e^{-2}=0.135$. This example suggests that, given the small number of cases and the small size of the cohorts, we could have easily observed frequencies very different from those that were really observed.

In order to answer the question if the observed frequencies of leukemias depart significantly from a linear dose-response relationship, we have adopted the following strategy. Making reference to the Standardized Mortality Ratio (SMR) values and the levels of exposure reported in our Table 2 we have compared four different equations, describing three different types of dose-response relationship, with the observed data:

$$y = e^{ax} \qquad (1)$$

where x is the average level of exposure in ppm-years for each different cohort, as

$$y = 1 + ax \qquad (2)$$

$$y = (1 + ax)^2 \qquad (3)$$

$$y = (1 + ax) \cdot (1 + bx) \qquad (4)$$

reported in our Table 2, and y is the expected SMR according to each one of the four equations.

Equation (1) relates to the type of function preferred by Rinsky et al.[4] (conditional logistic regression). One possible biological interpretation of this relationship is that benzene, in the interval of dosages considered, is both a promoting agent (or a stimulator of cell proliferation) and a genotoxic agent causing, for instance, chromosomal damage. In this case benzene could both increase the number of preneoplastic cells (clonal expansion) or the number of cell generations in clones at risk (cell proliferation) and, at the same time, increase the probability of rearrangement of some genes relevant in the process of leukemogenesis. There could be some sort of multiplicative effect of the two types of epigenetic and genetic factors combined. Empirically, within a given interval of dosages, an exponential function could fit closely the observed data for a given range of exposures.

Equation (2) relates to the standard conservative hypothesis for a linear extrapolation to low doses.

Equation (3) relates to the biologically possible case of some active metabolite of benzene affecting both alleles of the same recessive gene.

Finally, Equation (4) relates to the biologically possible case of some active metabolite of benzene affecting two different dominant genes. Obviously, other more complex multistage models are also possible. Here, we are interested in the general question, if multistage models tend to fit the results better than a single stage linear model.

We have considered the observed numbers of leukemias for each exposure cohort as the most likely μ of a Poisson distribution for a large population of which our cohorts are small samples.[15] According to this reasoning the highest possible probability that, for a given μ, we will obtain the observed number of cases is, in general:

$$P(\mu) = \frac{\mu^\mu}{\mu!} e^{-\mu} \qquad (5)$$

where μ is always a positive integer. In our case, the highest possible product of probabilities for the four levels of exposure is:

Table 2

Cumulative Exposure (ppm years; Pliofilm cohort)

Range	0-40	40-200	200-400	>400	Total
Range Cntr	20	120	300	540*	69
Leukemia Cases	2	2	2	3	9
Exp. Leukemia Deaths	1.83	.62	.17	.04	2.66
SMR	1.09	3.22	11.86	66.37	3.37
Extrapol. Cohort Sizes	799	272	74	20	1165

* Based on the average exposure of the 3 cases in this class; rounded data from Tables 2 & 4 of ref. 4
** Under the assumption that the age-distribution of the exposed cohorts are identical.

$$\frac{2^2}{2} \cdot e^{-2} \cdot \frac{2^2}{2} \cdot e^{-2} \cdot \frac{3^3}{3!} \cdot e^{-3} = 4.44 \cdot 10^{-3} \qquad (6)$$

Each "a" and "b" coefficient, in equations (1), (2), (3) and (4), determines an expected SMR (y), for the four levels of exposure. From each expected SMR we can obtain an expected value of μ according to the following Equation:

$$\frac{SMR_{exp}}{SMR_{obs}} = \frac{\mu_{exp}}{n_i} \qquad (7)$$

where n_i is the number of observed leukemias for exposure level i. From each expected μ value (for each level of exposure) we obtain a P(2), P(2), P(2) and P(3), respectively. With an iterative computer program[16] we can find the "a" and "b"coefficients for equations (1), (2), (3) and (4) respectively, maximizing the product P(2)·P(2)·P(2)·P(3). In this way we will have found the best "a" and "b" coefficients for the four equations according to the principle of maximum likelihood.[17] As a result of the procedure described above, equations (1), (2), (3), and (4) now become:

$$y = e^{7.886 \cdot 10^{-3} x} \qquad (8)$$

$$y = 1 + 3.88 \cdot 10^{-2} x \qquad (9)$$

$$y = (1 + 9.73 \cdot 10^{-3} x)^2 \qquad (10)$$

$$y = (1 + 9.73 \cdot 10^{-3} x) \cdot (1 + 9.73 \cdot 10^{-3} x) \qquad (11)$$

The results of our using equations (8), (9), (10) and (11) are reported in Tables 3 and 4 and illustrated graphically in Fig. 1. Because for equations (10) and (11) we have found identical coefficients, equation (11) can be adsorbed in equation (10).

The data of Table 3 were submitted to statistical analysis (χ^2 test), as shown in Table 5.

Table 3

Exposure ppm-yrs x	SMR$_{obs}$	SMR$_{exp}$ Eq. 8	Eq. 9	Eq. 10
20	1.09	1.171	1.776	1.427
120	3.22	2.576	5.656	4.698
300	11.86	10.65	12.64	15.36
540	66.37	70.70	21.95	39.11
ppm	n$_{obs}$		n$_{exp}$	
20	2	2.149	3.259	2.618
120	2	1.6	3.513	2.918
300	2	1.796	2.132	2.59
540	3	3.196	0.9922	1.768

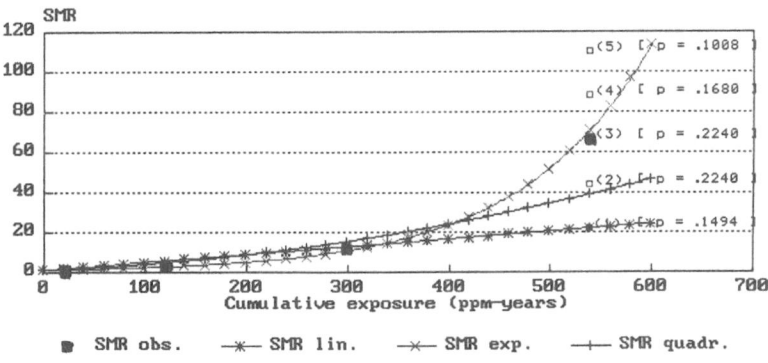

Fig. 1 Dose response curves for Eq. (8), Eq. (9), and Eq. (10). White squares refer to the hypothetical SMRs for 5, 4, 2, 1 leukemias observed in the cohort with the highest level of exposure (540 ppm)

Table 4

Cum. Exposure ppm-yrs	p^*_{max}	From Eq. 8	p From Eq. 9	From Eq. 10
20	.271	.269	.204	.250
120	.271	.258	.184	.230
300	.271	.268	.270	.252
540	.224	.223	.060	.157
$P_{20} P_{120} P_{300} P_{540}$		$4.458 \circ 10^{-3}$	$4.157 \circ 10^{-3}$	$6.108 \circ 10^{-4}$	$2.275 \circ 10^{-3}$
$(P_{20} P_{120} P_{300} P_{540})/p_{max}$	1	.9325	.137	.5103

*Obtained from Eq. 6

Table 5.

Model	x^2	p
Exponential	1.455×10^{-1}	
Linear	5.21	$.25 > P > .1$
Quadratic	1.428	

The x^2 or the linear regression equation is the closest to a statistical significance (.25 > p > .10). Larger cohorts for each level of exposure would be needed to decide whether to reject the linear interpolation.

The models referred to Eq. (2) and Eq. (4) are hierarchical (model (4) becomes model (2) if b=0), and a statistical analysis based on a likelihood ratio L_i test is therefore possible:[17]

$$x^2 = -2 \cdot \ln \frac{L_2}{L_4}$$

$x^2_1 = 2.63;$ $0.25 > p > 0.1$

We conclude that the exponential model and the two-stage model fit the data better than the linear model, but the number of leukemias available for our calculations is too small to permit a decision whether to reject the hypothesis of a linear relationship with benzene doses. If, at the highest level of exposure (540 ppm), one instead of three leukemias had been observed, a linear dose response relationship would fit the data very well.

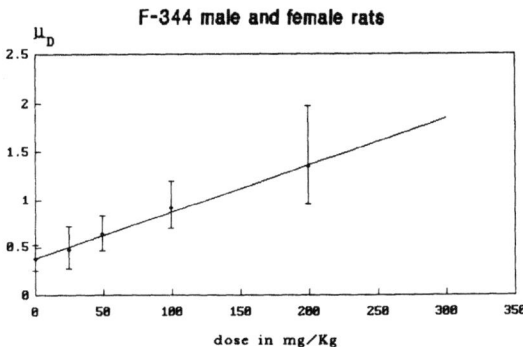

Fig. 2 Average number of malignant tumors per animal plotted against dose (D) in mg/Kg. Observations taken at 2 years after 2 years of treatment. Data fitted to $\mu_D = \mu_0 + axD$. Vertical bars: Standard Deviation for each experimental point.

LONG TERM CARCINOGENICITY EXPERIMENTS IN RODENTS

We have examined the results available in the literature, concerning long-term carcinogenicity experiments in rodents. To our knowledge only three experiments can be utilized for an analysis of the dose-response relationship.

A careful research in the existing literature, using both on line data banks[18,19] and the database of Gold et al.,[20-22] produced only two papers on long-term experiments in rodents that could be utilized in the investigation of dose-response relationships for benzene. The papers in question were from NTP[23] and from Maltoni et al.[24]

In the NTP study,[23] groups of male and female B6C3F$_1$ mice and female F344 rats were treated with daily doses of 25, 50 and 100 mg/kg, while groups of male F344 rats were treated in the same way, but with doses of 50; 100 and 200 mg/kg. Each control and dosed group was made up of 50 animals at the beginning of the experiment and, as deducible from Appendix Tables A and B of the paper, all the animals were necropsied and examined practically for all the different types of tumors at the end of the experiment. The treatment was performed by gavage 5 days/wk, for two years, that is for the duration of the experiment. For homogeneity with our second source of data, adjustments for intercurrent mortality, as reported in Appendix Table 5, were not considered.

In the second source of data,[24] results from experiments BT 901 and BT 902 were utilized. These were the only experiments useful for our purposes. Groups of 30 to 50 male or female Sprague-Dawley rats were used as controls or animals treated with benzene by gavage with daily doses of 50 and 250 mg/kg (Expt. BT 901) or 500 mg/kg (Expt. BT 902). The treatment was performed for 4-5 days/wk and lasted for 52 wks in the first experiment and for 104 wks in the second. The duration of both experiments was the animal life-span (more than two years).

In terms of malignant tumor bearing animals, using the experiments mentioned

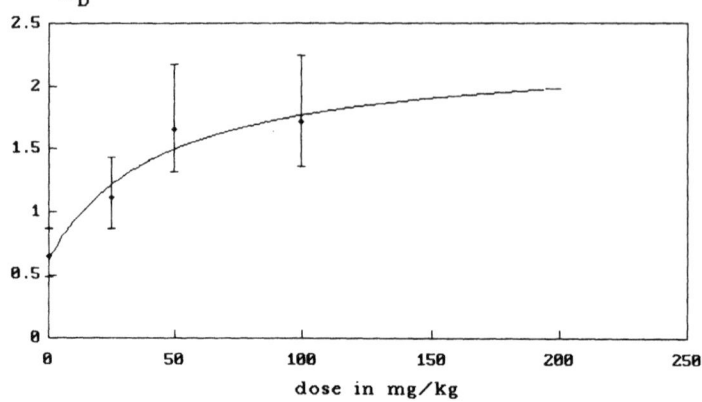

Fig. 3 Average number of malignant tumors per animal plotted against dose (D) in mg/Kg. Observations taken at 2 years after 2 years of treatment. Data fitted (least squares) to $\mu_D = \mu_0 + \{axD/(b+D)\}$. Vertical bars: Standard Deviation for each experimental point.

above,[23,24] we were able to collect six series of tumor incidence values. The selection of malignant tumors allowed for a more homogeneous treatment of the data reported in the two papers. As described above, the series of incidence values concerned both sexes of $B6C3F_1$ mice, F-344 rats and Sprague-Dawley rats. Essentially, all incidence values concerned more than 15 tumor bearing animals per group, and this size, as reported in Ref. 15, allows for treatment of the data according to a normal distribution.

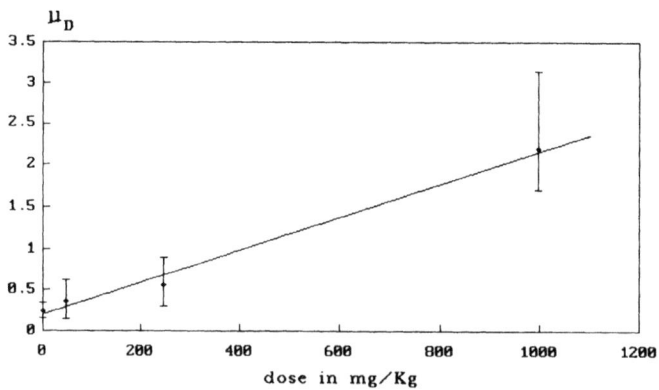

Fig. 4 Average number of malignant tumors per animal plotted against dose (D) in mg/Kg. A daily dosage of 500 mg/Kg for two years was considered equivalent to daily dose of 1000 mg/Kg for one year. Data fitted to $\mu_D = \mu_0 + axD$. Vertical bars: standard deviation for each experimental point.

Because no significant differences in incidence between males and females were observed, the data are reported together, in order to obtain larger sets. In addition, for the experiment on Sprague-Dawley male rats, the results related to the controls and the two lowest dosages of experiment BT 901 were not considered because tumor incidence in controls and treated animals was very low and different from all others experimental sets.

The results concerning F344 rats, B6C3F$_1$ mice and Sprague-Dawley rats are reported in Figs. 2, 3 and 4 respectively. As reported instead of tumor incidence (I) we have utilized the average number of tumors per animal (μ) for graphing the dependent variable. This kind of variable is much more correct when tumor incidences are relatively close to one. It is expressed by the equation:

$$\mu = -\ln(1-I) \qquad (12)$$

From the results of Figs. 2, 3 and 4 we can observe a linear dose-response relationship for rats and a saturation of the dose response relationship at high dosages for mice. It could be noteworthy that the possibility of a saturation of benzene metabolism at high dosages is strongly suggested by many short-term experiments and metabolic studies.[10]

CONCLUSIONS

In this paper we have examined the possible information about dose-response relationships obtainable from epidemiological data and long term experiments in rodents.

At low levels of exposure only epidemiological data can offer potentially useful information. As noted by Rinsky et al.[4] there is a tendency in the direction of a concave dose-response relationship. However, the size of the cohorts, in this most informative study,[4] is still too small to allow us to test the hypothesis of a linear dose-response relationship. Studies on DNA adducts seem to suggest a linear dose-response relationship at low and very low dosages. However, from these studies we know nothing about DNA repair, linearity of an initiation effect or possible promotion effects. Therefore, we can not exclude from these studies the possibility of non-linear dose-response relationships.

For high levels of exposure both epidemiological data[6] and long-term experiments in rodents are available.[23,24] Human data and long-term experiments in mice seem to suggest a flattening up of the dose-response relationship. However, we have to admit that the evidence of saturation of benzene metabolism is much more clear in the short-term studies and in the metabolism studies in rodents.[10]

Concerning the results obtained in rats we can not exclude the possibility that a promotional effect tends to cause a concave dose-response curve, while saturation of metabolism tends to cause a convex dose-response curve. The two opposite effects cancel each other out, generating a quasi-linear dose-response relationship.

ACKNOWLEDGEMENTS

We are grateful to Dr. Suresh Moolgavkar for his useful suggestions concerning the statistical analysis of the data. Supported by: 1) Grant from MURST: Interlab Project (IST-CIRC-Ansaldo); 2) Grants from MURST (60% and 40% to S.P.)

REFERENCES

1. S. Grilli, W.K. Lutz, and S. Parodi, Possible implications from results of animal studies in human risk estimation for benzene: non linear dose-response relationship due to saturation of metabolism, J. Cancer Res. Clin. Oncol. 113: 349 (1987).
2. S. Parodi, W.K. Lutz, A. Colacci, M. Mazzullo, M. Taningher, and S. Grilli, Results of animal studies suggest a nonlinear dose-response relationship for benzene effects, Env. Health Perspect. 82: 171 (1989).
3. M.G. Ott, J.C. Townsend, W.A. Fishbeck, and R.A. Langner, Mortality among individuals occupationally exposed to benzene, Arch. Env. Health 33: 3 (1978).
4. R.A. Rinsky, A.B. Smith, R. Hornung, T.G. Filloon, R.J. Young, A.H. Okun, and P.J. Landrigan, Benzene and leukemia. An epidemiologic risk assessment, New Engl. J. Med. 316: 1044 (1987).
5. S.N. Yin, G.L. Li, F.D. Tain, Z.I. Fu, Y.J. Chen, S.J. Luo, P.Z. Ye, J.Z. Zhang, G.C. Wang, X.C. Zhang, H.N. Wu, and Q.C. Zhong, Leukaemia in benzene workers: a retrospective cohort study, Br. Ind. Med. 44: 124 (1987).
6. E.C. Vigliani, Leukemia associated with benzene exposure, Ann. N.Y. Acad. Sci. 271: 143 (1976).
7. P.F. Infante, R.A. Rinsky, J.K. Wagoner, and R.J. Young, Leukemia in benzene workers, Lancet ii: 76 (1977).
8. R.A. Rinsky, R.J. Young, and A.B. Smith, Leukemia in benzene workers, Am. J. Ind. Med. 2: 217 (1981).
9. S.M. Brett, J.V. Rodricks, and V.M Chinchilli, Review and update of leukemia risk potentially associated with occupational exposure to benzene. Env. Health Perspect. 82: 267 (1989).
10. S. Grilli, S. Parodi, M. Taningher, and A. Colacci, Comparative metabolism and genotoxicity data on benzene: their role in cancer risk assessment, This volume (1991).
11. G.G. Bond, E.A. McLaren, C. Baldwin, and R.R. Cook, Update of mortality among chemical workers exposed to benzene, Br. Ind. Med. 43: 685 (1986).
12. M. Aksoy, Different types of malignancies due to occupational exposure to benzene: a review of recent observations in Turkey, Env. Res. 23: 181 (1980).
13. O. Wong, 1983, An Industry-Wide Study of Chemical Workers Occupationally Exposed to Benzene, Report to the Chemical Manufacturers Association, Environmental Health Associates, Berkeley, CA.
14. S.N. Yin, G.L. Li, F.D. Tain, Z.I. Fu, C. Jin, Y.J. Chen, S.J. Luo, P.Z. Ye, J.Z. Zhang, G.C. Wang, X.C. Zhang, H.N. Wu, and Q.C. Zhong, A retrospective cohort study of leukemia and other cancers in benzene workers, Env. Health Perspect. 82: 207 (1989).
15. G.W. Snedecor, and Cochran. Statistical Methods. The Iowa State Univ. Press., pp. 199-227, 1967.
16. BMDP Statistical Software. Cork Technology Park, Model Farm Rd. Cork, Ireland (1990).
17. D.G. Kleinbaum, L.L. Kupper, and K.E. Muller, "Applied Regression Analysis and Other Multivariate Methods," PWS-KENT Publ. Comp. 483 (1988).
18. Medline, Toxline and Cancerline (National Library of Medicine, Bethesda, MD, USA).
19. DIMDI (Deutsches Institut fur Medizinische Dokumentation und Information, Cologne, Germany).

20. L.S. Gold, C.B. Sawyer, R. Magaw, G.M. Backman, M. de Veciana, R. Levinson, N.K. Hooper, W.R. Havender, L. Bernstein, R. Peto, M.C. Pike, and B.N. Ames, A carcinogenic potency database of the standardized results of animal bioassays. Env. Health Perspect. 58: 9 (1984).
21. L.S. Gold, M. de Veciana, G.M. Backman, R. Magaw, P. Lopipero, M. Smith, M. Blumenthal, R. Levinson, L. Bernstein, and B.N. Ames, Chronological supplement to the carcinogenic potency database: standardized results of animal bioassays published through December 1982, Env. Health Perspect. 67: 161 (1986).
22. L.S. Gold, T.H. Slone, G.M. Backman, R. Magaw, M. Da Costa, P. Lopipero, M. Blumenthal, and B.N. Ames, Second chronological supplement to the carcinogenic potency database: standardized results of animal bioassays published through December 1984 and by the National Toxicology Program trough May 1986, Env. Health Perspect. 74: 237 (1987).
23. NTP Technical Report No. 289, Toxicology and Carcinogenesis Studies of Benzene (CAS No. 71-43-2) in F344/N Rats and B6C3F1 mice (Gavage Studies). U.S. Dept. of Health and Human Services, 1986.
24. C. Maltoni, A. Ciliberti, G. Cotti, B. Conti, and F. Belpoggi. Benzene, an experimental multipotential carcinogen: results of the long-term bioassays performed at the Bologna Institute of Oncology, Env. Health Perspect. 82: 109 (1989).
25. M. Mazzullo, S. Bartoli, B. Bonora, A. Colacci, S. Grilli, G. Lattanzi, A. Niero, M.P. Turina, and S. Parodi, Benzene adducts with rat nucleic acids and proteins: dose-response relationship after treatment in vivo, Env. Health Perspect. 82: 259 (1989).

A THRESHOLD FOR BENZENE LEUKEMOGENESIS

Bruce Molholt

Environmental Resources Management, Inc.
855 Springdale Drive
Exton, PA 19341

INTRODUCTION

Benzene is a very common environmental contaminant. Due to its proven epidemiologic association with acute myelogenous leukemia (AML) in exposed workers, this aromatic chemical has been classified as one of the few *known human carcinogens* (EPA Class A, IARC Class I). If human risks from background exposures to benzene are assessed according to EPA's linearized multistage model, a lifetime leukemia risk of 1-in-10,000 (10^{-4}) derives (1). Yet, given its mode of specific myelotoxicity, it is unlikely that there is any leukemogenic activity of benzene at background levels of exposure. In this paper I outline the evidence that there is a level of benzene exposure below which myelotoxicity and leukemogenesis fail to occur and suggest an experimental approach for definition of this leukemogenic threshold.

BENZENE GENOTOXICITY AND MYELOTOXICITY

Unlike most initiating carcinogens, benzene is not genotoxic in the majority of short-term test systems evaluated (2). Short-term tests which are negative for benzene include mutagenesis in microorganisms (*e.g.*, Ames test), invertebrates (*e.g.*, *Drosophila*) and mammalian cells in culture. On the other hand, after *in vivo* exposure to benzene, chromosomal aberrations, sister chromatid exchanges and micronuclei are all seen in bone marrow-derived cells (3). Hence, it appears that chromosomal effects of benzene exposure are limited to cells derived from bone marrow. Indeed, it has been known for nearly a century that benzene exposure adversely affects bone marrow as the target organ, resulting in a constellation of hematopoietic effects, including aplastic anemia, pancytopenia and leukopenia.

Are the leukemogenic effects of benzene exposure linked to these pervasive effects on bone marrow in general? It seems likely. A hint of what may be happening is found in B-cell leukemias in which the normal joining of variable and constant genes has gone awry and incorporates an oncogene (4,5). As will be seen below, chromosomal translocations are the hallmarks of leukemia. There is the thorny problem of cancers at additional organ sites which arise when animals are exposed to high concentrations of benzene (6), but this apparently anomalous result may be explained in terms of immunosuppression of natural killer lymphocytes which are active in cancer surveillance.

Leukocytes are produced in bone marrow from a pluripotent stem cell which differentiates in dedicated pathways according to specific cytokines elicited by stromal cells of the marrow matrix. These cytokines include interleukins (ILs), colony stimulating factors (CSFs) and hematopoietin. Interleukins and other cytokines attach to specific leukocyte receptors and transmit their differentiation signals to the nucleus via transmembrane tyrosine kinases. This is the same differentiation signal which goes awry during leukemogenic transformation. The transformation to chronic myelogenous leukemia (CML) must occur at the pluripotent stem cell precursor level, since CML cells may differentiate to either granulocytic progeny or lymphoblasts during blast crisis (7). A spectrum of cell lineages is also seen in acute myelogenous leukemia (AML), which is the leukemia most frequently associated with occupational exposure to benzene (8).

Table 1. Oncogene Translocations Of Human Leukemias (15)

Leukemia	Translocation	Oncogene(s)
CML, B-ALL	9q34 x 22q11	abl/bcr
AML (M2 or M4)	9q34 x 6p23	dek/can
APL	15q22 x 17q11	pml/rara
Pre-B ALL	1q23 x 19p13	e2a/pbx
B-CLL	14q32 x 11q13	bcl1 (prad1?)
	14q32x 1q13	bcl3
	8q24 x 12q22	btg1
T-ALL	14q11 x 8q24	myc*
	14q11 x 1p32	tcl5
	14q11 x 10q24	tcl3
	14q11 x 11p15	rbnt1
	14q11 x 11p13	rbnt2

*T-ALL oncogenes are juxtaposed with T-cell receptor gene tcr.

The pluripotent effects of benzene exposure on bone marrow may be understood in terms of reduced production of cytokines by stromal and other supportive cells of the bone marrow matrix. Benzene metabolites, including hydroquinone, have been reported to inhibit cytokine synthesis by bone marrow stromal cells (9-11). It is postulated that imbalance of this highly orchestrated cascade of growth factors disrupts normal leukocyte ontogeny, leading to the inadvertent rare translocation of oncogenes, an event which is normally confined to immunoglobulin loci of differentiating lymphocytes. It is probable that many other illicit chromosomal translocations are also induced in developing leukocytes by benzene exposure at doses which inhibit stromal cells. However, in humans, only leukemogenic translocations have been noted in that these patients become diseased and amenable to clinical analysis.

BENZENE EPIDEMIOLOGY

In order to understand how benzene may be affecting bone marrow to induce leukemia, it is helpful to look at that spectrum of leukemias seen among exposed workers. There is a common misperception that acute myelogenous leukemia (AML) is the sole malignancy seen among benzene workers. Although AML is certainly prominent (and the only malignancy which by itself is statistically significant), many other leukemias have been reported to be elevated among workers exposed to benzene, including chronic myelogenous leukemia (CML), lymphocytic leukemias, monocytic leukemia, multiple myeloma and lymphomas (for reviews see 12-14).

CHROMOSOMAL TRANSLOCATIONS IN LEUKEMIA

Although AML is more strongly associated with human exposure to benzene, I discuss CML here due to the stunning molecular genetic evidence regarding its genesis. The rationale is that similar genetic mechanisms, *i.e.*, chromosomal translocations, may control the onset of AML, however, the CML story is better understood, and therefore will be used as a prototype for benzene leukemogenesis. The onset of CML is associated with a chromosomal marker of diagnostic significance, the Philadelphia chromosome (Ph'). This pattern of an elongated chromosome 9 and shortened 22 is virtually diagnostic for CML. Ph' is also seen in other leukemias, such as acute lymphoblastic leukemia (ALL) and sometimes AML, although other translocations, such as 6p23 x 9q34 are more common to AML (15).

A closer examination of the 9x22 translocation by chromosomal banding in CML patients shows that the terminal band on the long arm of chromosome 9 (9q34) replaces most of the long arm of chromosome 22 (22q12-13) in the Philadelphia chromosome (Ph'). Many ALLs are also Ph'+ and it is believed that lymphoblastic transformation ("blast crisis") of CML is essentially a conversion of CML cells to ALL (16). It appears as if chromosomal translocations involving oncogenes are common mechanisms of leukemogenesis, including induction of AML, as shown in Table 1.

DNA analysis of the 9 x 22 Ph' translocation reveals juxtaposition of an oncogene (*abl*), which originates on chromosome 9, with a break cluster region (*bcr*) on chromosome 22. This *abl/bcr* fusion results in a new gene which now produces a larger, autocatalytic tyrosine kinase (p210) of 210,000 daltons molecular weight (17).

Transmembrane tyrosine kinases normally mediate the action of intercellular growth factors, including cytokines. Their activation by stromal-elicited growth factors is required for normal leukocyte differentiation, including the generation of antibody diversity in lymphocytes. The genetic activation of latent tyrosine kinase genes is frequently involved in oncogenic transformations. Activation of the p150 tyrosine kinase product of the normal *abl* oncogene (*c-abl*) requires that it bind its cognate growth factor on the exterior side of the leukocyte plasma membrane. This extracellular signal is transmitted via the transmembrane p150 kinase, resulting in its ability to phosphorylate tyrosines of phospholipase C inside the affected leukocyte. Once thus activated, phospholipase C in turn forms inositol triphosphate, that intracellular messenger responsible for activation of differentiation genes in the nucleus. Two modified *abl* oncogenes are known which, unlike *c-abl*, encode enzymes which permanently behave as if they are activated, *i.e.*, they always possess tyrosine kinase activity. One of these is *v-abl*, the gene carried by Abelson Murine Leukemia Virus, which induces leukemia in rodents. The other is the *abl/bcr* fusion gene characteristic of CML. (Table 2)

These observations concerning benzene exposure and the molecular genetic mechanisms of leukemogenesis, may be summarized as in Table 3.

The hypothesis that benzene leukemogenesis results from the indirect induction of chromosomal translocations in developing leukocytes may be tested directly. A useful test system has already been partially defined in that infection of irradiated mice by retroviruses carrying the human *abl/bcr* gene fusion develop CML (18). Furthermore, this system may be used to study the dosimetry of benzene leukemogenesis, as will be outlined further below. It is anticipated that benzene will behave in a transgenic mouse system to promote the leukemogenic *abl/bcr* translocation in a dose-responsive manner similar to that of other carcinogenic promotors. This system will allow detection of that threshold concentration below which benzene fails to impair stromal cell synthesis of cytokines and, hence, promote leukemogenesis.

THRESHOLDS FOR CARCINOGENIC PROMOTORS

Carcinogenesis may be defined in three stages: Initiation, promotion and progression. Whereas initiation and progression have been shown to require genetic alterations and are thus sensitive to genotoxic carcinogens, promotion is epigenetic and responsive to nongenotoxic carcinogens (promotors). The ability of the immune system to survey and reject tumors also may be impaired by nongenotoxic immunosuppressors. An overview of available dose-response curves for various classes of carcinogenic chemicals reveals that, whereas initiators/progressors yield carcinogenic risk which is linear with dose, by contrast, promotors/immunosuppressors display carcinogenic thresholds.

Table 2. Molecular Genetics of the Ph' Translocation in CML

- 9q34 (*abl*) x 22q11 (*bcr*) ⇒ 9q+ x 22q- (Ph', *bcr/abl*)
- The *abl* oncogene encodes a tyrosine kinase
-

Gene	Kinase	Phosphorylation Activity
v-abl	p160	autophosphorylates
c-abl	p150	no autophosphorylation
bcr/abl	p210	autophosphorylates

Benzo(a)pyrene [B(a)P] is a prototypic carcinogenic initiator. Its diol epoxide mutates DNA by binding to guanine (G) and causing a guanine-to-thymine (G -> T) transversion. Upon topical application to mouse skin, a clearly linear dose-response curve has been measured for papilloma initiation by B(a)P over four orders of magnitude (19). Some polycyclic aromatic hydrocarbons, act as both initiators and promotors of carcinogenesis. For example, 2-acetylaminofluorene (2-AAF) initiates cancer of the liver when fed to mice, but promotes bladder carcinogenesis, presumably due to the presence of required activation enzymes in the liver. The dose-response curves show that liver carcinogenesis by 2-AAF is linear (20), whereas a clear threshold exists for bladder carcinogenesis (21).

Table 3. Unifying Principles in Benzene Leukemogenesis

- Benzene is nongenotoxic to other than bone marrow cells
- Benzene metabolites indirectly promote chromosomal translocations in developing leukocytes
- Chromosomal translocations in leukocytes are occasionally leukemogenic

Nonlinear threshold-containing dose-response relationships are also seen in the promotion of bladder carcinogenesis by nitrilotriacetic acid (NTA) and of hepatocarcinogenesis by phenobarbital. Although clearly a threshold effect, the dose-response curve for NTA induction of bladder cancer has been attributed to a 28-hit curve (22). A similar threshold containing dose-response curve is seen for promotion of liver carcinogenesis by phenobarbital in partially hepatectomized rats initiated with diethylnitrosamine (23).

The anticipation is that, given a sensitive transgenic system for the measurement of leukemia-specific chromosomal translocations, benzene will show a dose-response curve similar to that for 2-AAF and NTA in urinary bladder carcinogenesis and phenobarbital in liver carcinogenesis. Through use of a sensitive test system, not only will it be possible to measure a benzene leukemogenic threshold, but that threshold level in turn, given adequate safety factors for interspecies conversion, may be utilized for the establishment of safe levels for human exposure to benzene.

A TRANSGENIC SYSTEM IN MICE FOR MEASUREMENT OF THE LEUKEMOGENIC THRESHOLD FOR BENZENE

The fact that mice transfected with a retrovirus carrying the human *abl/bcr* gene fusion develop CML defines a potential transgenic system for recreating this leukemogenic translocation in response to benzene treatment. An ideal system was described by D. Ornitz during this NATO Workshop (see ref. 24 for a similar transgenic construct). The Ornitz experimental design requires the establishment of two transgenic mouse strains, one of which in this construct would carry the human *abl* gene, the other the human *bcr* gene. Either or both human genes could be further marked with reporter genes, such as *lacZ*, or selective genes, such as *neo^r*, which are only activated by gene fusion and enhance the ability to detect and enumerate leukemogenic translocations in single cells. Mating of *abl* and *bcr* transgenic mice will produce offspring (F_1 hybrids) in which leukemogenic translocations can be measured as a function of the level of benzene exposure.

REFERENCES

(1) Hattemer-Frey, H.A., C.C. Travis and M.L. Land (1990) Benzene: Environmental partitioning and human exposure. Environ. Res. 53, 221-232.
(2) Parodi, S., D. Malacarne, S. Grilli, A. Colacci and M. Taningher (1992) Dose respons relationships for benzene: Experimental and human carcinogenicity data. This volume.
(3) ATSDR (1988) Toxicological profile for benzene. Agency for Toxic Substances and Disease Registry, Atlanta, GA.
(4) Tsujimoto, Y., J. Gorham, J. Cossman, E. Jaffe and C. Croce (1985) The t(14;-18) chromosome translocations involved in B-cell neoplasms result from mistakes in VDJ joining. Science 229, 1390-1393.
(5) Haluska, F.G., S. Finver, Y. Tsujimoto and C. Croce (1986) The t(8;14) chromosomal translocation occurring in B-cell malignancies results from mistakes in V-D-J joining. Nature 324, 158-161.
(6) NTP (1986) Toxicology and carcinogenesis studies of benzene (CAS No. 71-43-2) in F344/N rats and B6C3F1 mice (gavage studies). National Toxicology Program, Research Triangle Park, N.C.
(7) Greaves, M.F. (1986) Differentiation-linked leukemogenesis in lymphocytes. Science 234, 697-704.
(8) Keinanen, M., J.D. Griffin, C.D. Bloomfield, J. Machnicki and A. Chapelle (1988) Clonal chromosomal abnormalities showing multiple-cell-lineage involvement in acute myeloid leukemia. New Engl. J. Med. 318, 1153-1158.
(9) Gaido, K. and D. Wierda (1984) In vitro effects of benzene metabolites on mouse bone marrow stromal cells. Toxicol. Appl. Pharmacol. 76, 45-55.
(10) Gad-El-Karim, M.M., V.M.S. Ramanujam, A.E. Ahmed, and M.S. Legator (1985) Benzene myeloclastogenicity: A function of its metabolism. Am. J. Indust. Med. 7, 475-484.
(11) Gaido, K.W. and D. Wierda (1987) Suppression of bone marrow stromal cell function by benzene and hydroquinone is ameliorated by indomethacin. Toxicol. Appl. Pharmacol. 89, 378-390.
(12) Crump, K.S. and B.C. Allen (1992) Risk assessments for benzene-induced leukemia: A review. In this volume.
(13) Florida Petroleum Council (1986) Benzene in Florida groundwater: An assessment of the significance to human health. API, Washington, D.C.
(14) Goldstein, B.D. et al [eds] (1989) Benzene metabolism, toxicity and carcinogenesis. Envir. Health Persp. 82, 3-307.
(15) Solomon, E., J. Borrow and A.D. Goddard (1991) Chromosome aberrations and cancer. Science 254, 1153-1160.
(16) Desforges, J.F. and K.B. Miller (1986) Blast crisis - reversing the direction. New Engl. J. Med. 315, 1478-1480.

(17) Ben-Neriah, Y., G.Q. Daley, A. Mes-Masson, O.N. Witte and D. Baltimore (1986) The chronic myelogenous leukemia-specific p210 protein is the product of the *bcr/abl* hybrid gene. Science 233, 212-214.

(18) Kelliher, M.A., J. McLaughlin, O.N. Witte and N. Rosenberg (1990) Induction of a chronic myelogenous leukemia-like syndrome in mice with v-abl and bcr-abl. Proc. Natl. Acad. Sci., U.S. 87, 6649-6653.

(19) Burns, F.J. and R.E. Albert (1981) Relationship of benign papillomas to cancer induction in mouse skin. Cancer Detect. Prev. 4, 99-107.

(20) Littlefield, N.A., J.H. Farmer, D.W. Gaylor and W.G. Sheldon (1980) Effects of dose and time in a long-term, low-dose carcinogenic study. J. Environ. Pathol. Toxicol. 3, 17-34.

(21) Littlefield, N.A., D.L. Greenman, J.H. Farmer and W.G. Sheldon (1980) Effects of continuous and discontinued exposure to 2-AAF on urinary bladder hyperplasia and neoplasia. J. Environ. Pathol. Toxicol. 3, 35-54.

(22) Zeise, L., R. Wilson and E.A.C. Crouch (1987) Dose-response relationships for carcinogens: A review. Environ. Health Perspec. 73, 259-308.

(23) Goldsworthy, T., H.A. Campbell and H.C. Pitot (1983) The natural history and dose-response characteristics of enzyme-altered foci in rat liver following phenobarbital and diethylnitrosamine administration. Carcinogenesis 5, 67-76.

(24) Ornitz, D.M., R. Skoda, R.W. Moreadith and P. Leder (1992) Regulating gene expression in mammalian cell culture and transgenic mice with yeast GAL4/UAS control elements. This volume.

SOME POINTS FOR DISCUSSION FROM THE ONCOGENE WORKSHOP

Bruce Molholt

Environmental Resources Management, Inc.
855 Springdale Drive
Exton, PA 19341

When risk assessors meet molecular biologists with the aim of reaching consensus concerning mechanisms and mathematics of carcinogenesis, there is bound to be a language problem. Fortunately, in Vouliagmeni this potential source of misunderstanding was kept to a minimum. However, rather than being problematic in terms of our mutual understanding, because each side demanded precise terminology if we were to get anywhere at all in our discussions, the need for a clear definition of some well-accepted terms was refreshingly constructive. For example, should *carcinogenic promotion* be defined in terms of two stage model systems such as rat liver or mouse skin, in terms of increased tumorigenesis by initiators overall, in terms of specific mechanisms such as activation of protein kinase C or, more simply, as proliferation of stem cells? The best definition requires an overriding mechanistic understanding of promotion. And, concerning initiation, just what is the precise definition of an *activated oncogene*?

There is no question that oncogene research has played a dominant role in our understanding of the molecular mechanisms of carcinogenesis over the past decade. Perhaps the best understood two-step mechanism of carcinogenesis, however, remains the ablation of both copies of oncogene suppressor genes, such as *RB-1* in retinoblastoma and *p53* in colon cancer and other tumors. As Knudson correctly predicted 20 years ago, individuals may acquire such tumors either sporadically, through two genetic changes in appropriate target cells, or familially, through inheritance of one suppressor gene deficiency in the germline plus an additional genetic event in the appropriate target cell. We now know that this second genetic event in heritable carcinogenesis is often homozygosity of the inherited tumor suppressor gene defect.

Less understood is the role of point mutations in oncogenes of the *ras* family. Whereas transitions or transversions of codons 12, 13 or 61 are frequently found in *Ha-ras*, *K-ras* and *N-ras* of tumor cells, there is little agreement as to whether these are initiating mutations or whether, conversely, they occur after initiation and accelerate proliferation of initiated cells (promotion?). Although *ras* oncogenes are certainly important in carcinogenesis, I believe it was felt by most attendees that their role may be less pivotal than had been assumed 4-5 years ago.

How many steps are required for carcinogenesis? We can say one thing for sure, carcinogenesis is a multistep process. In retinoblastoma, it would appear that two steps are sufficient. However, Vogelstein's model of colon carcinogenesis embodies five steps or more. If each stage of a multistep process is rate-limiting and amenable to activation/inactivation by exogenous agents, risk assessment for specific chemicals or radionuclides must be placed within this context. In a practical sense, because the U.S. Department of Energy is just beginning to assess risk from mixed (chemical + radioactive) wastes, there will be an immediate currency for the application of mathematical models which account for multiple cellular interactions. We were fortunate in Vouliagmeni to have the presence of Suresh Moolgavkar, who not only presented an excellent paper in which he postulated a promotion role for familial adenomatosis polyposis (FAP) in colon carcinogenesis but also provided continual discussion with great relevance for modernization of risk assessment.

One of the most vexing goals of the risk assessor is to calculate risks to exposures which are orders of magnitude less intense than those found in the workplace or in experimental studies with rodents. At present, the EPA model assumes that all carcinogenic risks are linear according to dose down through infinite dilution. Whereas this simplified linear model may suffice to describe risks in the 10^{-7} to 10^{-4} range for genotoxic carcinogenic initiators, it is hardly suited to nongenotoxic carcinogenic promotors. Unlike initiators, promotors appear to have thresholds below which they fail to cause those physiologic changes required for promotion. For some promotors, such as 2,3,7,8-TCDD, this threshold would appear to be defined by the critical number of cellular receptors which must be occupied before the target cell dedifferentiates. Since most regulated chemicals are nongenotoxic and presumably promotors, the present EPA linear risk assessment model vastly overestimates risks from subthreshold doses and, hence, overly taxes industry by way of excessive abatement or cleanup costs which are not protective of public health.

Experimental data for carcinogenic dosimetry is limited by the statistical significance of observed effects over background incidence. Significant data for several decades of the dose-response curve may be obtained by either vastly increasing the number of experimental animals per test dose (the so-called Megamouse experiment, which Constantine Zervos wished to more aptly name *Kilomouse* in that 24,000 mice were involved) or by the utilization of target tissues such as mouse skin or rat liver in which multiple tumors may be counted. In all three systems, initiators increase carcinogenesis linear with dose whereas promotors exhibit threshold responses.

Two well-chosen examples of carcinogenic chemicals were presented as case studies at the Workshop. These prototypes of the marriage between risk assessment and molecular biology were methylene chloride, which by inhalation causes rodent lung tumors as well as liver tumors, and benzene, which is a known human leukemogen. Although experimental data are plentiful for both methylene chloride and benzene, we are still short of those complete mechanistic understandings which would enhance the accuracy of risk assessments for these carcinogens.

Anomalously, whereas methylene chloride is only genotoxic in short-term tests when activated by mixed function oxidases (MFOs, cytochrome P450s), this pathway is not associated with carcinogenicity *in vivo*. Rather, it is the glutathione (GSH) pathway, which metabolizes methylene chloride after saturation of the MFO pathway, that is associated with carcinogenic activation. Rory Conolly presented evidence that protein-DNA crosslinks are found proportionately to GSH-activated methylene chloride and speculated that possibly a sulfur-containing DNA alkylation mediates this activity as in the case reported for 1,2-dibromoethane.

Perhaps more is known about the epidemiology and genotoxicology of benzene than any other chemical carcinogen. Kenny Crump summarized the available epidemiologic studies of leukemia in workers with occupational exposure to benzene. Uncertainties in the exposure analysis and limited numbers of worker leukemias render

Table 1. <u>Carcinogenic Dose-Response Relationships For Initiators And Promoters</u>.

Assay System	Initiator/response	Promotor/response	Ref.
Mouse liver	2-AAF LINEAR		1
Mouse bladder		2-AAF* THRESHOLD	2
Mouse skin	B(a)P LINEAR		3
Mouse skin		Phorbol esters THRESHOLD	4
Rat liver	Diethylnitrosamine LINEAR		5
Rat liver		Phenobarbital THRESHOLD	5

*2-AAF is not activated to its ulimate carcinogen in bladder.

dosimetry models for prediction of risks by inhalation in the range of 1-10 ppm difficult. Ironically, there is no good model for benzene leukemogenesis in laboratory animals. Silvio Parodi summarized hundreds of genotoxic tests for benzene, the vast majority of which in prokaryotes and simple eukaryotes are negative. Benzene, however, is genotoxic *in vivo* for selected mammalian tissues, mainly for those hematopoietic cells which arise in bone marrow. Bruce Molholt summarized evidence that benzene metabolites inhibit elicitation of lymphokines by marrow stromal cells and speculated that the chromosomal translocations associated with many leukemias may derive from illicit genetic events normally required for generation of immunoglobulin diversity and perhaps other transgenetic events in hematopoietic cells.

One of the aims of the Workshop was to derive molecularly-based test systems which may aid risk assessors. To this end, two potentially useful systems for the measurement of benzene-mediated effects in bone marrow at low doses were discussed:

- Formation of micronuclei using the CREST assay (S. Parodi). This assay depends upon the presence of centromeres on fragments in micronuclei which may be detected fluorometrically.
- Transgenic construction of human *abl* and *bcr* regions in mouse such that translocation of genes may be monitored via formation of the *abl/bcr* fused gene. David Ornitz believes that this construct could be established with the appropriate genetic markers such as to visualize one *abl/bcr* fusion per 10^7 leukocytes.

The availability of powerful molecular genetic techniques for the measurement of extremely rare events in carcinogenesis should prove a boon to risk assessors. Whereas in the past we have had to rely upon data from human epidemiology or animal experimentation, both limited in statistical power for quantification in the low dose range, today we may proceed more swiftly, less expensively and with greater resolution through use of cell-based systems. However, utilization of the cell rather than the individual animal as a measure of dose-response presents an interesting enigma in terms of endpoint.

If one were to mutagenize Chinese hamster ovary (CHO) cells with diethylnitrosamine and select *hgprt* mutants on HAT medium, it would be possible to quantify mutagenic dose-response relationships in the range of 10^{-7} to 10^{-4} response. Is this dosimetry, however, comparable to the frequency of ovarian carcinoma initiation events in hamsters treated with diethylnitrosamine? Some insight into this question may be gained from Moolgavkar and Knudson's careful investigation of cellular kinetics as related to tumor incidence in retinoblastoma. The *RB-1* mutation which is inherited in familial retinoblastoma is said to be *dominant* in that 95 percent of progeny which inherit this mutation develop retinoblastoma by age 5. However, as a tumor suppressor gene, *RB-1* is clearly a *recessive* mutation at the cellular level since in heterozygotes the wild-type allele provides sufficient suppressor protein to keep retinoblastoma from developing. The 3-4 retinoblastomas which evolve spontaneously to yield mainly bilateral tumors in familial cases are relatively rare mutations ($\sim 10^{-5}$). Hence, in retinoblastoma, carcinogenic initiation events on a cellular basis which are $\sim 10^{-5}$ manifest themselves at 0.95 (five orders of magnitude higher) at the level of the metazoan individual. These unicellular/metazoan differences become even more exacerbated in tissues where the stem cell population is larger than several hundred thousand or where more than two steps are required for carcinogenesis.

Since introduction of the human *abl/bcr* fused oncogene into mouse myeloblasts results in the induction of murine chronic myelogenous leukemia (CML), it may be that one genetic event is sufficient for carcinogenic initiation in this system. In support of this notion, most humans with CML display a single chromosome 9 (*abl*) x chromosome 22 (*bcr*) translocation (one elongated 9 + a single Ph'). Comparing cellular and metazoan translocation events for CML, a single *abl/bcr* translocation in one cell may be sufficient for the onset of metazoan leukemogenesis. This simplified model does not take into account potential intervening events, such as the requirement for a second genetic event or immunosuppression of *abl/bcr*-containing myeloblasts in circulation.

REFERENCES

1. Littlefield, N.A., J.H. Farmer, D.W. Gaylor and W.G. Sheldon (1979) Effects of dose and time in a long-term, low-dose carcinogenic study. J. Envir. Pathol. Toxicol. 3, 17-34.
2. Littlefield, N.A., D.L. Greenman, J.H. Farmer and W.G. Sheldon (1979) Effects of continuous and discontinued exposure to 2-AAF on urinary bladder hyperplasia and neoplasia. J. Envir. Pathol. Toxicol. 3, 35-54.
3. Burns, F.J. and R.E. Albert (1981) Relationship of benign papillomas to cancer induction in mouse skin. Cancer Detect. Prev. 4, 99-107.
4. Van Duuren, B.L., A. Sivak, A. Segal, I. Seidman and C. Katz (1973) Dose response studies with a pure tumor-promoting agent, phorbol myristate acetate. Cancer Res. 33, 2166-2172.
5. Goldsworthy, T., H.A. Campbell and H.C. Pitot (1983) The natural history and dose-response characteristics of enzyme-altered foci in rat liver following phenobarbital and diethylnitrosamine adminstration. Carcinogenesis 5, 67-76.

CONCLUDING REMARKS, FINDINGS AND RECOMMENDATIONS

C. Zervos

Washington, DC

INTRODUCTION

The ARW delved into the intersection of three types of activities, three domains of endeavor, so to speak, shown in Fig. 1: Cancer Risk Assessments (CRAs), Oncogene research (Oncogenes) and Transgenics research (Transgenics). The papers included in this Proceeding are offered as the record of the events at the ARW.

The ARW was somewhat unusual and so are the Proceedings. Unlike most activities of the kind, they have more than a single scientific focus. They direct attention instead on three subjects two of which, Oncogenes and Transgenics, are closely related; the other, CRAs, is rather remote. The glue that hopefully unites all three together is societal need, the need to find ever better ways of reaping the benefits of technical and scientific development with minimal risks to health and the environment.

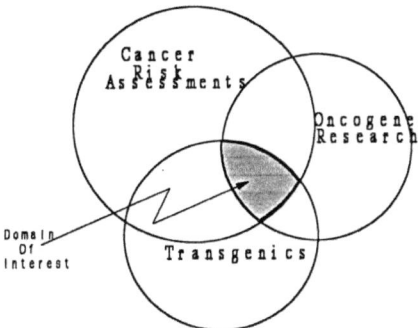

Fig. 1. The Three Domains Of The ARW

Single-focus Scientific workshops are usually interactions among scientists with interests characterized by great degree of homogeneity. Communication among such persons is, therefore, efficient. To a large measure, their interactions are productive because they share a common language and common assumptions. This ARW was planned with full awareness that multifocal activities engender high risks of communications failure. It was known from the beginning that bringing together three different groups of scientists would likely create communication and mindset problems, to say the least. It was felt however very strongly that given the current state of affairs at the intersection of the three domains of Fig. 1, the risks were well worth taking. May the reasons for this feeling become apparent below.

The reasons for this ARW were long-standing and strong convictions that: (i) that developments and discoveries in the basic biological sciences are extremely slow to influence the means and techniques used to predict the potentially adverse effects of technical development; (ii) that the societal consequences of this phenomenon are significant at every level, health, economic, societal structure, etc.; and (ii) that part of the problem is due to interdisciplinary communication barriers which are potentially removable.

This, the last paper of this proceedings constitutes the summary, findings, and recommendations of the ARW.

SUMMARY

In his introduction and charge to the participants, the author of this summary and editor of these Proceedings presented a brief and introductory description of the CRA. It included: a definition of the process; consideration of its origins; the force that drives its evolution; its evolutionary context, and its current societal significance. He stressed that in the past 25 years the uses of CRA have been expanding exponentially without a concomitant improvement in their scientific foundations.

To establish a common ground and to the extent possible a common language for the discussions to follow the author defined CRA as **a sequence of reasoning steps - some inductive some deductive - which permit the assessor to use accepted knowledge, available information, and a circumscribed set of data to provide the foundation for cancer risk related decisions.**

Dr. Anderson followed with a description of today's basic CRA procedure. In her description and discussion she was guided by the NAS well known U.S. NAS pamphlet. She identified a number of well known problems associated with: Hazard Identification; Dose-Response Assessment; Exposure Assessment; and Risk Characterization, the four steps in the procedure. She also discussed the success the procedure has had and identified some improvements that have been the result of hard work and perseverance on the part of dedicated individuals. Dr. Anderson's presentation and the subsequent discussion set the stage for the presentations that followed by focusing on the past efforts of mostly the regulatory community to incorporate into CRA's biological and biochemical information other than that obtained from the standard NCI/NTP bioassay.

Dr. Moolgavkar who followed Dr. Anderson delved deeper into the details, the problems and the prospects of the quantitative aspects of CRAs. Over the past 10 or so years his research efforts have focused on mechanistically based exposure-cancer incidence curves incorporating elements from the kinetics of cell replication, differentiation and death and the chemically caused genetic events that might affect such processes. He has made significant contributions to the multistage theory of carcinogenesis.

Dr. Moolgavkar's work constitutes an important link between carcinogenesis bioassays in experimental animals and epidemiological studies of cancer. His, however, is not the so called allometric approach. Instead he has attempted to model carcinogenesis in terms of fundamental processes, e.g., in terms of identifiable stages of the cell in its transition from normalcy to malignancy, in terms of the transition rates between the stages and the chemically induced genetic or epigenetic evens that affect such rates.

Dr. Moolgavkar's earlier work on the epidemiology and cellular/molecular genetics of retinoblastoma has given substance and foundation to the multistage theory of carcinogenesis. As a consequence there is a strong move to adopt it for regulatory purposes. This position was very apparent in several of the review/improvement efforts that have been under way.[1] Yet, Dr. Moolgavkar's introduction to his presentation and the two examples in the presentation itself should serve as a warning against hurried decisions. His analysis of the epidemiology of colon cancer and of the experimental data on radon induced lung cancer in mice point clearly to the fact that although the multistage theory of chemical carcinogenesis is probably the correct framework to work with, a one-model-fits-all approach is most likely to be a significant mistake. It also points to the fact that any excursions away from the simple and simplified models currently in use will cost significantly in terms of the data it will require.

At the end of Dr. Moolgavkar's presentation and the discussion that followed the major problems with the current procedure for CRA had been discussed fairly well. Those with indirect knowledge of these problems had increased awareness of them even though in the discussions a dichotomy between those in basic research and their regulation counterpart was still obvious.

With a reasonably good focus on, and understanding for, the CRA process and its many parts, Dr. Scherer discussed initiation by genotoxic agents and a number of the aspects of carcinogen-DNA interactions. This has been a long standing problem of all efforts to introduce the mechanistic features of chemical carcinogenesis into the CRA process.

Dr. Scherer discussed the intuitive proposition that the carcinogenic properties, including potency, of genotoxic agents must be somehow related to the type and quantity of DNA adducts formed during exposure to the carcinogenic agent of concern. He noted however, that many variables influence that relation in complex ways that have made its straight forward use in CRA impossible. He discussed some of these variables the effects they are likely to have on the adduct-potency relation and means to overcome any problems they might create.

Among the patterns that seem to have emerged from studies of chemically induced point mutations is that a number of them are specific both with respect to the chemical and with respect to the DNA sequence where they occur. Thus, even though on first sight the picture of chaos emerges from attempts to correlated degree of adduct formation with specific parts of the carcinogenic process, this picture is deceptive in that patterns do indeed exist, patterns which although more difficult to discern than originally thought, can none-the-less be found and used for CRA.

With respect to tissue, again there is no *prima facea* correlation between the degree of adduct formation and the potency of a carcinogen. But, according to Dr. Scherer, 'Combined analysis of cell-specific DNA modification, persistence of DNA adducts, and DNA replication is responsible for tissue-specific tumor formation.???"

Correlation difficulties also arise from the fact that a chemical can affect different stages of the carcinogenesis process in different ways. To state it succinctly, a chemical can initiate and promote, initiate but not promote and only promote. Further a specific mutation in and by itself can be an exclusively initiating event, an initiating and promoting event, or an exclusively promoting event. Dr. Scherer cited examples of all these possibilities.

The upbeat part of Dr. Scherer's presentation comes from his brief review of the results of efforts to map one-to-one if possible specific phenotypic events in a number of carcinogenesis processes onto specific genotypic alterations (mutations) in genomic DNA in general and specific proto-oncogenes and tumor suppressor genes in particular. Thus, the association of specific chemicals and their specific structure and reactivity features with specific types of mutations, i.e., transitions or transversion, constitutes proof of the existence of potentially useful pattern underlying the apparently chaotic adduct formation. So does the strong association between specific oncogenes and types of cancer.

Delving further into the major problematic aspects of CRA, Dr. O'Driscoll discussed extensively a very important aspect of CRA, the role of promoters. Specifically he addressed those that act by modifying the biochemical properties of Protein Kinase C and its role in mitogenesis and differentiation signal transduction. He placed the question of cancer promotion both in its scientific perspective in terms of the multistage theory of chemical carcinogenesis, and in its regulatory perspective pointing to the policy and economic implications of the distinction between initiators and promoters.

Cancer promotion as distinguished from cancer initiation, has been a phantom concept, rooted in operational distinctions that have created nothing but problems for the process of CRA and those who have been using it for regulatory purposes. Work in Dr. Weinstein's and other laboratories - including that of Dr. Castagna's which will be summarized below - reviewed and analyzed by Dr. O'Driscoll, seems to have gone a long way towards establishing an accepted basis for cancer promotion in molecular biology and biochemistry. Even though many fascinating and baffling details remain incomprehensible, a major part of it has been brought to heel under the overarching carcinogenesis theory of "progressive disorder in signal trunsduction." Further three major points have become apparent: (a) The different isoforms of PKC are intimately involved in cancer promotion; (b) The diverse experimental effects observed in relation to cancer promotion cannot be explained under the assumption of "... simple linear pathways of signal transduction."; and (c) "... the signal transduction pathways .... constitute a dynamic and highly interactive network, that includes considerable "cross-talk" and both positive and negative feedback control mechanisms."

It is possible that Dr. O'Driscoll's presentation might have given risk assessors and their clientele the uneasy feeling that as unraveled by research, the complexity of the carcinogenesis process is so great that prospects for help are indeed dim. Yet it seems that in the very complexity of the process Dr. O'Driscoll found the potential for experimental methods to identify cancer promoters and distinguish them from initiators based on powerful molecular biology and molecular genetics principles. He described mammalian cell culture lines promising to do just that.

Dr. Spandidos discussed the current techniques and the origins of oncogene research and covered in more detail the ras oncogene superfamily and their likely role in human neoplasia.

Dr. Dragani touched upon a major issue concerning CRA and its utility. An objection often raised against the use of certain strains of mice or rats is that such strains are 'susceptible' to the hepatocarcinogenic effects of chemicals and that as a consequence the results obtained therefrom have no meaning for the target population, i.e. humans, who are not genetically predisposed to develop hepatocellular carcinomata. This issue has been a bone of contention for many years with no obvious solution apparent at the physiology/pathology level. Dr. Dragani's presentation and the subsequent discussion suggested some progress, albeit inchoate in nature. Inbred strains of mice susceptible to liver cancer appear to inherit a genetic defect which is specific to the liver, dominant, and predisposes them to spontaneous or chemically tumors (HCC, etc). The presence of this defect is associated with higher - relative to insensitive strains - of Ha-*ras* mutation frequencies at codon 61.

There are human subpopulations genetically predisposed to liver tumors but the genetic damage associated with each of a number of syndromes is a risk factor acting epigenetically through liver cirrhosis which is very rare among mice. Human liver tumors are not characterized by high frequency of Ha-*ras* mutations.

The findings do not completely rule out the existence of inherited sensitivity among individuals or groups of individuals sensitivity that could be very hard to identify. They suggest however that the sensitive murine strains might not be appropriate for liver cancer bioassays.

In her introduction, Dr. Castagna provided a brief but clear description of current though concerning the roles of oncogenes in cancer induction. She covered a number of reviews concerning specific parts of the process. She emphasized the multistage nature of chemical carcinogenesis and focused in on promotion and the role of PKC.

She examined promoters in the light of whether and how they might interact with PKC. The central theme of her paper was that "promoters impair the expression of PKC isoforms present in a given tissue, thereby generating a cellular phenotype which prevents cells from returning to the quiescent state."

Studies on rhabdomyosarcoma-derived cell lines provided evidence that the PKC isoform gene and the c-sis gene are correlated and also related to malignancy potential. The studies also suggest, but not prove, that correlation of PKC isoform, c-sis expression - which is related to release PDGF - and malignancy potential might be connected with an autocrine circuit. It concludes that: "... changes in the expression of individual isoforms of PKC appear to play a major role in determining tumor cell phenotype. It is hypothesized that alterations in the expression of PKC isoforms result from adaptation process in response to chronic exposure of cells to agents which damage the activity of this key enzyme." It suggests that the relative levels of expression of PKC isoforms [in various tissues?] may allow to assess the carcinogenicity of some chemicals with potential promoting activity." The clear suggestion again being that it is possible to move away from the current operational definition of promotion to one rooted in and supported by molecular biology and molecular genetics.

Dr. Rhim summarized and described in considerable detail the directions of current research in oncogenes and tumor suppressor genes and what is currently thought to be their role in human and mammalian tumor induction. He focused on the uses of mammalian cell culture for the study of the different steps in the transition from normalcy to malignancy. Along with previous and subsequent presentations, Dr. Rhim's presentation and the discussion that followed conveyed the sense that with the uses of mammalian cell cultures it is now possible to learn a considerable deal about the process of cancer induction by chemicals at reasonable cost. Specific suggestions as to how to put that knowledge to use for CRA were considerably harder to come by. It is not possible however to suppress the feeling that once the specific aspects of the various stages of chemical carcinogenesis become known to some detail their potential uses will not be late in coming.

Dr. Pellicer discussed in vivo model systems for the study of *ras* oncogene involvement in carcinogenesis. In his introduction, he gave a good summary of pertinent information concerning the *ras* family of oncogenes, their potential role in signal transduction, and thereby their potential for being targets for carcinogenic agents.

Dr. Pellicer, in his presentation and during discussions, raised a number of questions which can be posed as scientific programs for research and would improve understanding of the carcinogenic process. He considered transgenic models fundamental in future research to unraveling the steps of carcinogenesis; the value of tissue and cell cultures potentially great; and comparative metabolism as the premier means to answer a number of questions of extrapolating from animals to man.

He described studies of specific oncogenes for their carcinogenic properties *in vivo* using techniques for tumor DNA isolation, vector preparation, introduction into cell culture, monitoring for transforming potential, blotting and identification of transforming genes, re-introduction into different cell cultures, isolation of transformed cells, and introduction into nude mice for determination of carcinogenic potential. He described the prevailing theory of carcinogenesis in terms of its three stages;[1] and because *ras* mutations are specific and are thought to be important he tacitly accepted that 'Many authors view as a fact *ras* oncogene involvement in human tumorigenesis ...' even though specific questions about it still need to be answered.

He described the mouse thymic lymphoma induction by MNU and radiation as a model system for the study of the role of *ras* oncogenes in carcinogenesis. The K- N-*ras* oncogene are the targets of MNU and mutations occur in all three labile mutations. He describe transgenic mice with a MMTV/N-*ras* construct and the pathological abnormalities observed in them.

One of Dr. Pellicer's important finding is that introduction of the N-*ras* oncogene using the MMTV promoter did not produce the thymic lymphoma from which the oncogene was isolated.

Dr. Haliassos (Spandidos) presented primary work and data indicating that it is now possible to obtain adequate amounts of genetic material (DNA) from fixed pathology slides to determine whether specific oncogenes have undergone a specific point mutation in malignant cells. The approach they described appears to overcome the problem created by the fact that tumors and especially advanced cancers are heterogeneous collections of different cells and

---

[1] He includes progression and the acquisition of the potential to metastasize as a distinct stage

it is therefore difficult to detect specific point mutations born by sells which comprise less that 5% of the cell population. The technique which is based on the use of the polymerase chain reaction (PCR) opens great possibilities for mapping known oncogene activation for different tumor tissues. If it is now possible to determine easily what mutation is induced by a given chemical reactivity and what mutation is associated with particular stages in the carcinogenesis process then there exists a direct experimental connection between the chemical and the disease. The CRA consequences of this are many and potentially very significant and should be explored.

Dr. Cordaro made a wide ranging and pace-setting presentation of background information concerning transgenics and oncogene research. He expressed the very hopeful attitudes that both types of research will eventually contribute significantly to the development of substantially better CRA procedures.

An extensive overview of the topic of transgenic animal production and potential utility in many fields of biomedical research was given by Dr. Christopher Cordaro. He discussed in brief the two principle methods - retroviral vectoring and naked DNA microinjection - of introducing foreign genetic material into the cells of a developing embryo. He also analyzed the advantages and disadvantages of the two methods and under what experimental conditions each might be advantageous compared to the other.

Concerning the relation between the construction and utilization of transgenic animals and cancer, Dr. Cordaro delved into two areas which are directly or indirectly related to CRA: the design of antineoplastic agents and the design of test systems for safety evaluation. The description of his strategy for rational drug design is of CRA interest because it is based on an overarching theory of chemical carcinogenesis, one that assumes a multistage process for cancer and associates each of the stages with specific genetic alterations involving the activation of one or more members of the proto-oncogene families or one or more members of the tumor suppressor or anti-oncogene families. His strategy of rational drug design consists of suggestions for interventions designed to take advantage of current knowledge of oncogene biochemistry and physiology for the purpose of reversing the effects of the various stages theorized to comprise the carcinogenesis process. Although cancer therapeutics was not in the topics to be considered in the workshop, it is of interest to note that theories about the fundamental carcinogenesis process were, especially one that affords to experimentalists the opportunity to understand the process and to design measurements and data collection protocols which can put the bioassay results into biochemical and molecular biologic perspective. Thus, Dr. Cordaro's strategy for rational drug design lead to considerable discussions concerning experiments that might use transgenic animals carrying spectra of oncogene and anti-oncogene combinations designed to dissect the carcinogenesis process into its component parts. The potential benefits to derive from such experiments would be in the area of hazard identification. Appropriate data could strengthen or weaken the confidence attaching to the inter-species extrapolation step. But such benefits that might be derive would be of the qualitative rather than the quantitative type. The ARW participants made no suggestions that imminent improvements were forthcoming from the use of transgenic animals in the quantitative aspects of risk assessments save perhaps the notion that some estimates for the rate constants necessary for the use of the Moolgavkar model might now be possible under specific conditions of experimentation.

What Dr. Cordaro's called "Strategy for Carcinogenesis and Mutagenesis" included similar proposal for the construction of animals bearing combinations of activated oncogenes and antioncogenes for the purpose now of elucidating the component part of the process of chemical carcinogenesis. Dr. Cordaro's strategies are based on what is thus far known about the activation of oncogenes and inactivation of tumor suppressor genes in human neoplasias. However he described only one transgenic animal construct that had been already in use. In the discussions that followed Dr. Cordaro's proposals it became apparent that his strategies were believed to be potentially profitable at some time in the future but at the moment it was felt that there were too many knowledge gaps to provide any real support in the quantitative aspects of CRA.

In a *tour de force*, Dr. Ornitz described the potential powers of the new techniques in dissecting the process of carcinogenesis and studying its details. Often, the introduction into the mouse genome of a stably expressible mutant allele of even a single gene whose product plays a significant role in the transduction of extracellular signals for growth or differentiation can have a devastating effect on the health of the transgenic animals. If such effects are on reproduction capabilities it is difficult or impossible to establish and maintain the founder strains of transgenic animals. Binary systems have been described before. Appropriate activator sequences are introduced into a transactivator strain of mice and the target gene is introduced into a target strain. The latter apparently provides for tissue specificity of the activator. The other strain provides the target gene which remains essentially silent in the absence of the trans activator. Upon crossing of the two phenotypically normal strains of mice the pathology expected from the activation of the target genes in the tissues dictated by the transactivator gene is observed.

Most of the work with binary systems is based on viral activators which introduce the potential problems expected from their class. Dr. Ornitz described work with the yeast activator GAL4 and, *inter alia*, oncogenes involved with murine mammary carcinogenesis. The described results are extremely encouraging and point up several facts. The day is obviously not very far when it will be possible to introduce into experimental animals combinations of genes thought to be involved in species-specific processes of interest. The design of experiments will then be limited only by what is known about the appropriate gene combinations, and although knowledge about combinations responsible for genetic diseases is accumulating very rapidly analogous knowledge about chemically induced cancer does not. Spade work designed to identify and correlate for instance oncogenes with particular experimentally induced tumors and chemical reactivity with specific mutation is not advancing at the rate one would hope for.

And as things appeared to assume some manageable shape and form, Dr. Tavitian informed the ARW participants that they are really not as simple as they appear. He described the *ras* superfamily of genes and its three subfamilies *ras*, *rab* and *rho*, their appearance and distribution among different species from yeast to man, and the distribution of some of their protein products in different tissues and subcellular components. Although it appears that, functionally, the products of these genes are related, one way or another, to intercellular growth and differentiation signal transduction and perhaps specifically to the small G proteins, their true functions and their scope and extend are still largely unknown. As to the contribution to carcinogenesis by their mutant forms a similar state of knowledge prevails. Thus, as we lift the lid of the black box, known as chemical carcinogenesis, we discover a level of complexity that we could not have anticipated. The prevailing feeling however appeared to be that practical advantage should always be taken of the knowledge available at the moment always being aware that, but not deterred by, the certainty that ultimately things will turn out to be more complex than they appear at the moment.

Dr. Ahn gave a brief but detail-packed description of current knowledge about: oncogenes and tumor suppressor genes; how they are thought to be involved in the process of carcinogenesis, spontaneous or chemically induced; of the methods currently in use for the introduction of oncogenes into mice specifically designed for the qualitative and perhaps quantitative study of chemical carcinogenesis; of the problems and advantages associated with these methods; and of the methods and attendant problems of producing transgenic animals with inactivated tumor suppressor genes. Certainly, from the view point of Dr. Ahn, and judging from the context of the discussion that followed his presentation, in the near future transgenics, and specifically transgenics carrying appropriate combinations of oncogenes should make the process of CRA, be it qualitative or quantitative a much more rigorous exercise.

Dr. Polet described a specific transgenic animal example showing clearly the power of the new techniques to dissect the process of carcinogenesis and to study the individual steps in considerable detail to the advantage of cancer therapy and perhaps cancer prevention measures such as CRA. Deregulation of the c-*myc* protooncogene has been associated with a number of tumors including lymphocyte tumors in which it is translocated into the heavy Ig locus and comes under the influence of the Ig promoter. Transgenic mice carrying the enhancers either of the light or heavy Ig chains linked to the mouse c-*myc* developed B cell lymphomas confirming the role of c-*myc* deregulation in tumor development. Dr. Polet extended the work in outbred rabbits with interesting results. Rabbits carrying the protooncogene linked to the short chain enhancer developed polyclonal acute lymphoblastic leukemias, similar to childhood leukemias, and those carrying the same gene linked to the heavy chain enhancer developed monoclonal tumors of a variety of tissues. The transgenic animals constructed by Dr. Polet and her colleagues is likely to prove valuable both in the study of the development of ALL and in the study of leukemogenic agents such as benzene and of the ways they produce their insidious effects.

Dr. Conolly changed gears as scheduled and presented a brief but substantial report on the CRA for methylene chloride (MC). This review was scheduled because the CRA of MC represents a significant turning point in CRA method development and acceptance. Dr. Conolly discussed the idea of dissecting the process of chemical carcinogenesis along the model of Moolgavkar et al. and of dealing with the two parts, the pharmacokinetics and pharmacodynamics, separately using experimental data for modeling. He focused on the evolution, development, and partial acceptance of the Physiologically Based Pharmacokinetic (PBPK) models and their advantages and remaining uncertainties. He identified areas where data and information collection efforts should focus to permit the ideas of PBPK models to be applied with all the rigor that current experimental science will permit.

He next focused on the more difficult part of the CRA process, i.e., toxicodynamics. He sought a bridge between the Moolgavkar model parameters and experimental observations much like Dr. Moolgavkar proposed and referred to during the various periods of discussion. The implication of Dr. Conolly's presentation and discussion was indeed that oncogene and transgenic techniques can and should provide the necessary knowledge and information to unravel how MC causes mutations - DNA-protein crosslinks? what is the carcinogenic significance of such mutations , and how can we account for the inter- and intra-species differences is such mutations and their consequences. He also touched on a very significant but seldom discussed aspect of CRA the contributions to estimated uncertainties made by polymorphisms either in the metabolic machinery of the members of each species or strain, or in the targets to a toxicant such as MC .

Ms Charnley complemented the presentation of Dr. Conolly focusing on the toxicodynamics of methylene chloride and guided by the Moolgavkar two stage model of cancer induction. She accented the fact that even in such well studied topics of chemical carcinogenesis as the mouse liver preneoplastic lesions there is no agreement concerning their meaning and their identity and it is difficult to identify specific stages in the process with phenotypic alterations whose physiological and biochemical parameters can be determined experimentally for the purpose of estimating the constants required by the Moolgavkar model. Much like the case of identification of the specific enzymes involved in the activation of MC and in identifying the specific target of the activated species, Ms Charnley speculates that mapping of the activated oncogenes and judicious use of the resulting knowledge to construct transgenic animals to test hypothesis about their actions and function in chemical carcinogenesis will eventually eliminate considerable portion of the uncertainty currently associated with specific forms of the Moolgavkar model.

What appears to be disheartening about the MC case is that work guided by the requirements for the application of the Moolgavkar model has not been forthcoming in the past 2-4 years. In the eyes of those interested there appears to be no incentive to further pursue the consequences of the application of the PBPK model. The 50- to 100-fold gain obtained through the use of the latter appears to have exhausted the impetus for serious experimental inquiries into the mechanisms of MC carcinogenesis even though the Moolgavkar model combined with development in oncogenes and transgenics would appear to promise considerable pioneering progress.

Dr. Crump introduced the benzene (BZ) case study with a review of the numerous risk assessments that have been prepared and published over the past 10 years. BZ was selected as a case study because unlike MC, the various risk assessments were exclusively based on epidemiology data.

For BZ epidemiology data appear more abundant than any other case that comes to mind. Risk assessors are often reminded to use epidemiology data whenever and wherever they might be found available because, the argument goes, they eliminate the problems that arise from the need for interspecies extrapolations of cancer probabilities.

The BZ case is characterized by an additional feature. Occupational data were used to estimate probabilities for the occupational setting. Thus, the CRA process did not require intraspecies interpolation either. It is therefore considered a straight forward case of looking into the problems that might arise with the CRA process under optimal circumstances and of the potential role oncogenes and transgenics knowledge might play in minimizing the impact of such problems.

And a number of serious problems were indeed identified by Dr. Crump's thorough presentation and the discussions that followed. The major group of problems, as could be expected, has its origins in the need to determine exposures in detail retrospectively before a dose-response curve can be constructed. In a *tour de force*, Dr. Crump presented a very vivid picture of such problems and the effects they might have on the probability estimates needed for risk management. Assistance with the problem of retrospective determination of exposure patterns and their toxicological significance was suggested by the discussion along the lines of the Pereira group at Columbia U. which is making perhaps the first attempts in the direction of testing for activated oncogenes in asymptomatic workers. It would appear however that the value of such efforts will depend strongly on how well they are planned and coordinated.

A smaller but very important set of problems arises from the need to identify the types of lesions to be considered in the CRA. Leukemia is a group of malignancies. One could argue that BZ is a general leukemogen or that it is specific. Epidemiology data are unlikely to provide definitive evidence one way or the other. However, the molecular biology and molecular genetics of human cancer could.

Dr. Crump's discussion also pointed out to other problems with lasting and outstanding effects on CRA and its results: the healthy worker problem, a phenomenon that is not limited to BZ; and the problem of unanticipated findings.

The healthy worker phenomenon is rather common one that complicates CRA often. The explanations that have been offered are empirical. As a result definitions of the resulting problem are circular. It would help considerably if the phenomenon acquired a sound basis in molecular biology and genetics.

Similarly, the observation in the chinese study that BZ is a lung carcinogen has parallels in the oncology literature. Whether, the observation is the result of sampling statistics or an artifact or whether it has more real and serious causes is a very important question. Plausible genetic or environmental causes for the observation abound and it certainly seems that currently available theories and techniques - those used in oncogene and transgenic research - should help narrow down the possibilities.

Drs. Grilli and Parodi reviewed the experimental evidence concerning the metabolism and toxicity of BZ in a number of species including humans. The central point of his review was that there is a wealth of data of all kinds which when used with the PBPK models for a CRA yields cancer probability increments comparable to those obtained from analytical epidemiology studies. The same data however casts great doubts on the wide range validity of simple calculations even in the presence of apparent agreement between the two approaches. The uncertainties in both approaches are such that one cannot rule out the possibility that the agreement is simply a coincidence. Simply stated both approaches are shaky and they must be substantiated by additional evidence collected independent of the bioassay or the analytical epidemiology study.

Saturation of BZ metabolism in a variety of species is an intriguing feature of the data reviewed by Dr. Grilli. It is difficult to understate its implications for risk assessments based on leukemia, or generally cancer, probabilities observed at saturating levels of exposure.

A lot of metabolic data was presented and discussed in the context of what is the toxic metabolite of benzene or what DNA adduct is responsible for the induction of leukemia or other malignancies. This questions however remain unanswered even though it must be assumed that the techniques are available to approach these questions. The toxicodynamics of BZ are a feasible project of research one that can yield data that is likely to place the CRA of BZ on a sounder than hitherto basis. Comparative metabolism, the genetics of metabolizing enzymes and the carcinogenic species and its target, if identified, would strengthen the base of CRA no matter what the database used.

Dr. Molholt Continued with a summary of the information currently available concerning the molecular events of the process of leukemogenesis and of the potential role of benzene and its metabolites. He pointed out that chromosomal translocations are probably at the heart of all leukemias: "It appears as if chromosomal translocations involving oncogenes are common mechanisms of leukemogenesis, including induction of AML, ..." He also suggested that the benzene induced translocations are likely to be the result of inhibition of the production of cytokines by bone marrow stromal cells which control the development and differentiation of pluripotent hematopoetic stem cells. Dr Molholt's implication is that benzene is not an initiator - it does not behave as an initiator in the classic initiation tests - but as a promoter and that therefore a threshold could and should be determined for the leukemogenic effects of benzene. Finally he suggested an experimental means to determine such threshold using a binary transgenic animal system with the two protooncogenes *abl* and *bcr*.

RECOMMENDATIONS

In their presentations, some workshop participants made specific suggestions for the improvement of CRAs offering novel experimental procedures for the collection and interpretation of data based on the principles and methods of molecular biology and molecular genetics (cf. Pellicer, Ornitz, Polet, Molholt, Crump, Conolly.) Others made more general suggestions (Cordaro, Ahn, Rhim.) All presentations elicited extensive discussions which at times tended toward the anticipated dichotomy mentioned earlier. However, this difficulty was managed by repeatedly posing to the participants the central questions of the workshop:

- What other data or information could be obtained before, during or after the standard cancer bioassay, information that is comparatively cheap to obtain, and can aid in lifting the fog that surrounds the usual bioassay results? and
- Given the ARW discussions about their current status and what ARW participants know or heard concerning oncogenes and transgenics, is it possible to recommend any type of experimentation, short term or long, that will strengthen the science used to back CRAs?

Removed from specific objects of regulatory attention and controversy, e.g., MC, BZ, etc., such inquiries did not elicit consensus responses pointing to specific directions. From the discussion however it appeared that differences in acceptance standards constitute one of the major reasons why concepts and techniques from rapidly advancing research disciplines are not incorporated expeditiously into regulatory and control activities such as CRA. Different acceptance criteria apply in the two domains. Research disciplines can accept, and as a matter of fact are usually based on, techniques that have not as yet emerged from what recently has been termed their "heroic" phase.[2] DNA microinjection, PCR, and the various types of blotting are among the latter. Regulatory activities on the other hand, by nature, must be based on proven techniques that can be performed by anyone who is adequately trained and has access to modern state-of-the-art laboratory facilities. Thus, systematic and well planned evaluation and validation would appear to constitute ways to expedite the transfer of technique and method from the basic biological disciplines to the regulatory domains in general and CRA specifically.

Along this line, on several occasions, the ARW participants were asked for opinions about the techniques of oncogene and transgenics research as tools for regulatory CRA activities. Two examples were specifically mentioned to give impetus to the discussion.

In cancer bioassays, it is often difficult to associate causally exposure to the agent or agents of concern on one hand, and the appearance of malignancy on the other. As a consequence, interminable disputes and arguments ensue with respected scientists on either side of the causation issue. It would therefore be both prudent and economical to find ways to decide on more definitive grounds the question of causality. Could we, for instance, request those who conduct cancer bioassays to determine the state of activation of a complement of oncogenes in malignant tissues along with the determination of the pathology of the same tissues. Would such requirement by the regulatory agencies be reasonable in terms of time and resources? would the required experimental techniques be adequate? would the information be useful?

Although a number of reservations were expressed concerning the ease and cost of identifying activated oncogenes in malignant and normal tissues of experimental animals the general view was that such tests are becoming common place very rapidly. It was argued that a bigger problem with the idea was the fact that although intuition says that such information would be useful in more than one ways, little specific potential for utility could be identified in the absence of a systematic effort to establish specific and detailed correlations in the chain of events:

Chem. Reactivity ⇒ Type of Mutation ⇒ Transformation Potential

---

Some very good and promising work is currently done by the Balmain group in Glasgow[3] to correlate the chemical reactivity of carcinogens and the type of point mutation they elicit. Much, much more of the same type of work is required to validate the ideas that will permit molecular biology and molecular genetics data to be required for premarketing approvals and CRAs.

Work on the correlation between the type of mutations and transformation potential, i.e., on mapping the pattern of activated oncogenes onto the nature and strength of transformation is also going on at many places. But it is not driven by the needs of cancer prevention activities such as CRA. Impetus comes from cancer therapeutics instead. Thus, patterns of oncogene activation in human neoplasias are often determined for purposes of prognosis and for the determination of therapeutic regiment. Consequently, questions specific to CRA are answered only incidentally and do not receive the design attention they deserve

Similarly, although talk abounds, little specific attention has been directed to developing transgenic animals specific for carcinogen testing purposes. Development of transgenic species or strains of test animals will be a difficult task to be accomplished in two steps before use for CRA. First they must be constructed and then validated. At the moment, however, there is very little construction and even less validation underway. Currently, transgenics are constructed almost exclusively for therapeutics. Yet, the opportunities in the test animal development area abound. During the discussions of the MC BZ cases, for instance, it was suggested that transgenic species/strains be developed and used to narrow the uncertainties associated with the interspecies metabolism of the substance and those pertaining to the Glutathione Transferase isozyme that activates it. A similar approach was later suggested for the elimination of the uncertainties associated with the CRA for benzene. The idea received some consideration.

REFERENCES

1. Cf., for instance, the recent efforts of the US Federal Coordinating Council for Science, Engineering, and Technology.

2. Grease, R.P. (1992). History of Science. The Trajectory of Techniques: Lessons from the Past, Science, 257:350-353.

3. Bailleul, B., Brown, K., Ramsden, M., Akhurst, R.J., Fee, F. and Balmain, A. (1989). Chemical Induction of Oncogene Mutations and Growth Factor Activity in Mouse Skin Carcinogenesis, Environ. Health Persp. 81:23-27.

INDEX

κ-chain enhancer 214, 215
μ-chain enhancer 214, 215,111, 122, 124, 125, 139, 141,126
129/J 68, 116, 13-16,
1,4-benzoquinone 15, 17-22, 25, 27-29, 31, 34, 35, 187, 145, 185, 188-191 193, 270, 275-277, 287,
1AT 72
2-AAF 36, 38, 41, 43, 44, 46, 49, 52, 53, 242, 270, 275, 278, 279, 282-284, 286, 295, 297, 299, 301, 306, 307, 309-311, 315, 317, 329, 331
5-azacytidine 191
5'flanking regions 74, 86, 89, 119, 53-55, 58-60, 59, 103, 155, 157, 156,
7-meG 31, 33, 71-75, 77-79, 89, 96, 115, 122, 127, 157, 170,
a-bomb survivors, leukemia latency pattern 250
AAF 31, 33, 36, 307, 309, 313, 315
Abelson Murine Leukemia Virus 105, 308
aberrant development 190
aberrations 46, 145, 146, 149, 150, 187, 193, 269, 270, 274, 288, 292, 305, 310
abl 96, 100, 141, 144, 148-150, 152, 306-311, 316, 330
abnormalities 46, 73, 78, 103, 107, 120, 139, 187, 194, 269, 310, 324
absolute risk model 247, 250, 256
absorption 5, 73, 270, 276, 282, 290
AC3 70
acid rain 11
acids, fatty 284, 287, 301
acinar cells 37, 36, 40, 41, 169, 171, 189, 216
activated rho 178
activation 23, 32, 35, 37, 42, 45, 48, 49, 51, 54-59, 61, 64, 77, 80-85, 88, 89 95, 96, 98-100, 102, 103, 105, 109, 112, 113, 115, 117, 123, 125-128

activation (continued) 130, 131, 143, 147, 150, 151, 156, 166, 169, 170, 186-188, 191, 193-195 214, 236, 263, 275, 280, 286, 307-309, 313-315, 325, 326, 328, 331
activator 57, 84, 89, 100, 134, 156, 166, 169, 172, 326
acute lymphoblastic leukemia 199, 306
acute leukemias 256
acutely transforming retroviruses 61
adaptation 88, 323
additive or relative risk model, 258
adduct, DNA 5, 31-34, 40, 41, 114, 229, 277, 285, 321, 322, 329
adducted 36
adducts, DNA 31-34, 36, 38-40, 45, 52, 59, 114, 117, 148, 150, 192, 226, 263, 270, 274, 275, 277, 280, 282, 287-290, 301, 303, 321
adenocarcinoma 37, 41, 80, 120, 126,130, 171, 192
adenoma 74, 102, 108, 130, 189, 194, 234, 236, 237
ADP ribosylation 177
AFB1-modified DNA 32
aflatoxin 31, 38, 39, 42, 71
agencies 3, 6, 232, 239, 331
aggressive stage of cancer 64
agonist 47
aguaninic sites 32
AIP 73
A/J 70
AKRxRF 116
alcohol 71, 74
alkyl-DNA 32
alkylated DNA 238
alkylation 238
ALL 206, 209, 210, 213-215, 242, 245, 246, 248-253, 255, 256, 269, 275, 276, 278, 299, 300, 305-307, 313, 314, 319, 322, 324, 327, 329, 330
allele 72, 105, 125, 128-130, 136-138

333

allele (continued) 142, 148, 316, 326
Alpha1-antitrypsin 72
ASO 128
alteration, DNA 40, 45, 50, 67-69, 75, 100,102, 114, 121, 122, 151, 233, 234
altered growth control 186
alveolar cells 221, 233
ambient air quality 10
Ames tests 21, 52, 53, 145, 146, 302, 303, 305
AML 193, 305-307, 330
ampicillinase gene 34
Amplification, gene 58, 64, 88, 96, 95, 100, 127-129, 131, 133, 134, 141, 145, 170-172, 188, 193, 194
amplified genes 64, 71, 87, 127, 129, 131, 159, 188
analytical epidemiology 283, 284, 329
anaplastic cells 186
anemia 131, 149, 256, 305
animal genome 188
animal life-span 297
anti-diolepoxides 276
anti-rab sera 180
antibiotics 69
antibodies 33, 36, 51, 180, 181, 212
antioncogenes 176
antipeptide 57
AP-1 54, 57, 59, 64, 65, 64-66
AP1 50, 65
Apl-ras gene 174
aplastic anemia. 256, 305
Aplysia 173, 174
aplysiatoxin 83
aquifer 7
Armitage-Doll model 23, 25, 26, 260
arsenic 4, 8-10, 12, 13, 19, 241
artificial restriction sites 128, 129
ASARCO 6, 20
asbestos 16, 241
assessment of risk 1-10, 9-12, 11-15,17-22, 28, 31, 34, 35, 41, 43, 44, 46, 52,53, 71, 75, 77, 111, 122, 123, 124, 145, 182, 183, 193, 217-219, 223, 224, 226-229, 239,241, 242, 246-250, 252, 254-256, 258, 260-263, 283, 285, 291, 302, 310, 314, 320

association 19, 71, 74, 79, 115, 118, 233, 240, 246, 261, 262, 302, 305, 322
athymic mice 101
autocatalytic tyrosine 307
autocrine mechanisms 88, 323
avian leukosis virus 62
azaserine 37, 36, 40, 41

B-cells 96, 148, 150, 192, 214, 306, 310
B-cell lymphomas 148, 192
B-CLL 306
B-lymphoblastic leukemia 213
B-lymphocyte lineage 204, 210
B-galactosidase 166
B(a)P 309, 315
B2, probable human carcinogens 233
B6C3 hybrid 71
B6C3F1 76-78, 146, 217, 218, 228, 231, 233, 279-281, 286, 290, 299, 300, 303, 310
B6D2F1 119
BAAM 12, 13
background mortality 250, 251
bacterial assays for point mutations 233
bacterial mutagenesis 192, 193
BALB/c mice 69-71, 76, 77, 140, 285
base pairing 32
bcl-2 96, 100
Bcl1 304
bcl3 304
bcr 304-309, 314, 328
benign prostatic hyperplasia 162
benign tumors 44, 54, 105, 113, 120, 162, 185, 233, 311, 317
Benzene 5, 8, 10, 11, 13, 20, 83, 89, 149, 230, 241-246, 248, 247-252, 260-263, 269, 270, 274, 275-278, 277-294
benzene, epidem. studies 242, 292
benzene exposure 5, 243-246, 248, 252, 254, 256, 261, 278, 290, 293-303, 305-310, 314-316
benzene exposure, short-term, high-level 257
benzene, human carcinogen 241
benzene metabolism, toxic pathways 283
benzene, occupational exposure 243, 245, 261, 262, 302

benzene potency parameter 251
Benzo(a)pyrene 309
Benzo(a)pyrene 5, 146-148
best available technology 11, 13
bigenic mice 156, 161-164, 166, 167, 169
bile 44, 55, 57, 83, 130, 233
bile duct fibrosis 233
bile salts 83
binary system 155-157, 159, 160, 166, 167, 171
biliary tract 233
bioaccumulation 14
bioactivation 36, 282
bioassay 2, 10, 13-15, 193, 217, 219-221, 233, 286, 320, 325, 329, 330
biological relevance 232
biological switches 180
biomedical research 11, 15, 76, 325
biopsy 73
birth rate 238
bladder 233, 309, 311, 315, 317
blast cells 306, 307, 310
blast crisis 306, 307, 310
blood disorders, exposure to benzene 241
blot 51, 85-87, 116, 121, 127, 131, 135, 136, 138, 140, 141, 159-160, 164, 203, 205, 206, 211, 212, 214
body burden 14
bone hyperplastic transgenic mice 192
BOP 31, 36, 37, 36-38
bounded log-logistic-like function 237
brain 56, 58, 73, 117, 147, 174, 180-184
breast 218, 228
breast cancer 193
BRL-ras 174, 178
bronchiolar cells 233
bryostatin 85
btg1 304
Burkitt's lymphoma 96, 97, 100, 105, 106, 148, 152, 187, 193, 201, 215

c-abl 141, 186, 187, 194, 306
c-erbB 86, 172, 186
c-fos 50, 54, 86, 186, 192
c-H-ras 48, 49, 99, 102, 189
c-Ha-ras 46, 48, 77, 86, 107, 193
c-jun 50, 54, 86
c-Ki-ras 42, 79, 86
c-myc 50, 55, 86, 96, 105, 122, 125, 126

c-myc (continued) 141, 145, 148, 149, 152, 171, 186, 187, 189, 192, 193, 199-201, 203, 204, 206, 213-216, 327
c-myc linked to an Ig enhancer 201
c-myc or N-myc gene hybrid 192
c-myc transgenic rabbits 199, 204, 206, 209, 213-215
c-neu 122, 124, 126, 171, 189
c-onc 85, 86, 106
c-sis 86-88, 105, 323
c-src 97, 104
C1 48, 51, 101, 235
C1 domain 48, 51
C3H/He 70
C3H-10T1/2-PKC4 50
C3H 48, 50, 57, 67-70, 76, 77, 140, 148
C3H-10T1/2 48, 50, 57
C3Hf 69, 71, 76
C57BL/6 116, 148, 274, 290
Ca_2^+-activated 83
Ca_2^+ 47, 55, 57, 59, 83, 89
CAG 11, 18
calpain 47, 56, 57
cancer 215, 217, 218, 220, 221, 223, 224, 226, 228, 229, 231-234, 236, 237, 239, 240, 242, 246, 250, 251, 260, 261, 263, 276, 278, 280, 283-289, 291, 302, 306, 309-311, 313, 317, 319-331
cancer, mouse skin model 187
cancer potency factor 233
cancer risk 1-3, 9, 14, 15, 17, 18, 20, 21, 27, 28, 31, 41, 46, 53, 217, 218, 122, 124, 223, 224, 229, 231, 232, 234, 239, 260, 263, 276, 278, 280, 283-285, 302, 319, 320
cancer risk assessment 2, 21, 28, 31, 41, 46, 122, 218, 223, 229, 263, 283, 285, 302
carbon tetrachloride 7, 16, 83, 283, 289
carboxy terminal domain 122
carcinogen 5, 10, 12, 11, 13, 15, 18-20, 31-35, 38, 39, 44, 45, 53, 58, 59, 67, 69, 70, 88, 98, 101, 102, 112, 114, 115, 117, 122-125, 147-151, 186, 192-194, 197, 217, 231, 233, 236, 237, 239-241, 260, 261, 280, 288, 303, 315, 321, 329, 331
carcinogen-modified bases 32

Carcinogen uptake 33
carcinogenesis 4, 11, 13, 15, 18, 19,21, 22, 24, 23, 28, 29, 31, 35, 36, 38-46,48-50, 52-59, 67, 68, 69, 70, 75-77, 81, 82, 88, 89, 91, 95, 98-105, 107-109, 111-114, 123, 125-128, 130, 133, 134, 145, 146, 150, 152, 153, 170,185, 186, 185, 188, 191-195, 197, 228, 229, 231, 232,237-240, 260, 263, 280, 282 284-287, 289, 290, 303, 308-311, 313, 314, 316, 317, 320-328
carcinogenesis, genotoxic theory 231
carcinogenic potential 185, 193, 194
carcinogenicity, long term experiments 298
carcinoma 2, 35, 37, 38, 42, 47, 49, 64, 67, 71, 72, 75, 78-80, 95, 96, 95, 96, 95-97, 102, 104, 105-108, 121, 126, 131, 152, 193, 194, 199, 205, 207, 209, 210, 214, 215, 234, 237, 240, 316
cartilage formation 209
Case control studies 242
catabolism 73
catalytic site 48
Caucasian 73
causal 123, 145
causative 41, 53, 111
CBA 67, 68
CBI 277
CC3 70
CD-1 119, 270, 286
CDC42 163, 171, 174, 177
cDNA 48, 50, 51, 55-57, 141, 160, 173, 174, 176, 181, 182
cell 199, 201-203, 205, 206, 209, 210, 212, 214, 216, 224, 226, 229, 231, 232, 236, 237, 238, 240, 242, 250, 254, 234,257, 270, 276, 278, 285-288, 291
cell kinetics 23, 28
cell morphology 48, 178
cell motility 173
cell proliferation 22, 28, 36-38, 95, 103, 117, 123, 143, 148, 229, 231, 232, 234, 236, 237, 295
cell rounding and disorganization 178
cell transformation tests 193

cellular architecture 234
Cellular localization 177, 180
cellular proliferation, differentiation 127
cellular promoter 191
centrilobular cells 33, 36
centroacinar cells 37
cervical cancer 130
chemical workers 245, 260, 262, 302
chemoprevention 53
childhood leukemia 199, 202, 213
chimeric transgenic mice 190
chlordane 5, 12, 14, 16
chloride-induced hepatotoxicity 237
chlorination 17
chlorobenzilate 12, 13, 17, 18
chloroform 7, 16, 83, 89, 128, 218, 228
chromatid 107, 146, 149, 192, 290, 305
chromosomal aberrations, breakpoints 187
Chromosomal abnormalities 107, 187, 310
chromosomal breakpoints 62, 187
chromosomal loss 188
chromosomal 200, 274, 291, 295, 305-311, 315, 331
chromosome aberrations 145, 187
chromosome 15q23-q25 73
chromosome 14q32.1 72
chromosome 1p34 73
chromosome 6p21.3 73
chromosome 11q23.2-qter 73
chromosome 5, 29, 55-57, 62, 72, 73, 97, 100, 102, 105, 106, 125, 134, 136-138, 144-146, 149, 150, 152, 153, 170, 186, 187, 212, 215, 270, 269, 270, 285, 289, 291, 307, 310, 316
Chromosome translocations 100, 269, 270, 289, 310
chromosomes 270
chronic leukemia 256
chronic myelocytic leukemia 62
chronic myeloid leukemia 187
chrysarobin 84
cirrhosis 67, 72-75, 78, 79, 323
cis-acting genes 64
clastogenic agents 33, 149, 269, 285
clastogenicity tests 269
Clean Air Act 10
clonal expansion 24, 25, 238, 295
clonal isolation of transgenic ES cells 190

clonal tumors 24, 25, 51, 81, 101, 102, 112, 124, 190, 238, 293, 308
clone 19, 83, 293
cloning 56, 57, 85, 104, 106, 113, 116, 133,134, 136, 160, 181, 183, 287
Clostridium Botulinium strains C and D 177
CML 187, 304-307, 314
co-carcinogenic mechanism 34
coding 38, 48, 50, 62-64, 72, 98, 115, 117, 119, 122, 137, 142, 146, 148, 171, 181, 182
codon 34, 36, 38, 40, 42, 67, 71, 77, 102, 117, 118, 126-130, 136, 148, 187, 323
codon 61 67, 71, 77, 117, 148
codon 12 34, 38, 40, 42, 128-130, 148, 187, 323
cofactors 44, 47, 83
cohort 78, 233, 239, 243-246, 248-251, 253-257, 258, 259, 261, 262, 293-297, 302
colon 22, 23, 25, 26, 25, 27, 29, 35, 38, 44-46, 49, 52, 55, 57, 63, 83, 95, 96, 95, 96-99, 103, 113, 130, 131, 148, 152, 193, 194, 233, 313, 314, 321
colon adenocarcinoma 130
colonic adenomas 128, 131
colonic neoplasias 128
colorectal 29, 41, 55-57, 59, 83, 97, 102, 109, 128, 131, 152, 187, 194
Comparative metabolism 263, 291, 302, 324, 330
competitor oligonucleotide 64
concave dose-response relationship 299
Conjugation 33, 134, 147, 217, 219-222, 224, 275, 291
consensus sequences 34
conservatism 4
conserved sequences 173, 216
construct 48, 119, 120, 135, 136, 143, 151, 160, 188, 190, 202, 206, 213, 243, 309, 316, 324, 326, 328
conversion 9, 41, 44, 62, 64, 81, 102,106, 114, 142, 143, 275, 307, 309
convex dose-response curve 301
convolution 37
cooperative effects 44, 48, 101

cost-effectiveness 43
covalent bonds 32
cross-talk 51, 320
crude lung cancer rate 246
crude rate of leukemia 243
CSFs 304
cumulative dose 250-254, 256
cumulative exposure 245, 247, 251-253, 255, 278, 296
cumulative exposure index 245
cumulative exposure, log transform 253
curve-fitting 232
Cys2 51
cysteine 48, 51, 58, 147, 175, 178, 288
cystic fibrosis 136, 162
cytokines 306-308, 330
cytomegalovirus 139
cytoplasmic IgM 199, 205
cytoplasmic oncogenes 187
cytosolic exchange factor 180

d-guanosine 270, 277
D-ras2 174
D-ras3 174, 176
DAG 47
damage, DNA 5, 37, 36, 39, 41, 46, 74, 88, 114, 123, 137 276, 280, 286, 288, 291, 295, 323
DCC 97, 102, 103, 194
DEAE-Sephacel 51
deaminase 73
deficit of leukemias 251
degradation 18, 47, 60, 151
deletions 35, 43, 45, 47, 50, 51, 56, 78, 99, 136, 148, 157, 171, 176, 186, 194, 256
DEN 7, 8, 31, 33, 35, 36, 39-41
deoxycholate 55, 82, 83
deregulated genes 194
dermal uptake of chemicals 257
detection of point mutations 127, 130, 131
detoxifying pathways 33, 280
developmental stages 124
diacylglycerol 45, 83
dichloromethane 228, 229, 240, 282
Dictyostelium discoideum 174
diet 67
dietary effects 38, 43, 46, 76
diethylnitrosamine 8, 70, 309, 315

337

differential susceptibility among species 193
differentiated cancer 186
differentiation 32, 49, 64
dimethylsulfoxide 50
diols 275, 307
dioxin 4, 6, 7, 16, 19, 52, 69, 77, 227
direct microinjection of DNA 188
DMGT 112, 116
DMN 8, 31, 33
DNA 5, 31-37, 36-41, 45, 46, 48, 52, 54-56, 61-64, 86, 91, 95-100, 102, 104, 107, 108, 112, 113, 114, 116, 117, 119, 120, 123, 125-131, 133-146, 148-152, 157, 164, 170, 171, 173, 187-193, 199, 200, 203, 205, 204-206, 210, 211, 212-215, 218, 221, 224, 226, 229, 231, 232, 233, 238, 258, 263, 269, 270, 274, 275, 277, 280, 285, 287-290, 301, 307, 309, 315, 321, 322, 324, 325, 327, 329, 330
DNA adducts 31-34, 36, 38-40, 52, 150, 226, 263, 275, 277, 281, 289, 301, 321
DNA binding domain 157
DNA BK virus with BK T antigens 192
DNA demethylation 191
DNA JC virus with JC T antigens 192
DNA methylation 55, 152, 191
DNA modification 32-36, 39, 56, 123, 319
DNA repair 40, 114, 123, 146, 192, 193, 280, 301
DNA size constraints 188
DNA, viral, single copy integration 189
DNA, viral, site specific integrations 189
DOC 82, 84, 83, 84
dominant 295, 313, 316, 323
dominant event 35
dominant gene 22, 35, 72, 73, 112-115, 146
donor DNA 112
dose-response model 4, 20, 231, 234, 237, 239
dose response slope 251
dose-response flattening 301
dot blot hybridization 127
double strand 32

down-regulation 47, 54, 60
Drinking Water Act 10
Drosophila 146, 170, 174, 176, 181
drug metabolism 69, 228
Dukes stages 130
duration of exposure 244, 253
DY 163, 162, 169
dysmyelination 192
dysplasia 128
DZ x RV 164, 166, 169

Eκ-myc 199, 202, 205, 206, 209, 213-215
Eμ-myc 199, 202-204, 206, 210, 213-216
Eμ-myc DNA construct 213
E1B/CAT 159
e2a 304
early B-lymphocytes 204, 213
Eco RI 86
effect modifiers 253
effector region 175, 180
EGF 84, 178
eight or sixteen cell morula stage embryos 190
EJ 97, 106, 107, 153
elastase 72, 160, 163, 166, 169, 171
electron-rich 32
electrophilic 32
elevated risk for leukemia 242
embryogenesis 119
embryonic carcinoma 199, 205, 207, 209, 210, 214, 215
emissions 6, 13
emphysema 72
endogenous factors 185
Endoplasmic Reticulum 74, 179, 180, 205
enhancer 57, 135, 139, 145, 148, 155, 159, 160, 170-172, 189, 192, 200, 201, 202, 206, 214-216, 325
environment, measures/protective devices 246
enzymatic DNA amplification 129
EPA 1-7, 8-14, 13, 15, 17-19, 219, 224, 229, 233, 234, 239, 252, 254, 260, 261, 270, 305, 314
Epidemiological data 16, 242, 247, 255, 256, 284, 293, 301, 307, 315, 316, 321, 328, 329
epidemiology 2, 4, 19, 44, 52, 53, 58, 123, 228, 229, 261, 283, 284

epididymis 159, 160, 162, 164, 166
Epigenetic mechanisms 295
epithelial cells 48, 49, 55, 56, 99, 102, 104, 107, 108, 124, 162, 166, 171, 183, 192, 229, 236
epoxides 147, 228, 309
error-correcting processes 32
ES cell, injection into blastocyst 188, 190
esophageal cancer 35, 37, 41, 108, 148
esophagus 35, 37, 38, 99
essential regulatory processes 180
estimates of lifetime risk of cancer 251
ethylation 36
ethylene oxide 8, 241
evolution 40, 64, 76, 104, 141, 173, 231, 320, 327
excretion 33, 274, 278, 286, 290
exo enzyme C3 177
exon 62-64, 66, 112, 136, 200
exposure profile 243
expression of an inserted oncogene 190
expression of hybrid oncogenes 191, 192
expression, oncogene 45, 120, 121, 166, 169, 180, 189, 191, 323
extrapolation 2, 9, 10, 12, 14, 15, 17, 21, 43, 44, 49, 52, 53, 193, 217, 219, 222-224, 226, 227, 232, 255, 263, 276, 278, 282, 284, 280, 292, 294, 295, 325
extrapolation, out of range 255

F1 69, 70, 76, 77, 139, 140, 142, 148, 310
F21 85-87
F21-T 85-87
F344 rats 290, 299, 300
FAH 72, 73
familial retinoblastoma 190, 316
FAP 22, 25, 29, 44, 57, 97, 102, 127, 130, 194, 128, 314
farnesylation 173, 175
fatty acids 84, 89, 175, 233
fatty acylation 175
fatty degeneration 233
FBJ osteosarcoma virus LTR hybrid 192
feedback 51, 218, 322
fertility 120, 121
fibroblasts 34, 45, 49-51, 55, 56, 58, 59, 61, 65, 99, 101, 102, 106, 108 112

fibroblasts (continued) 162 , 183
fibroplasia 74
FIFRA 10, 13
finger-like regions 48, 51, 58
fission yeast 174, 181
FITC-labeled anti Ig antibodies 212
fixation 37, 123
flanking sequences 34
flattening of response 293
fly ash 7
focal hyperplasia, 236
foci of cellular alteration 35, 36, 40, 41, 112, 116, 192, 193, 233, 234, 236, 237, 240, 311, 317
focus 3, 17, 36, 37, 36, 66, 117, 147, 174, 234, 236, 242, 319, 321, 327
focus forming 117
follicular cells 62, 96
fos 49, 50, 54, 58, 59, 64, 66, 86, 144, 150, 186, 192
founder mice 119, 120, 135, 139, 140, 147, 151, 188, 190-192, 209, 215, 326
fractionation 22, 27
full likelihood approach 258
functions of the rho protein 178
fusion of vesicles 180
FVB mice 162

G proteins 64, 81, 112, 326
G→T 307
G25K 174, 177
g47.1 arms 117
G6Pase 74
GAL4/236 157, 159, 161, 162, 169
GAL4 155-157, 156, 157, 159, 160, 161, 160-163, 162, 164, 166, 169-172, 311, 326
GAP 175-177, 179, 183
gastrointestinal cancer 5, 29, 73
GDI activity 179
GDI 175, 177, 180, 183
GDP 64, 112, 115, 173, 175, 178
GDP-Dissociation Inhibitor 175
genes 22, 23, 25, 26, 25, 27, 31, 33-36, 35, 38, 42, 44-47, 50, 54-57, 59, 61, 63-65, 64, 65, 66, 67, 69-73, 75, 77-80, 86-88, 90, 91, 95, 97, 98, 100, 101, 102, 104-106, 108,112-120, 122

genes (continued) 124-127, 131, 133-137, 139, 141-143, 145, 146, 148, 149, 152, 153, 155-157, 159-163, 166, 167, 169-171, 173, 174, 176-178, 181-184, 186-192, 194, 201-205, 210, 215, 216, 269, 295, 306-311, 313, 316, 323, 326, 327
gene amplification 133, 134, 188
gene conversions 188
gene deletions 187
gene, dominant 295
gene products 122, 171, 183
Gene-specific 33
gene superfamily 173, 175
generalized 5, 6, 134
Genetic 293
genetic cancers 190
genetic constitution 67
genetically resistant mice 67, 68, 69, 71, 75
genetically susceptible animals 67-71, 75, 77
genetics 42, 44, 67, 68, 76, 78, 103, 106, 133, 134, 155, 306, 319-321, 327-329
genomic 31, 124, 127, 131, 135, 140, 141, 151, 166, 173, 190, 203, 206, 322
Genotoxic agent 295
genotoxic chemicals 52, 69, 185, 242
genotoxicity 114, 115, 133, 134, 145, 150, 193, 231, 233, 238, 263, 269, 274-276, 284-287, 289, 293, 302, 305
genotoxicity tests, in vitro 193
geranylgeranylation 175
germ-line 67-69, 74, 203, 205
germline 22, 44, 71, 119, 138, 156, 188, 190, 199, 200, 203, 205, 204-206, 209, 210, 215, 313
GH 73
gip 193
glands 33, 119-121, 160, 162, 278, 280
glucose-6-phosphatase 74, 79
Glycogen 72, 74, 79
glycoprotein 72, 137, 139
Golgi bodies 176, 177, 179-183
goodness-of-fit comparisons 234
granulocytes 270, 306
growth factors 38, 86, 89, 103, 169-171, 183, 205, 213, 308

growth hormone gene 160, 163, 171
GSH 33, 217, 219, 218-224, 226, 228, 275, 315
gsp 193, 194
GTP-binding 45, 181-183
GTP-GDP binding domain 178
GTP/GDP binding site 175
GTP 45, 64, 112, 115, 144, 173, 175-178, 181-183
GTP hydrolysis 144, 177
GTPase 64, 112, 115, 175, 176, 182, 183
GTPase activating protein 175, 182
GTPgS 180
guanine 32, 34, 39, 40, 103, 181, 277, 309
guanine-to-thymine transversion 309
guideline 278

H-*ras* 34, 35, 40, 48, 49, 56, 57, 61, 63, 95,96, 95, 96, 95, 96, 98, 99, 102, 105, 108, 112, 122, 144, 146-148, 173, 174, 186, 189, 192, 193
Ha-*ras* 46, 48, 57, 58, 67, 70, 71, 77, 81, 86, 88, 105, 107, 108, 126, 193, 313, 323
halogenated chemicals 52
hamster 36, 37, 36-38, 40, 41, 89, 145, 284, 316
Harderian gland 120, 121
hazard 2, 11, 289, 320, 325
HBV 71, 74
HCC 71-75, 323
health effects of benzene 241
health-protective course 232
HEM 6
hematopoietin 306
heme 73
hemizygous animals 190
Hemochromatosis 72, 73, 78
hemosiderosis 233
hepatic foci 234
hepatic lobular architecture 234
hepatitis B virus 71, 78
hepatocarcinogenesis 40, 67-71, 75, 77, 236-238, 309
hepatocellular vacuolization 233
hepatocyte proliferation 233
hepatocytes 234
hepatoma 78, 189, 199, 205, 206, 209

hepatoma (continued) 214
hepatomegaly 74
heptachlor 5, 12, 14, 16
Herpes Simplex Virus 166
heterodimers 64
HFE 72, 73
hGH 160, 163
HIV 133, 139-143, 153, 166, 170
high-dose experiments 3, 151
high stringency conditions 127
histopathologic changes 234
HL-60 49, 50, 55, 57, 59, 97
HLA-A 73
homologies, sequence 149, 151, 173
homologous DNA 63, 64, 100, 112, 113, 122, 124, 149, 171, 173, 181, 182, 188, 190
homologous recombinations 188
homology 47, 61, 64, 97, 112, 151, 174-176, 186, 216
homozygosity 190, 215, 311
hormones 160, 162, 163, 171
hot spots 98, 115, 149
HSV-1 166
HT29 49, 55, 83
HTLV 1 166
human cells 101-103, 105, 123
human cancer risk 217, 276, 280, 283, 284
Human T-Cell Leukemia Virus 166
hybrids 63, 68-71, 119, 143, 151, 189, 191, 192, 194, 311
hybrid proteins 63
hydrocarbons 147
hydrolase 73, 228
hyperlipidemia 74
hydrophobic 83, 136
hyperplasia 157, 162
hyperplastic 162
hyperuricemia 74
hypoglycemia 74

IARC 41, 78, 111, 125, 263, 269, 275, 286, 288, 305
Ig κ-chain enhancers 199
Ig-enhancer 214
Ig gene rearrangement 202
Ig gene rearrangment, diversification 202
Ig promotor/enhancer 201

IgH enhancer 192
indole 83
immature blast-like cells 213
immortalization 99, 101-103, 185, 186
immune system 46, 138, 139, 141, 144, 308
immunoblotting 51
immunocytochemistry 33, 35, 37, 36, 180
immunofluorescent analysis 204
Immunoprecipitation 50
immunosuppressed patients 128
immunosuppression 306, 316
immunosuppressors 308
in situ hybridization 180
inactivation/deletion of tumor suppressor genes 186
inactivation 35, 45, 81, 89, 91, 100, 103, 141, 147, 186, 188, 190, 314, 326
inbred animals 68, 69, 76, 113, 201, 323
inbred strains 68, 69, 76, 323
incidence 2, 10, 22, 23, 25-28, 37, 46, 53, 68-71, 76-79, 120, 152, 193, 200, 215, 219, 218, 221, 223, 224, 228, 233, 240, 243, 277, 280, 290, 299, 300, 314, 316, 320
induction 18, 32-35, 37, 38, 40, 41, 50, 52, 67, 105, 107, 113, 114, 116, 118, 120, 122, 123, 126, 144-148, 157, 166, 171, 192, 193, 231, 233, 239, 240, 261-263, 269, 270, 274, 277, 284, 286-291, 307-309, 311, 314, 317, 323, 324, 328-330
industrial standards 241
infancy 73
Inflammation 74
ingestion 4, 7, 9, 15, 19
inhalation exposures 4, 18-20, 219, 218, 222-224, 227-229
inheritable changes 242
inhibition of transcription 191
initiating activity 35, 44, 45, 69, 115, 263, 277, 305, 313, 322
initiation 6, 35, 70
injury 82, 229
inositol 47, 56, 308
inositol triphosphate 308
Insertion 140, 141, 147, 149- 151, 163, 169, 171, 176, 177, 190
insertional mutations 188

341

insertional mutagenesis 61, 62, 141, 149, 151, 190
Int-2 96, 155, 159-163, 162-164, 163, 166, 167, 169-172
integration site 135, 151, 190, 191
inter-species variations 21, 325
inter-individual variations 44
interactions 38, 44, 46, 96, 124, 133, 145, 150, 314, 319, 321
interleukins (ILs) 306
internal dose 276, 278, 283
interspecies extrapolation 222, 227, 269, 281, 283, 309, 328, 332
interview/questionnaire responses 242
intron 62, 64, 112, 216
inverse dose-rate effect 28
IRIS 8, 14, 19
IRLG 2, 3, 19
iron 73, 78
isoforms 43, 48, 49, 51, 52, 58, 86, 88, 142
isoprenylation 176
isozymes 36, 54, 56, 57

J558L 159, 160
job categories 243, 258
Job histories 243
jun-B 86
jun-D 86

K-*ras* 35, 38, 42, 61, 63, 75, 80, 95, 96, 95-97, 102, 105, 112, 116, 118, 126-131, 144, 147, 148, 152, 173, 174, 176, 187, 193, 313
K-*ras* point mutations 128, 130
K-*ras* allele 130, 148
Kaposi sarcoma 130
karyotypic abnormalities 194
keratinocytes 89, 98, 99, 101, 106, 107
kidney transplant patients 130
kidney transplantation 128, 130
kidney 37, 58, 99, 108, 127, 128, 130, 131, 147, 160, 164, 192, 202, 203, 210, 211, 280
killer cells 150, 306
kinase 34, 43, 46-48, 51, 53-60, 65, 81, 82, 84, 88, 89, 95, 145, 148, 307, 308, 313, 322
kinases 45, 55, 103, 143, 178, 186, 306

kinases (continued) 307
kinetic 18, 39, 228, 229
kinetics 22, 23, 25, 27, 28, 35, 40, 83, 116, 221, 237, 283, 316, 320
Kirsten 106, 107, 112, 125, 173
K_m 220, 283
Kupffer cells 33

lac Z 157, 159, 166
lacrimal gland 119
lactic acidosis 74
large litter size 188
large T antigen 96
latency 115, 116, 120, 122, 139-141, 151, 189, 192, 194, 195, 250, 251, 254, 258
latency distribution 258
later embryonic stages 190
leaky expressions 166
LET 22, 27, 28
leukemia 95, 96, 95, 96, 95, 96, 187, 241, 245, 248, 294, 305-307, 316
leukemia, link to benzene 242
leukemia incidence 243
Leukemia Inhibitory Factor 157, 159
leukemia dose response, shape 253
leukemia risk and benzene exposure 245
leukemogenesis 105, 125, 295, 305, 307-310, 315, 316, 330
leukemogenic agents 20, 260, 305-310, 327, 330
leukemogenic cellular transformation 258
Leukocytes 306-309, 316
leukopenia 305
LIF 157, 159
light chain, Ig 199, 201, 204, 205
likelihood 2, 32, 124, 258, 296, 298
likelihood ratio 296
linear dose response model 258
linearized multistage model 232, 276, 283, 305
lipids 46, 162
litter 162, 188
localization of the rab proteins 180
locus PI 72
locus 136-138, 166, 169, 177
low stringency annealing 174
low-dose extrapolation 2, 3, 5, 20, 21, 124, 151, 231, 239, 311, 317

LTR 61, 62, 124, 119, 143, 157, 159-161, 189, 190, 192
LTR, proviral 190
lung cancer and benzene exposure 246
lymphatic and hematopoietic cancers 245
lymphoblastic leukemia 199, 213, 216, 307, 327
lymphoblasts 203, 306
lymphocytes 128, 139, 140, 143, 145, 146, 150, 192, 199, 202-206, 213, 270, 277, 285, 289-291, 306, 307, 310
lymphocytic leukemia 96, 98, 201, 202, 206, 216, 260, 307
lymphoma 62, 96, 97, 105, 106, 115, 116, 119, 122, 124, 125, 145, 148, 187, 189, 193, 199, 201, 205, 210, 211, 214-216, 324
lymphomas 95, 117, 192, 307
lyngbyatoxin 83

M-MuLV as an insertional mutagen 190
M-MuLV expression, not permissive 189
M-MuLV 189, 190
macrophages 275
male pronucleus 119, 134, 188
malignancy 78, 88, 96, 98, 100, 103, 104, 107, 115, 128, 152, 187, 190, 215, 307, 320, 323, ,331
malignant tumors 21, 23, 24, 23, 29, 28, 35, 38, 41, 44, 46, 50, 55, 61, 63, 80, 81, 88, 91, 99, 101, 102, 105, 108, 113-115, 117, 123, 126, 144, 171, 186, 185, 226, 234, 245, 254, 299, 300, 324, 331
malignant neoplasms 113, 171, 245
mammalian mutagenesis 192, 193
mammary gland 10, 62, 99, 98, 105, 107, 119-121, 124-126, 148, 149, 157,
mammary gland (continued) 159, 162, 166, 167, 169, 170-172, 189, 192, 216, 233, 277, 280, 326
map 322
mapping, gene 78, 100, 105, 157, 325, 328, 331
marked blood changes 257
marrow 203, 205, 216, 270, 274-277, 280, 281, 283, 285, 286, 289-291, 305-307, 309, 310, 315, 316, 330

mature B-lymphocytes 204
mechanism of action 2, 22, 34, 38, 48, 50, 52, 54, 72, 89, 95, 96, 95, 96, 95, 96, 100, 114, 123, 135, 138, 139, 144, 182, 185, 194, 195, 224, 226, 228, 229, 231-234, 238, 239, 284, 285, 287, 290, 313
mechanistic data 5, 218, 226, 232, 233
membrane 46, 47, 57, 64, 74, 89, 112, 144, 171, 173, 175-177, 179, 180, 270, 308
membrane ruffling 173
MEN-1 194
mesenchymal differentiation 209
messenger RNA 45, 47, 83, 89, 308
metabolic activation 32, 37, 115, 147, 263, 275, 280, 286
metabolism 18, 33, 35, 56, 69, 78, 89, 123, 145, 146, 149, 150, 152, 217, 220-222, 226, 228, 229, 239, 263, 269, 274, 275, 277, 281-285, 287, 288, 290, 291, 302, 310, 324, 329, 330, 332
metalloprotease 50
metastasis in human cancer 193
metastatic cancer 85, 86, 90, 114, 115, 119, 120, 122, 148, 149
meteorological data 6
methylated DNA 36, 37, 191
methylation, DNA 34, 37, 40, 55, 102, 137, 152, 176, 191
methylene chloride 5, 7, 83, 217, 219, 218-224, 226, 228, 229, 231-234, 236-240, 289, 314, 315, 327, 328
mezerein 85
MHC class I 192
Michaelis constant 220, 283
Michaelis-Menten equation 275, 283
microinjected DNA 119, 203, 206
microinjection into blastocysts 190
micronodular structure 74
micronucleated cells 277
mispairing 32, 34
mitogen 50, 89
mitogenesis 52, 53, 59, 322
mitogenic agents 57, 103, 238
mitotic index 96, 145, 146, 190, 290
MMTV 62, 119-122, 124-126, 157, 159-162, 169, 189, 192, 170, 171

MMTV (continued) 324
MMTV LTR 119, 157, 159-161, 189, 192
MMTV-int-2 169
Model, linear 295
Model, multistage 295
model, tumor development/growth 215
modeling, dose-response 2-4, 6, 8, 19, 39, 218, 219, 222, 224, 227-229, 231, 232, 234, 236, 285, 291, 327
models, BBDR 21, 232
Modified primers 129
modulators 46
molecular basis of cancer 53, 103, 151
molecular epidemiology 52, 53, 123
Moloney murine leukemia virus 189
monitoring 3, 6, 7, 14, 15, 59, 249, 257, 289, 290, 324
monoclonal tumors 33, 51, 205, 213, 214, 327
Moolgavkar, Venzon, Knudson 232
morphologic data 51, 73, 75, 236
morphological evidence 234
morphology 48, 49, 51, 78, 79, 178, 210
mortality from benzene 250
mRNA 47, 57, 135, 136, 139, 143, 148, 149, 160, 166, 169, 183
Msp I enzyme 129
muconaldehyde 270, 275-277, 279, 287, 291
muconic acid 270, 274-276, 279, 282, 283
multifactorial process 53, 81, 91
multigene family 64, 173
multiple myeloma 245, 307
multiplicative effect 295
multiplicity of tumors 277
multistage model 16, 15, 18, 22, 28, 29, 31, 40, 43-45, 48-50, 52, 53, 59, 67, 91, 111, 113, 114, 217, 231-233, 247, 250, 251, 260, 276, 283, 285, 295, 305, 320-323
multistep process 31, 34, 68, 76, 81, 91, 95, 99-101, 103, 108, 113, 123, 125, 126, 144, 152, 185, 186, 314
murine heavy chain enhancer probe 203
mutagens 31, 38, 40, 81, 145, 146, 149, 190, 232, 234, 238
mutagenesis 31, 32, 39, 40, 48, 53, 58, 61, 62, 105, 141, 149-151, 190, 192, 193, 276, 286, 287, 305, 325

mutation 24, 27, 32, 38, 71, 95, 96, 102, 130, 331
mutation pattern 32, 34, 36, 38
mutations by genotoxic chemicals 185
mutators 23
mutH E.coli 34
myc 49, 50, 54, 55, 61, 64, 86, 96, 97, 99-101, 105, 122, 125, 126, 131, 141, 144, 145, 148, 149, 150, 152, 157, 159, 171, 186-189, 192, 193, 199-206, 209, 210, 213-216, 306, 327
myelodyplastic syndromes 256
myelogenous leukemia 96, 100, 148, 149, 193, 305-307, 311, 316
myeloma 245, 307
myelotoxicity 287, 305

N-*ras* 61, 63, 79, 95, 96, 95, 96, 97, 105, 106, 111, 112, 116-122, 124, 125, 144, 147, 173, 174, 176, 187, 189, 192, 193, 313, 324
N-*myc* 49, 54, 64, 96, 145, 148, 186, 188, 189, 192, 193, 216
N-nitrosamines 33, 36
N-substituted 45
N-alkylation 32
NAS 1, 2, 9, 11, 19, 250, 255, 261, 320
nasal epithelia 33
NDEA 70, 77
necrosis 74, 233
neoplasms 74, 125, 185, 192, 100
neoplastic tissues 35, 38, 40, 41, 68, 75, 91, 98, 100-104, 106-109, 112, 115-117, 123, 127, 144, 145, 181, 186, 199, 202, 226, 233, 234
neoplastic lymphocytes 199
neoplastic nodules 233
nervous system 73, 192
neu 38, 42, 61, 95, 122, 124, 126, 148, 157, 159, 166, 169, 171, 172, 189
neuroblastoma 96
neurofibromatosis type 1 locus 177
neutrons 116
neutrophils 72
NF1 194
Ni-induced tumors 85
NIH 3T3 cells 61, 112, 113, 116, 194
nitrilotriacetic acid 14, 309
nitrogen oxides 10

NMBzA 31, 35, 37, 38
NMU 31, 34, 111, 116, 117, 122
no-effect dose or threshold 193
non-genotoxic agents 21, 43-46, 52, 69, 115, 124, 286
nodules, hyperpalstic 35, 36, 40, 70, 209, 210, 233, 239
non-linear models 252
non-lymphoid tumors 199
non-mutational process 45
non-randomness 34
non-specific protein binding 64
non-target tissues 34, 40
noncoding exon 66
nonneoplastic effects 74
nonrandom chromosome alterations 186
nonthreshold 2, 3, 5
Northern blot 51, 85-87, 120, 121, 135, 136, 141, 143, 148, 149, 159-160, 164, 169, 180
NTA 14, 15, 17, 309
NTP 126, 219-221, 228, 233, 240, 276, 278, 286, 299, 303, 310, 320
O-alkylation 32
O-alkyl-modified bases 32
O^6-meG 31, 32, 34, 37, 38
occupation as a surrogate for exposure 242
occupational exposure 241, 243, 247, 249, 251, 252, 255, 260-262, 285, 286, 302, 306, 315, 328
occupational benzene standards 249
ODC 50
odds ratio 253, 256
oligonucleotides 63, 65
Oligonucleotide primers 129
onc 85, 86, 99, 105, 106
oncogene 35, 41, 42, 45, 46, 48-50, 54, 56-58, 61, 62, 64, 77, 80, 81, 88, 90, 91, 95-100, 103-108, 111, 113, 115-131, 143, 144, 147, 148, 150, 152, 155, 159, 163, 170-172, 176, 178, 181, 183, 186-195, 197, 199, 200, 214-216, 236, 238, 306-308, 313, 316, 319, 322, 324, 325, 327, 329, 331
oncogene activation 61, 80, 96, 98-100, 115, 123, 127, 128, 131, 186, 188, 193, 236, 325, 331

oncogene expression 41, 48, 88, 107, 118, 122, 191, 192, 194, 195, 238
oncogene product 50, 57, 181
oncogene, enhancer, promoter fusion 200
oncogenes 35, 38-40, 42, 43, 45, 46, 49, 51, 52, 54, 56, 58, 61-64, 76, 77, 79, 81, 88, 91, 95, 96, 97, 98, 97-101, 103-108, 111-119, 122-127, 131, 133, 134, 141, 144, 146, 150, 152, 153, 159, 170, 173, 175, 181, 185, 186, 188, 189, 191, 192, 194, 195, 200, 214-216, 231, 239, 306-308, 313, 319, 322-324, 326-331
oncogenes, identification in specific cells 188
oncogenes, tissue-specific expression 188
oncogenes, deregulated expression 200
oncogenic agents 31, 35, 38, 58, 64, 95, 104, 107, 173, 185, 213, 307
oncogenic signals 213
oncoprotein action 192
oncoprotein synthesis, consequences of 192
one-hit model 14, 16, 15
ontogeny 304
organ specificity 35, 125, 193
organotropism 36
ornithine 50, 55
OSHA 241, 243, 248, 261
OSTP 2, 3, 20
oval cells 237, 240
overexpression of genes 49-51, 54, 56, 86, 88, 178

p150 305, 306
p21 65, 98, 112, 125, 170, 181, 183
p210 307, 308, 311
P450 36, 147, 275, 290
p53 35, 38, 41, 42, 75, 80, 97, 101-103, 109, 141, 148-150, 152, 194, 313
PAH 16, 45, 147, 287, 309
palytoxin 84
pancreatic cancer 36-38, 41, 42, 233
pancreatic ducts 33
papilloma 35, 107, 309
parotid 162
parsimony 22
partially transformed phenotype 178
pathogenesis 41, 64, 67, 69, 74, 75, 79

pathogenesis (continued) 91, 98, 103, 113, 140
PB-PK models 283, 284
PB 228, 283, 284, 289
PB 9, 69, 129
PBG deaminase 73
PBGD 72, 73
PBPK 217-219, 221-224, 226, 227, 228, 282, 327-329
pbx 306
PC12 49, 50, 54
PCE 269
PCR 42, 51, 71, 80, 127-129, 133, 134, 140, 141, 149, 152, 325, 330
PCR amplification and K-*ras* 128
PCT 73
PDGF 86, 88, 323
peak exposures 258
pedigree 22
peripheral chromatin condensation 205
periportal tissues 33
pesticides 9, 12, 13
pGEM GAL4 159
Ph' 307, 308, 316
phage library of genomic DNA 203
pharmacokinetics 5, 9, 18, 152, 217, 218, 221, 227-230, 269, 276, 282, 283, 289, 327
phenobarbital 69, 84, 315
phenotype 112, 114, 157
phenyl glucuronide 278
phenyl sulfate 278, 279, 282
phenylguanine 277
phenylmercapturic acid 278, 290
pheochromocytoma 49, 183
Philadelphia chromosome 105
phorbin 50, 55
phorbol 43, 46-48, 51, 53-55, 57-59, 65, 81-84, 88, 89, 108, 187, 315, 317
phosphatase 74, 79
phospholipase C 308
phospholipid 46, 47, 54, 56, 57, 59, 83, 88, 171
phospholipid-binding 83
phospholipid-dependent 46, 54, 56, 57
phospholipid-dependent 59, 83,88
Phylogenetics 68
PK model 283,284
PK 152, 228, 269, 283, 284, 289

PKC 43, 45-52, 53-60, 65, 81, 82, 84, 83-88, 89, 322, 323
plasmacytomas 201, 206, 215
plasmid 65, 108, 159
pleiomorphic 72
pleiotropic effects 51, 59, 236
pluripotent cells 306, 330
pluripotential cells 209
pml 306
pMV7 48, 51
point mutations 35, 38, 42, 79, 95, 96, 102, 105, 117, 118, 126-128, 173, 187, 193, 233, 324, 331
Poisson 250, 258, 294, 295
Poisson regression 250, 258
pollutants 8, 10, 12
poly(A+) RNA 85, 86
polychlorinated biphenyls 52
polyclonal tumors 205, 214
polygonal cells 210
polylinker 160, 163
Polymerase Chain Reaction 325
polynuclear aromatic hydrocarbon 277
polyoma middle T antigen 97
polyposis 22, 128, 312
polyps 23
porphobilinogen 73
Porphyria 72, 73, 78, 79
post translational modifications 47, 178
postpartum 163, 169
potency 276, 302, 303, 321
prad1 306
pre-B-lymphocytes 204
Pre-B ALL 306
precancerous changes 35, 36, 41
predictor of leukemia 253
predisposition to cancer 53, 67, 76, 106, 122, 125, 139
preimplantation embryos, retroviral infection 189
preneoplastic cells 238, 295
prerequisite for gene expression 191
prevention of cancer 29, 43, 44, 52, 53, 58, 59, 76, 88, 91, 111, 123, 128, 139, 141, 202, 209, 287, 289, 327, 331
principle of maximum likelihood 296
priority setting 11
probes, molecular 85-87, 113, 117, 119

probes, molecular (continued) 128, 164, 174, 181, 190, 200, 203, 205, 204, 210, 212
progeny 120, 121, 134, 135, 138-141, 146, 151, 192, 306, 316
prognosis 64, 75, 131, 331
prognostic marker, 128
progression 35, 68, 70, 185
progression of familial polyposis 130
proliferation 33, 201, 295
proliferative stage 102
promotion 85, 114, 308, 309, 322
promoters 52, 81, 84, 113
promoting agent 295
promotion tests, in vitro 193
promotor 35, 201, 308, 309, 314, 315
Promutagenic 33, 34, 36, 37
prostate 99
prostatic gland cancer 130
protease 56, 72, 137, 139, 142, 143
protein product 95
protein kinase C 313, 322
proteolysis 47, 56
proto-oncogenes 45, 61-64, 77, 91, 95, 98-100, 103, 122, 123, 141, 143, 170, 175, 181, 322
proviral sequences 190, 191
proviral sequences, stochastic activation of 191
pseudo-pregnant mice 188
pSV2NEO 159
pulmonary tumor induction 193

qualitative risk assessment 2, 11, 17, 28, 43, 176, 282, 325, 327
quantitative risk assessment 2, 3, 6, 11, 13-15, 17, 18, 20, 28, 40, 43, 58, 67, 69, 70, 77, 152, 176, 185, 193, 224, 226, 241, 242, 246, 257, 258, 260-262, 282, 286, 288, 320, 325- 327
quasi-linear dose-response relationship 301
quiescent genes 85, 148, 191, 323
quinol 282

R-*ras* 174, 176, 181
R6-PKC3 48-50, 57
R6 48-51, 57
R9-T 85-87

R9 85-87
rab/YPT family 174
rab proteins in vesicular traffic 180
rab 173-176, 178, 179, 180, 182, 183, 326
rab effector region 180
rab family 175
*rab*1 173, 174, 182
*rab*2 in Sezary syndromes 181
*rab*3B,5 and 6 174
*rab*7 174, 178
*rab*8, 9, 10 and 11 174, 178
*rac*1 174, 177
*rac*2 174, 177
racemic mixtures 276
radiation 2, 11, 22, 27-29, 84, 107, 111, 115, 116, 126, 228, 231, 246, 256, 261, 324
radon 22, 27-29, 28, 29, 321
ralA 174, 176, 181, 183
random insertional mutagenesis 190
*rap*1A 174, 176, 177
*rap*2 174, 176, 177, 181
*rar*a 306
ras 34, 36, 35, 38-40, 42, 45, 46, 48, 49, 54-59, 61, 63-65, 67, 70, 71, 75, 77, 79-81, 86, 88, 95, 96, 95, 96, 95, 96, 98, 97-99, 98, 101, 102, 101-108,111, 112, 115, 116, 117-131, 144, 152, 153, 157, 159, 171, 216, 313, 322-324, 326
ras gene 34, 36, 35, 38, 42, 54, 65, 67, 70, 71, 75, 77, 79, 105, 106, 112, 115-119, 124, 125, 146-150, 173-178, 180-184, 186, 187, 189, 192, 193
ras gene family 77, 106, 116, 125, 174, 186
ras oncogenes in tumorigenesis 111
ras protein 122, 174, 182
ras proteins, subcellular localization of 176
ras-related small G-proteins 180
ras subfamilies 174
ras superfamily 173-176, 180, 326
ras-like genes in neoplasia 180
rate-limiting step 23, 49, 118, 119, 312
rates of spontaneous mutation 232
Rb gene 97
Rb1 194

rbnt1 304
rearranged VDJ gene 199, 205
rearranged JH genes 212
rearrangement, gene 5, 32, 46, 95, 98, 100, 114, 136, 138, 144, 190, 193, 202, 204, 205, 210, 212, 215, 295
rearrangement of the flanking sequence 190
recall bias 242
receptor 38, 45, 47, 53, 58, 81, 84, 89, 95, 103, 139, 143, 147, 149, 171, 186, 306, 314
recessive traits 72-74, 114, 136, 146, 190, 295, 316
recessive mutations of suppressor genes 190
recombinant retrovirus construct 190
recombination 61, 100, 140, 141, 144, 145, 148-151, 188, 190
recombination, homologous, site-directed 190
regeneration, tissue 74, 261
regression analysis, conditional logistic 253
regulatory elements 119, 156, 166, 191, 194, 195
regulatory sequences 61, 64, 100, 142, 155
relative risk 71-73, 247, 250-252, 255, 256, 258, 293
relative risk model 247, 250-252, 256, 258, 293
relative risk, quadratic model 252
Repair, DNA 28, 31-34, 36, 37, 39, 40, 114, 123, 146, 149, 151, 192, 193, 229, 280, 301
replication 21, 22, 25, 27, 32, 34, 37, 96, 139, 140, 142, 143, 148-151, 191, 224, 226, 320, 321
replicative phase 34
replicative incorporation 34
reporter gene 157, 159, 160
residual risk 11, 13, 127
respiratory absorption 270
restriction enzyme digestion, DNA 204
ret 186, 189, 193, 194
reticulum, endoplasmic 74, 179, 180, 205
retina 22, 25
retinoblastoma 22, 25, 29, 97, 153, 190, 194, 218, 313, 314, 316, 321

retinoic acid 50, 57, 88, 147
retroviral infection 48, 51, 56, 61, 91, 96, 97, 99, 134, 135, 151, 186-191, 325
retroviral transduction 187
retroviral oncogenes 91, 96, 97, 186
retrovirus 107, 135, 139, 189, 190, 309
retroviruses 99
RF/J 116
RFAGT1-1 cells 65
RFLP 129, 138
RFP 69
RFP-induced 69
Rhabdomyosarcoma 81, 85, 88, 90, 323
rho-GAP 177
rho 173-179, 178, 181-183, 326
rho, interaction with actin 178
RHO1, 174
RHO1 174
RHO2 174
rhoA 174, 177, 178
Rifampicin 69, 76
Risk, upper-bound 3, 5, 7-10, 12, 11, 12, 14, 13, 15, 17
risk 1-14, 217-219, 223, 224, 226-229, 231-234, 239, 241-243, 245-248, 252, 254-258, 260-263, 276, 278, 280, 283-286, 289, 293-295, 301, 302, 305, 308, 310, 313-316, 319, 320, 322, 323, 325, 328, 329
risk assessment in standard setting 241
risk model 247, 250
risk of benzene-induced leukemia 256
RMS 9-4/0 85, 86
RNA 50, 64, 85-87, 120, 121, 139, 142, 143, 159, 164, 270, 280, 287
RNase 120, 159-161, 166, 169
rough endoplasmic reticulum 205
RV line 159, 161, 166
*ryh*1 174, 181
*ryh*2 174

S4-T 85-87
S4 85-87
saccharomyces cerevisiae 156, 170, 174, 291
safety 2, 241, 243, 260, 261, 286, 309, 325
safrole 17
salivary gland 96, 120, 159, 160, 162

salivary gland (continued) 164, 192, 233
sarcoma 58, 88, 97, 101, 105-108, 112, 124, 125, 130, 139, 140, 143, 153, 173
*sas*1 174, 175
*sas*2 174
saturation of metabolism 269, 301, 302
SCE 149, 269
SEC4 174, 175, 179, 180, 182, 183
second signal for transformation 215
selection pressure 178
selective oligonucleotide hybridization 71
selective *in vitro* amplification 128
selectivity of oncoprotein actions 191
self-regulation 115
sensitivity of detection 127
sensitivity analysis 258
sequence-specific effects 33, 215
sequencing 42, 80, 98, 114, 134, 138, 140, 141, 149
serine 46, 82-84, 177
serum 72, 143, 178
short gestation period 188
signal transduction 43, 45-49, 51-54, 59, 64, 84, 88, 95, 99, 103, 118, 144, 152, 159, 205, 213-215, 306, 308, 322, 324, 326
silent transgene 155
single amino acid substitutions 98
single base mutations 32, 98
Single cell zygotes 203
sis 58, 86-88, 105, 186, 323
site-directed 48, 190
Site-specific 6, 8, 32, 39
skin tumor induction in mice 193
small cell lung carcinoma 64, 97
smelters 6
smg 25A, B and C 174
smokers 246
SMR 233, 244-246, 294-296
sodium dodecyl sulfate 51
soil micropores 8
somatic cell alterations 21, 22, 31, 46, 98-100, 105, 114, 215, 285, 291
somatic hypermutation 215
southern blot 42, 51, 86, 113, 116, 119, 135, 136, 138, 140, 141, 148
Sp1 64

specific antibodies 36, 180
spermatogenesis 122
spleen 116, 119, 160, 164, 202, 209, 210, 212-214, 270
spleen 277, 280
splenomegaly 203
splice 38, 161, 160, 163
splicing 47, 55, 135-137, 142
spontaneous tumors in animals 193
spontaneous transformation 101
squamous cells 96, 101, 107, 149, 193, 276
staining 33, 36, 37, 212
standard mortality ratio (SMR) 244
Standardized Mortality Ratio 292
stem cell 23, 186, 188, 190, 234, 236-238, 240, 270, 306, 313, 316
step-wise analysis 35
stereological measurements 70
stimulator of cell proliferation 295
stomach cancer 96
stromal cells 169, 270, 291, 306-308, 310, 315, 330
subcarcinogenic doses 123
suppressor genes, model for inactivation 190, 313, 316, 322, 323, 325-327
supralinearity 280
surface Ig 204, 205
surface IgM 212
surface immunoglobulin 199
survival 4, 78, 96, 135
susceptible cells 21, 24, 61, 67-71, 75, 77, 148, 191, 193, 236, 237, 269, 274, 282, 323
SV40 viral DNA 188
SV40 T 97, 108, 122
SW 80 cell line 128
Syn- 85-88
Syn+ 85-88
synchronous colon cancers 130
synergistic effects 44, 126, 277, 285, 289
syngeneic 85, 86, 88
Syrian hamster 36, 42, 284

T-ALL 306
T24 62-65, 64, 65, 108
tagging 113
tandem head-to-tail arrays 188
target organ/tissue 194

349

target site 283
target strain 155, 324
target tissue 5, 9, 34, 40, 115, 194, 217-219, 218, 219, 222-224, 226, 238, 287, 314
TC10 174, 177
TC21 176
TCDD 20, 69, 238, 314
TCE 3, 6, 7
tcl3 306
tcl5 306
teleocidin 83
Tetrachlorodibenzo-p-dioxin 69, 77, 84, 227
thapsigargin 84
thermal cycled fusion 51, 56
thioguanine mutants 270
threonine 46, 84, 89, 175, 176
threshold 2, 15, 193, 253, 305, 308, 309, 314, 315, 330
Thy-1 122
thymic tumors 111, 115-119, 122, 125, 126, 324
thymidine kinase 34, 145
thymus, 113
TIMP-1 50
tissue-specific activation 191
tissue culture 112, 155, 160, 166, 188
TLV 253
toluene 7, 83, 89, 242, 286, 288
toxic gene 166
toxic effects 4, 11, 237, 263, 270, 275
toxicity 4, 11, 18, 124, 224, 227, 229, 231, 233, 234, 239, 240, 285-288, 290, 291, 310, 329
TPA 43, 45-52, 54, 55, 57, 65, 84
transactivation 157, 159
transactivator 155-157, 159-163, 166, 167, 169, 326
transcripts 86, 121, 126, 135, 137, 148
transcription 45, 49, 50, 54, 57, 59, 64, 65, 81, 100, 135, 139, 141-145, 155, 156, 161, 170, 191
transcriptional activator 156, 166
transcriptional elements 185
transcription factor 45, 49, 50, 54, 57, 59, 81, 155, 156, 161
transduction 43, 45-49, 51-54, 59, 61, 84, 88, 95, 103, 112, 134, 144, 152, 187

transduction (continued) 322, 324, 326
transfection 34, 57, 65, 81, 95, 97, 101, 104, 107, 108, 126, 157, 186, 190, 194
transfection assays 186, 194
transformants 116, 117
transformation 20, 21, 23, 28, 34, 48, 49, 51, 54-56, 63, 68, 91, 96, 97, 99, 101-104, 106-109, 112, 113, 115, 117, 123,125, 134, 144, 145, 173, 178, 181, 191-193, 201, 213-215, 258
transformation, molecular mechanisms 213
transforming genes 101, 104-106, 124,
transgene 159, 204
transgene expression, factors of 191
transgenic 51, 67, 69, 76, 81, 111, 117-126, 133-153, 155-157, 159-161, 160-163, 166, 167, 169-172, 185, 186, 188, 189, 188-192, 194, 195, 199-206, 209, 210, 213-216, 308-311, 316, 324-332
transgenic animal models 188
transgenic mice 21-126, 133, 134, 136-153, 155, 157, 159-161, 162, 163, 166, 169-172, 185, 186, 188-192, 194, 195, 201, 206, 215, 216, 310, 311, 324, 327
transient transfection 157
transition 38, 128, 232, 234, 238, 254, 320, 323
transition rates 232, 234, 238, 320
translocase 74, 79
translocation 62, 96, 99, 306, 316
translocations 62, 100, 137, 144, 148-150, 187, 306-310, 315, 330
Transmembrane molecules 42, 55, 112, 136, 137, 306-308
transversions 32, 117, 145, 147, 148, 313
trichloroethylene 7, 9, 290
truncation 47, 143
tubules 37
tumor 5, 20, 22, 24, 26, 25, 29, 28, 31-38, 40, 41, 43-46, 48, 49, 51-56, 58, 59, 61, 62, 65, 66-71, 75, 76, 81-86, 88, 89, 91,95-97, 96, 97, 100, 102-104, 108, 109, 111, 113, 114, 115-119, 121-126, 128, 130, 131, 133, 141, 143, 144, 146-150, 153, 157, 170, 171, 173, 175, 176, 186

tumor (continued) 185-190, 192-195, 212-215, 217, 199, 201, 207, 209, 210, 219, 221, 223, 224, 226, 231-233, 237-240, 277, 280, 283, 299, 300, 313, 316, 317, 321-327
tumor etiology, 31
tumor progression, molecular marker of 128
tumor suppressor gene 35, 38, 42, 45, 49, 52, 56, 81, 91, 97, 96, 97, 100, 103, 104, 123, 133,141, 143, 144, 146,-147-150, 153, 176, 182, 186, 188, 190, 194, 313, 316
tumor suppressor genes, inactivation 45, 100, 188, 326
Tumor multiplicity 70
tumor, undifferentiated 210
tumorigenesis 67, 68, 71, 76, 88, 100, 102, 107, 111, 114, 115, 117, 118, 122-125, 148, 152, 159, 170, 171, 231, 233, 238, 239, 313, 324
tumorigenicity 86
tumors of B-lymphoid origin 214
turnover 74, 137, 149, 226
TWA 241, 243, 245, 249
two-mutation model 22, 24, 23, 25, 27-29
two-stage model 4, 41, 218, 228, 231, 232, 234-236, 238, 298
tyrosine 73, 148, 175, 178, 186, 306-308
tyrosine kinase 148, 178, 186, 304, 306-308
tyrosinemia 72, 73, 78

U.S. EPA Risk Assessment 252
UAS 155, 157, 159, 160, 166, 169, 171, 311
undifferentiated cancer 186, 207, 209, 210
Undifferentiated B-Lineage 210
unit risks, geometric mean 252
Unscheduled DNA synthesis 62, 63, 146, 290
uptake due to dermal contact 257
urethan 70, 71, 76
urethan-induced tumors 71, 76
UROD 72, 73
uroporphyrinogen 73

v-*fps* 178, 183
v-*fos* 49
v-H-*ras* 99, 122, 189
v-*abl* 308, 311
v-*myc* 49, 99, 200, 202
v-*myc* probe 200, 202
V3 47
variable tandem repeat 62, 64
vas deferens 162
VDJ genes 199, 215
vector 48, 51, 56, 113, 135, 151, 160, 163, 190, 191, 324
vesicular traffic 180
vesicular nuclei 209, 210
VH1 215
VHS 6, 19
vinyl chloride 8, 9, 11, 13, 17, 283, 286
viral promoter 156
viral vector 191
Von Gierke's 74
VP16 157, 166
VTR 62, 64

WAP gene 192
waste sites 11, 18
water, nationwide quality criteria 15
WBC 203
weight-of-evidence 11-13
weighted cumulative dose 250, 252, 254
well-defined cohort 243
Whole mount preparations 162, 167, 170
Wilm's tumor 190, 194
window dose 250
WLM 28, 29
workers, morbidity/mortality 257
WT1, 194

x-ray irradiation 101

yeast 64, 145, 155-157, 170-174, 177, 180-182, 288, 311, 326
Yin and Yang 91
YPT 173, 174, 178, 180-182

zero order kinetics 283
zinc 9, 48, 51, 58
Zymbal 278, 277, 280, 284, 287-289

351

If you have any concerns about our products,
you can contact us on
ProductSafety@springernature.com

In case Publisher is established outside the EU,
the EU authorized representative is:
**Springer Nature Customer Service Center GmbH
Europaplatz 3, 69115 Heidelberg, Germany**

Printed by Libri Plureos GmbH
in Hamburg, Germany